Ecological Risk Assessment

Ecological Risk Assessment

Editor: Francis Lawrence

www.callistoreference.com

Callisto Reference,
118-35 Queens Blvd., Suite 400,
Forest Hills, NY 11375, USA

Visit us on the World Wide Web at:
www.callistoreference.com

ISBN: 978-1-63239-830-7 (Hardback)

The publisher's policy is to use permanent paper from mills that operate a sustainable forestry policy. Furthermore, the publisher ensures that the text paper and cover boards used have met acceptable environmental accreditation standards.

Trademark Notice: Registered trademark of products or corporate names are used only for explanation and identification without intent to infringe.

Printed in the United States of America.

Cataloging-in-publication Data

Ecological risk assessment / edited by Francis Lawrence.
   p. cm.
Includes bibliographical references and index.
ISBN 978-1-63239-830-7
1. Ecological risk assessment. 2. Ecology. 3. Environmental impact analysis. I. Lawrence, Francis.
QH541.15.R57 E26 2017
577.27--dc23

Table of Contents

Preface

Ecological risk assessment is defined as the monitoring of ecological resources to calculate changes and hazards faced by the ecosystems. Long-term damage can lead to larger sections of wildlife and flora being adversely affected. Models of risk assessment to deal with concrete effects that can be assessed as well as theoretical foresight that can be analyzed. This book on ecological risk assessment discusses the various models of ecological assessment and its relation to the severity of risk that is posed. From theories to research practical applications, case studies related to all contemporary topics of relevance of this field have been included in this book. It is a compilation of chapters that discusses the most vital concepts and emerging trends in the field of ecological risk assessment. As this field is emerging at a rapid pace, the contents of this book will help the readers understand the modern concepts and applications of the subject.

Every book is initially just a concept; it takes months of research and hard work to give it the final shape in which the readers receive it. In its early stages, this book also went through rigorous reviewing. The notable contributions made by experts from across the globe were first molded into patterned chapters and then arranged in a sensibly sequential manner to bring out the best results.

It has been my immense pleasure to be a part of this project and to contribute my years of learning in such a meaningful form. I would like to take this opportunity to thank all the people who have been associated with the completion of this book at any step.

Editor

Sources of Heavy Metals in Surface Sediments and an Ecological Risk Assessment from Two Adjacent Plateau Reservoirs

Binbin Wu[1], **Guoqiang Wang**[1]*, **Jin Wu**[1], **Qing Fu**[2], **Changming Liu**[1]

1 College of Water Sciences, Beijing Normal University, Key Laboratory of Water and Sediment Sciences, Ministry of Education, Beijing, China, 2 Chinese Research Academy of Environmental Sciences, Beijing, China

Abstract

The concentrations of heavy metals (mercury (Hg), cadmium (Cd), lead (Pb), chromium (Cr), copper (Cu) and arsenic (As)) in surface water and sediments were investigated in two adjacent drinking water reservoirs (Hongfeng and Baihua Reservoirs) on the Yunnan-Guizhou Plateau in Southwest China. Possible pollution sources were identified by spatial and statistical analyses. For both reservoirs, Cd was most likely from industrial activities, and As was from lithogenic sources. For the Hongfeng Reservoir, Pb, Cr and Cu might have originated from mixed sources (traffic pollution and residual effect of former industrial practices), and the sources of Hg included the inflows, which were different for the North (industrial activities) and South (lithogenic origin) Lakes, and atmospheric deposition resulting from coal combustion. For the Baihua Reservoir, the Hg, Cr and Cu were primarily derived from industrial activities, and the Pb originated from traffic pollution. The Hg in the Baihua Reservoir might also have been associated with coal combustion pollution. An analysis of ecological risk using sediment quality guidelines showed that there were moderate toxicological risks for sediment-dwelling organisms in both reservoirs, mainly from Hg and Cr. Ecological risk analysis using the Hakanson index suggested that there was a potential moderate to very high ecological risk to humans from fish in both reservoirs, mainly because of elevated levels of Hg and Cd. The upstream Hongfeng Reservoir acts as a buffer, but remains an important source of Cd, Cu and Pb and a moderately important source of Cr, for the downstream Baihua Reservoir. This study provides a replicable method for assessing aquatic ecosystem health in adjacent plateau reservoirs.

Editor: Jonathan H. Freedman, NIEHS/NIH, United States of America

Funding: This research was supported by Beijing Higher Education Young Elite Teacher Project (Grant No. YETP0275), the Program for New Century Excellent Talents in University (Grant No. NCET-12-0058) and the Fundamental Research Funds for the Central Universities (Grant No. 2012LZD10). The funders had no role in study design, data collection and analysis, decision to publish, or preparation of the manuscript.

Competing Interests: The authors have declared that no competing interests exist.

* Email: wanggq@bnu.edu.cn

Introduction

There is worldwide concern about heavy metal contamination because of the environmental persistence of these elements, biogeochemical recycling and the ecological risks that metals present [1,2]. Large numbers of anthropogenically generated heavy metals from urban areas, agricultural areas and industrial sites are discharged into aquatic environments where they are transported in the water column, accumulated in sediment, and biomagnified through the food chain [3], resulting in significant ecological risk to benthic organisms, fish and humans [4]. Sediments are the main sink for heavy metals in aquatic environments [5], and sediment quality has been recognized as an important indicator of water pollution [6]. However, heavy metals are not permanently bound to sediments [7], and they may be released into the water column when the environmental conditions change (e.g., temperature and pH) or when sediments undergo other physical or biological disturbances [8]. Furthermore, reservoir construction generally leads to an increase in residence time, resulting in high accumulations of heavy metals in sediments. Consequently, it is important to analyze sediments from reservoirs for heavy metals to support environmental management, particularly for sediments from drinking water reservoirs.

Understanding the sources of pollutants in aquatic sediments is important for pollution control. Statistical approaches, such as Pearson correlation analysis, principal components analysis (PCA), and cluster analysis, are considered to be effective tools for uncovering pollution sources and have been used successfully in many studies of heavy metal pollution in sediments [1,2,3,7,9,10,11]. Risk assessments of the environmental pollution are also critical for sediment analysis. The ecological risk of heavy metals in sediments differs for different receptors (e.g., sediment-dwelling organisms, fish or humans). The thresholds in sediment quality guidelines (SQGs) have been used to evaluate the potential adverse effects of heavy metals on sediment-dwelling organisms in freshwater systems [12,13,14,15]. However, few SQGs have been developed to assess the adverse effects of heavy metals in sediment on higher trophic levels (fish or other wildlife) [16,17]. The potential ecological risk index proposed by Hakanson [18] is based on heavy metal concentrations in sediment, and it is the simplest and most popular method for assessing the human health risk from fish consumption.

The rapid growth of urbanization and industrial development has resulted in increasing heavy metal pollution in the aquatic sediment of the Yunnan-Guizhou Plateau [11,19]. Several cascade hydropower stations have been built along the region's large rivers (e.g., Wujiang, Jinshajiang and Nanpanjiang) since the 1950s, and stations are still being built for electricity production today, leading to a continuous series of reservoirs along the rivers [20]. Previous studies have evaluated the carbon (C) cycle [21,22] and the mercury (Hg) balance [20,23] in adjacent plateau reservoirs, and have demonstrated how upstream reservoirs influence downstream reservoirs. However, little research has been conducted on other heavy metals in the adjacent reservoirs on this plateau. In addition, several decades after their construction, the functions of these reservoirs were changed, so they now supply drinking water to the human population, which has grown rapidly due to economic growth. Currently, these reservoirs are the main drinking water sources on the Yunnan-Guizhou Plateau. Although some pollution sources were closed or moved when the reservoir functions were changed, the residue of previous pollutants still remains in reservoir sediments. Furthermore, the Yunnan-Guizhou Plateau is famous for its karst landforms, and the hydrogen carbonate (HCO_3^-) concentration and pH are both high in the aquatic environment [24,25,26]. The alkaline environment favors heavy metal accumulation in sediment [27], while the karst landform promotes interactions between groundwater and surface water through fractures (sinkholes, conduits and caves) or carbonate bedrock [26,28,29]. These interactions can complicate heavy metal transport and increase the ecological risk of secondary pollution. Therefore, it is important and necessary to investigate the heavy metal pollution and to assess the associated pollution sources and ecological risks from reservoir sediments on the Yunnan-Guizhou Plateau, particularly for drinking water reservoirs. The objectives of this paper are to (1) identify the pollution sources of heavy metals in the sediment from two adjacent drinking water reservoirs on the Yunnan-Guizhou Plateau, (2) estimate the associated ecological risk by considering different receptors, and (3) discuss the influence of upstream reservoirs (as buffers or sources of heavy metals) on downstream reservoirs.

Materials and Methods

Study Areas

The Hongfeng and Baihua Reservoirs are two adjacent reservoirs on the Yunnan-Guizhou plateau, just northwest of Guiyang City, the capital of Guizhou Province, Southwest China (Fig. 1). These two reservoirs were constructed on the main channel of the Maotiao River, a branch of the Wujiang River in the Yangtze River Basin, in 1958 and 1960, respectively. The Maotiao River was one of the first rivers to be used for cascade hydropower in China. The Hongfeng is the first reservoir and the Baihua is the second of seven cascade hydropower stations along the Maotiao River. The Hongfeng Reservoir covers a water surface area of 57.2 km², while the Baihua Reservoir covers an area of 14.5 km². Both reservoirs are very deep, with each having a maximum depth of approximately 45 m. The Hongfeng Reservoir consists of the North and South Lakes (which have different flow directions, Fig. 1), and has five main inflows, two into the North Lake and three into the South Lake. The Maotiao River is the only outlet of the Hongfeng Reservoir, and also serves as the major inlet of the Baihua Reservoir. The Baihua Reservoir has eight additional minor inflows and one outlet. For their first 30 years, the reservoirs were mainly used for electricity generation and flood control. During this period, as industry, agriculture, tourism and fishery production were established and developed in

the basin, the water quality in both reservoirs declined. However, because of an increasing demand for water through the 1990s, the two reservoirs were designated as drinking water sources for Guiyang City. The major function of both reservoirs was changed to drinking water supply in 2000, at which point the government strengthened their environmental protection. The pollution sources have gradually decreased, but the sediments may still hold residue from earlier pollution.

Heavy metal pollution is one of the most prominent environmental problems in the study reservoirs, mainly owing to intense anthropogenic activities. Figure 2 shows the distribution of main point sources in Hongfeng and Baihua Reservoirs basin from the first China pollution source census in 2008 (provided by the Guiyang Research Academy of Environmental Sciences). There are many mining, smelting, mechanical manufacture, chemical and other industries (e.g., building material, food and pharmaceutical factories) in the catchment area of the Hongfeng Reservoir (1596 km²), which are major sources of heavy metals (Fig. 2). There is also a large coal-fired power plant (300 MW) situated on the southeast bank of the Hongfeng Reservoir [30], which is the main source of atmospheric deposition, especially for Hg and other heavy metals associated with coal combustion. The catchment area of the Baihua Reservoir is 1895 km², but pollutants are first transported into the Hongfeng Reservoir, which may serve as a buffer for the Baihua Reservoir. The Baihua Reservoir receives direct inputs from an area of only 299 km². Even so, there are many intense point sources in this small area, including various heavy and light industries (Fig. 2), which have resulted in serious heavy metal pollution in the Baihua Reservoir. In particular, the Baihua Reservoir is noted for its Hg contamination from the Guizhou Organic Chemical Plant (GOCP) [31], which is located in the upper reaches of the Baihua Reservoir and downstream of the Hongfeng Reservoir. The GOCP used Hg-based technology to produce acetaldehyde and discharged Hg-laden wastewater to the Baihua Reservoir via the Dongmenqiao River until 1997. The pollution caused by the GOCP persists to the present day.

Sampling and Analysis

Two field surveys were conducted in December 2010 and April 2012. Water and sediment samples were collected from 26 sites in the Hongfeng and Baihua Reservoirs (Fig. 1 and Table 1). The field studies were permitted by the Administration of Hongfeng, Baihua and Aha Reservoirs, and did not involve endangered or protected species. There were 13 sites in each reservoir, comprising inlets of main tributaries (sites 1–5 in the Hongfeng Reservoir and sites 14–22 in the Baihua Reservoir) and representative sites within both reservoirs (sites 6–13 in the Hongfeng Reservoir and sites 23–26 in the Baihua Reservoir). Surface water samples were collected in acid-washed polyethylene sample bottles and were acidified with 1:1 nitric acid: deionized water. Water samples were stored at 4°C immediately upon returning from the field. The upper 0–10 cm of sediment was collected, placed into pre-cleaned polyethylene bags, and taken to the laboratory. All sediment samples were freeze-dried and passed through a 2 mm nylon sieve to discard the coarse debris. A pestle and mortar was then used to grind the sieved sediments until all particles were fine enough to pass through a 0.147 mm nylon sieve. Sediment samples were digested in a microwave digestion system with a HNO_3-HF-$HClO_4$-HCl acid mixture solution before analysis for total heavy metal content. All water samples and the solutions of the digested sediment samples were analyzed by inductively coupled plasma atomic emission spectroscopy for Cr, Cu and Pb. Cd concentrations were determined by graphite

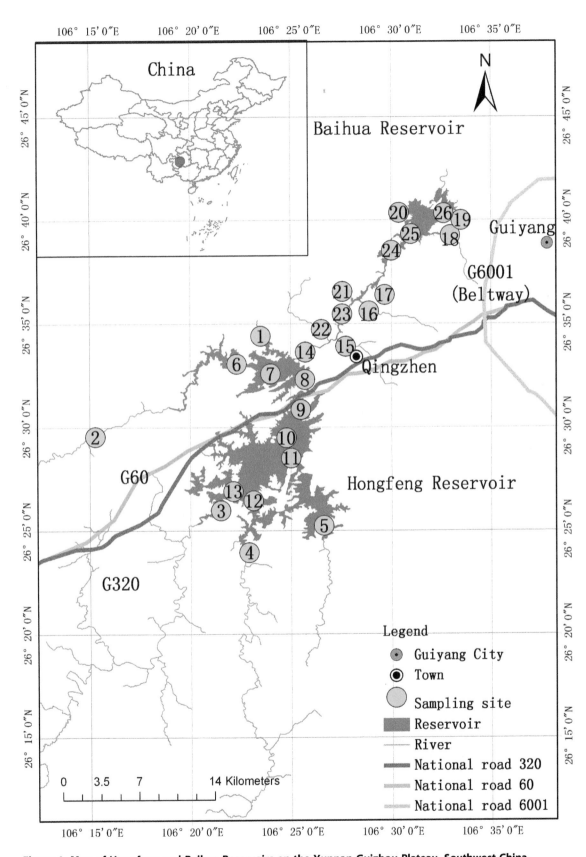

Figure 1. Map of Hongfeng and Baihua Reservoirs on the Yunnan-Guizhou Plateau, Southwest China.

furnace atomic absorption spectrophotometry. As and Hg were measured using atomic fluorescence spectrometry. Quality assur- ance and quality control of the analyses processes were assessed by duplicates, method blanks and standard reference materials [11].

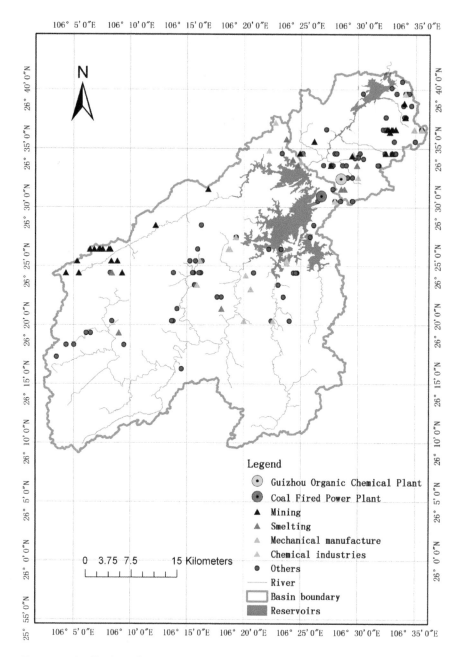

Figure 2. Distribution of main point sources of pollution in the Hongfeng and Baihua Reservoir basins.

Spatial and Statistical Analyses

Spatial and statistical analyses were performed by using Arc GIS 9.3 and SPSS 17.0 for Windows (SPSS Inc., Chicago, IL) to investigate the heavy metal pollution sources separately for the Hongfeng and Baihua Reservoirs. A one-way ANOVA was performed on heavy metal concentrations in sediment to determine whether the differences between the two field surveys were significant. Pearson correlation analysis was used to determine the relationships between the heavy metals in sediment. To obtain more reliable information about the relationships between the heavy metals, a PCA with Varimax normalized rotation was performed separately for the Hongfeng and Baihua Reservoirs. The PCA calculated eigenvectors to determine the common pollution sources, and components with eigenvalues greater than 1 were considered to be relevant [32]. Components

with factor loadings above 0.75, between 0.5 and 0.75, and between 0.3 and 0.5 were considered to be strong, moderate and weak, respectively [33]. Boxplot is a convenient way to depict the full range of data and compare the distributions among different datasets. In this study, the boxplot was used to compare the heavy metal concentrations at site 14, the outlet of the Hongfeng Reservoir and inlet of the Baihua Reservoir, with concentrations at other sites in the tributaries and at sites within the reservoirs, in order to discuss the influence of the upstream Hongfeng Reservoir on the downstream Baihua Reservoir.

Ecological Risk Assessment

We used two methods to assess the ecological risk of the heavy metals in surface sediments to benthic organisms and humans. First, we used the consensus-based SQGs for freshwater ecosys-

Table 1. Locations of sampling sites in the Hongfeng and Baihua Reservoirs on the Yunnan-Guizhou Plateau, Southwest China.

Reservoir	Site	Longitude	Latitude	Description	
Hongfeng Reservoir	1	106°23′30.64″	26°34′24.52″	At main tributaries	Maibao River
	2	106°15′3.58″	26°29′22.80″		Maiweng River
	3	106°21′30.89″	26°25′58.34″		Yangchang River
	4	106°22′54.88″	26°23′56.79″		Maxian River
	5	106°26′42.01″	26°25′4.80″		Houliu River
	6	106°22′11.10″	26°33′44.07″	Within reservoir	Taipingdi
	7	106°24′2.05″	26°32′34.80″		Center of the North Lake
	8	106°23′17.43″	26°32′27.13″		Junction of the North and South Lakes
	9	106°26′07.06″	26°30′58.03″		Houwu
	10	106°24′39.31″	26°29′19.07		Center of the South Lake
	11	106°25′04.17″	26°28′33.89″		Jiangjundong
	12	106°23′01.81″	26°26′26.73″		Yangjiajun
	13	106°22′01.22″	26°26′37.57″		Sanjiazhai
Baihua Reservoir	14	106°25′43.88″	26°33′37.65″	At main tributaries	Outlet of Hongfeng Reservoir
	15	106°27′19.60″	26°34′7.06″		Dongmenqiao River
	16	106°28′51.88″	26°35′36.16″		Maicheng River
	17	106°29′39.21″	26°36′23.43″		Dianzishanggou River
	18	106°32′56.81″	26°39′15.61″		Maixi River
	19	106°33′16.38″	26°40′6.00″		Banpochanggou River
	20	106°30′22.33″	26°40′21.27″		Maolizhaigou River
	21	106°27′33.58″	26°36′32.72″		Xiaohekou River
	22	106°26′47.46″	26°34′17.11″		Changchong River
	23	106°27′25.20″	26°35′27.60″	Within reservoir	Huaqiao
	24	106°30′4.42″	26°38′31.42″		Xuantiandong
	25	106°30′50.23″	26°39′1.37″		Tangerpo
	26	106°32′48.39″	26°40′18.16″		Chafan

tems that were proposed by MacDonald et al. [15], which included a threshold effect concentration (TEC) and a probable effect concentration (PEC). TECs are the concentrations below which adverse effects are not expected on sediment-dwelling organisms, while PECs are concentrations above which adverse effects are expected to occur frequently [34,35]. The mean PEC

Table 2. Ecological risk assessment criteria for the sediment quality guidelines (SQGs) and Hakanson index.

Method	C or E_r^i	Potential ecological risk for single heavy metal	m-PEC-Q or RI	Ecological risk for all Heavy metals
SQGs	$C<TEC$	Low	m-PEC-Q<0.1	Low (<14%)[a]
	$TEC <C<PEC$	Moderate	0.1< m-PEC-Q <1.0	Moderate (15–29%)[a]
	$C>PEC$	High	1.0<m-PEC-Q<5.0	Considerable (33–58%)[a]
			m-PEC-Q >5.0	Very high (75–81%)[a]
Hakanson index	$E_r^i<40$	Low	$RI<95$	Low
	$40<E_r^i<80$	Moderate	$95<RI<190$	Moderate
	$80<E_r^i<160$	Considerable	$190<RI<380$	Considerable
	$160<E_r^i<320$	High	$RI>380$	Very high
	$E_r^i>320$	Very high		

C: concentration of heavy metal in surface sediment.
TEC: threshold effect level; PEC: probable effect level [15].
m-PEC-Q: mean PEC quotient; [a]incidence of toxicity [36].

Table 3. Summary statistics for heavy metal concentrations in surface sediments from the Hongfeng and Baihua Reservoirs.

		Hg	Cd	Pb	Cr	Cu	As	
HongfengReservoir	Mean (Min, Max)	0.32(0.08, 1.03)	0.28(0.01, 0.85)	28.41(0.10, 89.20)	86.91(34.10, 141.00)	43.50(15.70, 93.60)	15.33(0.12, 45.54)	This study (2010.12)
	S.D.	0.26	0.36	27.55	29.65	25.09	14.92	
	CV(%)	82.86	130.78	96.97	34.12	57.67	97.31	
	Mean (Min, Max)	0.27(0.04, 0.56)	0.53(0.31, 1.37)	30.22(1.21, 89.20)	82.78(41.00, 141.00)	45.46(23.20, 73.80)	23.31(17.31, 29.61)	This study (2012.4)
	S.D.	0.17	0.27	25.63	23.45	17.72	3.98	
	CV(%)	62.33	51.27	84.78	28.32	38.99	17.07	
	2007.3	0.99		34.31		89.11	49.90	Huang et al. [40]
	2008.8			58.39	120.16	69.84		Zeng et al. [41]
	2008.10	0.66	0.77	35.91	87.98	91.85	29.74	Liu et al. [42]
	2009.5	0.46	0.65		118.00	88.00	40.80	He et al. [27]
Baihua Reservoir	Mean (Min, Max)	0.68(^{a}, 2.20)	0.58(0.01, 1.00)	27.28(0.10, 51.90)	76.24(30.80, 143.00)	36.71(0.36, 65.90)	29.95(7.95, 48.19)	This study (2010.12)
	S.D.	0.73	0.37	16.89	31.50	19.77	14.80	
	CV(%)	108.24	64.15	61.90	41.31	53.86	49.42	
	Mean (Min, Max)	0.45(0.01, 1.25)	0.61(0.23, 1.00)	27.84(0.10, 51.90)	76.38(30.12, 143.10)	43.16(9.38, 73.55)	26.23(7.95, 34.75)	This study (2012.4)
	S.D.	0.48	0.28	18.15	31.45	21.49	7.30	
	CV(%)	107.21	46.13	65.19	41.17	49.79	27.82	
	2007	18.90	0.88	16.05	59.75	74.97	53.34	Huang [39]
	2010.5		0.95	38.90	66.00	67.50		Tian et al. [43]
Natural Background Value		0.08–0.15	0.08–0.12	18.50–23.90	73.90–94.60	27.30–36.70	27.00–50.00	NEPA [44]

All concentrations are in mg/kg dry weight. ^{a}: not detected. S.D.: standard deviation; CV: coefficients of variation.

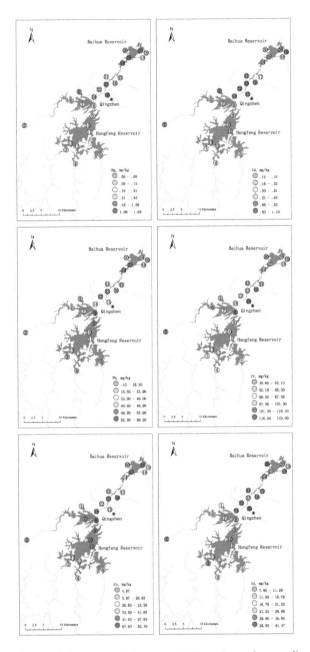

Figure 3. Heavy metal concentrations in surface sediments from Hongfeng and Baihua Reservoirs.

quotient (m-PEC-Q) [36] was also calculated for each sediment sample to assess the biological significance of the contaminant mixtures as follows:

$$m-PEC-Q= \frac{\sum_{i=1}^{n}(C_i/PEC_i)}{n} \qquad (1)$$

where C_i is the sediment concentration of compound i, PEC_i is the PEC for compound i and n is the number of compounds i. Four ranges of the mean PEC quotient were developed by Long et al. [36] for ranking samples in terms of toxicity incidence (Table 2).

We also used the Hakanson index, which reflects the risk to human health from fish consumption. This index is based on the assumption that the sensitivity of the aquatic system depends on its productivity [3,18]. The potential ecological risk index (RI) was introduced to evaluate heavy metal pollution in sediments by considering the toxicity of heavy metals and the environmental response. The RI is calculated as follows:

$$RI = \sum E_r^i \qquad (2)$$

$$E_r^i = T_r^i C_f^i \qquad (3)$$

$$C_f^i = C_0^i / C_n^i \qquad (4)$$

where RI is the total potential ecological risk index for multiple metals, E_r^i is the potential ecological risk index for a single metal, and T_r^i is the toxic-response factor for a given metal, considering both toxicity and the sensitivity. C_f^i is the contamination factor, C_0^i is the metal concentration in the sediment and C_n^i is a reference value for metals. In this study, because both reservoirs are moderately eutrophic [37], T_r^i was described as Hg (40) > Cd (30) > As (10) > Cu = Pb (5) > Cr (2), based on the assumption that the bioproduction index was 5 [18]. C_n^i was defined as the upper limit of the natural background value for a given metal in the study area (Table 3). Four ranges of the risk factor RI were suggested by Hakanson, based on eight metals (polychlorinated biphenyls (PCBs), Hg, Cd, As, Pb, Cu, Cr, and zinc (Zn)). PCBs and Zn were not considered in this study. Based on the different contributions of these elements to the ecological risk index RI, the adjusted evaluation criteria for RI based on the six metals in this study are listed in Table 2.

Table 4. Correlations between heavy metals in surface sediments from the Hongfeng and Baihua Reservoirs.

	Hg	Cd	Pb	Cr	Cu	As
Hg	1.000	0.090	0.321	−0.234	−0.038	−0.343
Cd	−0.422*	1.000	−0.540**	−0.168	−0.237	0.625**
Pb	0.511**	−0.506**	1.000	0.438*	0.595**	−0.524**
Cr	0.495*	0.013	−0.086	1.000	0.785**	0.046
Cu	0.241	−0.013	0.006	0.509**	1.000	−0.183
As	−0.414*	0.476*	−0.314	0.067	−0.333	1.000

Hongfeng Reservoir in the upper right corner (blod); Baihua Reservoir in the lower left corner.
Levels of significance: *p<0.05; **p<0.01.

Table 5. Principal Component Analysis (PCA) for heavy metals in surface sediments from the Hongfeng and Baihua Reservoirs.

Heavy metal	Hongfeng Reservoir			Baihua Reservoir	
	F1	F2	F3	F1	F2
Hg			0.973	−0.669	0.523
Cd		0.924		0.821	
Pb	0.605	−0.575	0.378	−0.800	
Cr	0.929				0.902
Cu	0.930				0.799
As		0.851	−0.334	0.705	
Variance (%)	35.29	32.24	21.35	37.84	29.32
Cumulative (%)	35.29	67.53	88.88	37.84	67.16

Factor loadings smaller than 0.3 have been removed.
Extraction method: PCA, Rotation method: Varimax with Kaiser normalization.

Results and Discussion

Heavy Metal Concentrations in Water Samples and Surface Sediments

Of the 6 heavy metals, only Hg and As were detected in the water samples in December 2010, while Hg, Cd, Cr (VI) and As were detected in April 2012. Their concentrations were similar in the Hongfeng and Baihua Reservoirs, with Cd, Cr (VI), and As concentrations lower than Class I as defined in the Chinese Environmental Quality Standards for Surface Water (GB3838-2002, <0.001 mg/L for Cd, <0.01 mg/L for Cr (VI), and < 0.05 mg/L for As) and Hg concentrations ranging from Class I (GB3838-2002, <0.00005 mg/L) to Class IV (GB3838-2002, 0.0001–0.001 mg/L) among different sites. The low heavy metal concentrations in water were primarily due to the accumulation in sediments because the alkaline environment in both reservoirs provides ideal conditions for adsorption and precipitation [27]. Moreover, the sediment accumulation rate in both reservoirs was quite high [38], contributing to the removal of heavy metals from the water column. A prior one-way ANOVA analysis was conducted to examine the variation in heavy metal concentrations in sediment between the two field surveys. None of the heavy metals in Hongfeng and Baihua Reservoirs displayed significant variation in means (p>0.05), although Cd in Hongfeng Reservoir and As in both reservoirs showed significant changes in their variances (p<0.05). The significant differences in variance for Cd and As are mainly because of their higher concentrations in the sites within the reservoirs and the reduced spatial heterogeneity in the second field sampling comparing to the first one. However, the general spatial patterns for Cd and As (with higher concentrations at sites in the tributaries than at sites within the reservoirs) were still similar between the two field surveys. Those results indicate that the pollution sources for the metals were relatively stable between the two surveys. The concentrations of heavy metals in surface sediments of both reservoirs from the two field surveys are summarized in Table 3. Heavy metal concentrations in sediment were much higher than those in water. In general, the mean Hg, Cd and As sediment concentrations in the Baihua Reservoir were higher than those in the Hongfeng Reservoir, while Pb, Cr and Cu were higher in the Hongfeng Reservoir. Comparison with the results of previous studies [27,39,40,41,42,43] shows that most of the heavy metal concentrations in both reservoirs have decreased, though by differing amounts (Table 3). In particular, the Hg concentrations in the Baihua Reservoir have decreased signifi-

cantly, indicating that measures taken in recent years have been effective and resulted in improvements. During the two field surveys, the mean concentrations of Hg, Cd, Pb and Cu, and the maximum concentrations of Cr in both reservoirs exceeded the upper limit of the natural background values for the study area [44], indicating anthropogenic sources. However, the concentrations of As (including the minimum and maximum values) in both reservoirs were well within the range of the natural background values [44], implying no significant anthropogenic impact and primarily lithogenic sources.

Heavy Metal Pollution Sources

To develop control strategies for environmental pollution, it is very important to identify its source. Spatial and statistical analyses were performed to identify the possible pollution sources for heavy metals in the Hongfeng and Baihua Reservoirs. The average concentrations of heavy metals in the sediments from the two field surveys were used to study the spatial distributions, while all data in the sediments from the two field surveys were used in a Pearson correlation analysis and PCA. The spatial distribution patterns of Hg, Cd, Pb, Cr, Cu and As in surface sediments of both reservoirs are shown in Figure 3. The Pearson correlation coefficients and the results of the PCA for the investigated metals are shown in Table 4 and Table 5, respectively. All of the results were generally consistent with each other.

Specifically, the PCA yielded three significant components for Hongfeng Reservoir and two significant components for Baihua Reservoir, accounting for 88.88% and 67.16% of the cumulative variance, respectively (Table 5). For the Hongfeng Reservoir, the first component (F1), explaining 35.29% of the total variance, had strong positive loadings of Cr and Cu, and moderate positive loading of Pb. Those three heavy metals exhibited similar spatial distributions in the Hongfeng Reservoir, with unexpectedly higher concentrations at reservoir sites than at tributary sites. In particular, site 8 (at the junction of the North and South Lakes) showed the highest concentrations for all of the three heavy metals, and site 9 (near Houwu) also showed relatively high concentrations (Fig. 3). In addition, those three heavy metals were highly correlated (Table 4, p<0.01 for Cr-Cu and Cu-Pb, p<0.05 for Cr-Pb), indicating their similar origins or comparable chemical properties [45]. This phenomenon might be caused by two possible reasons. Firstly, Bai et al. [46] found that traffic pollution was responsible for the high heavy metal concentrations (including comparable Cr, Cu and Pb concentrations with our study) along

Table 6. Results of ecological risk assessments for single heavy metal from two methods for the Hongfeng and Baihua Reservoirs.

	Hg	Cd	Pb	Cr	Cu	As
TEC	0.18	0.99	35.8	43.4	31.6	9.79
PEC	1.06	4.98	128	111	149	33
Hongfeng Reservoir						
% samples which exceeded TEC	69.23	7.69	38.46	92.31	61.54	84.62
% samples which exceeded PEC	0.00	0.00	0.00	15.38	0.00	15.38
% samples with $E_f^i<40$	30.77	7.69	100.00	100.00	100.00	100.00
% samples with $40<E_f^i<80$	23.08	53.85	0.00	0.00	0.00	0.00
% samples with $80<E_f^i<160$	38.46	23.08	0.00	0.00	0.00	0.00
% samples with $160<E_f^i<320$	7.69	15.38	0.00	0.00	0.00	0.00
% samples with $E_f^i>320$	0.00	0.00	0.00	0.00	0.00	0.00
Baihua Reservoir						
% samples which exceeded TEC	53.85	7.69	38.46	84.62	46.15	92.31
% samples which exceeded PEC	30.77	0.00	0.00	7.69	0.00	38.46
% samples with $E_f^i<40$	38.46	23.08	100.00	100.00	100.00	100.00
% samples with $40<E_f^i<80$	15.38	0.00	0.00	0.00	0.00	0.00
% samples with $80<E_f^i<160$	7.69	15.38	0.00	0.00	0.00	0.00
% samples with $160<E_f^i<320$	15.38	61.54	0.00	0.00	0.00	0.00
% samples with $E_f^i>320$	23.08	0.00	0.00	0.00	0.00	0.00

All concentrations are in mg/kg dry weight.

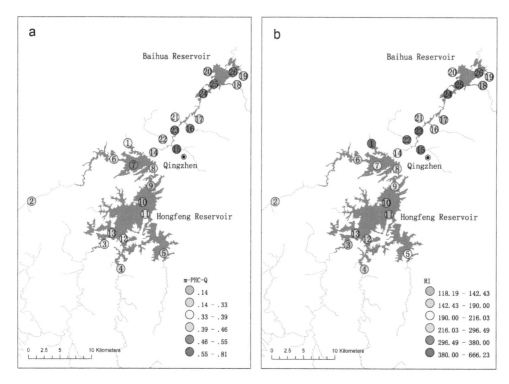

Figure 4. Mean PEC quotient (a) and potential ecological risk indexes (b) of heavy metals in sediments.

the roadside of National Road 320 in the Yunnan province (adjacent to Guizhou Province), and Zhu et al. [47] also found that road dust samples were severely polluted by Cr, Cu and Pb in another metal smelting/processing industrial city in Guizhou Province. In this study, the National Road 60 and the National Road 320 pass close to the junction of the South and North Lakes of the Hongfeng Reservoir (Fig. 1), which suggests that the traffic emissions, through atmospheric deposition and road runoff, could result in heavier pollution in sites near the roadway (site 8 and 9). Secondly, the high concentrations of Cr, Cu and Pb at reservoir sites are likely to be related to the residual effect from former industrial activities (e.g., mining, smelting, mechanical manufacture and chemical industry). The metals are more likely retained in the sediment of sites within the reservoirs rather than sites in the tributaries because heavy metal accumulation in lake sediments is generally higher than that in rivers [3]. Additionally, the complex hydrodynamic conditions at the junction of the South and North Lakes may affect the heavy metal distributions in sediment, which requires further research. Therefore, the first component (F1) might reflect mixed sources from traffic pollution and the residual effect of former industrial influence. The second component (F2), explaining 32.24% of the total variance, was dominated by Cd and As. Similar spatial patterns were observed for Cd and As, with higher concentrations at tributary sites than at sites within the reservoirs, indicating that they mainly come from the inflows (Fig. 3). As expected, significantly positive correlations were found between Cd and As (Table 4, p<0.01). However, Cd showed apparent anthropogenic origin, with most sites exceeding its natural background values, while As levels suggested natural origins, with all sites well within the natural background values (Table 3). Cd is closely related to industrial activities, such as smelting, electroplating and plastics production in the upstream areas. Hence, F2 may reflect the pollution through inflows from both industrial activities and natural weathering and erosion. The

third component (F3) had strong positive loading on Hg and a weak positive loading on Pb, accounting for 21.35% of the total variance. The highest Hg concentration in Hongfeng Reservoir was found at site 1 in the tributary of Maibao River, which has several smelting and chemical industries in its upstream (Fig. 2). Meanwhile, for both Hg and Pb, the North Lake were more polluted than the South Lake, and three tributaries in the South Lake showed concentrations well within the natural background values. Feng et al. [30] found that runoff due to soil erosion was the main source of Hg in sediment in the South Lake of the Hongfeng Reservoir. Thus, F3 may reflect the pollution from inflows from industrial activities in the North Lake and lithogenic origin in the South Lake. In addition, He [48] found that atmospheric deposition from coal combustion was also an important source of Hg in the Hongfeng Reservoir, which was not clearly distinguished by the PCA.

For Baihua Reservoir, the first component (F1), explaining 37.84% of total variance, showed strong positive loadings on Cd and As. Similar spatial distributions (with higher concentrations at tributary sites than at sites within the reservoirs) (Fig. 3) and positive correlations were also found between Cd and As (Table 4, p<0.05). As discussed above, F1 in Baihua Reservoir might be similar to F2 in Hongfeng Reservoir, including the pollution through inflows from both industrial activities and natural origin. The second component (F2) had a moderate positive loading of Hg and strong positive loadings on Cr and Cu. The highest Hg and Cr concentration in the Baihua Reservoir was found at site 15 in the Dongmengqiao tributary, which received wastewater from the GOCP and many other industries (Fig. 2). The highest Cu concentration in the Baihua Reservoir was at site 16 in the tributary of the Maicheng River, which has several industries (especially mining) in its upstream (Fig. 2). Strong associations were found between Cr and Cu (p<0.01) and between Cr and Hg (p<0.05) (Table 4). F2 obviously represented industrial activity

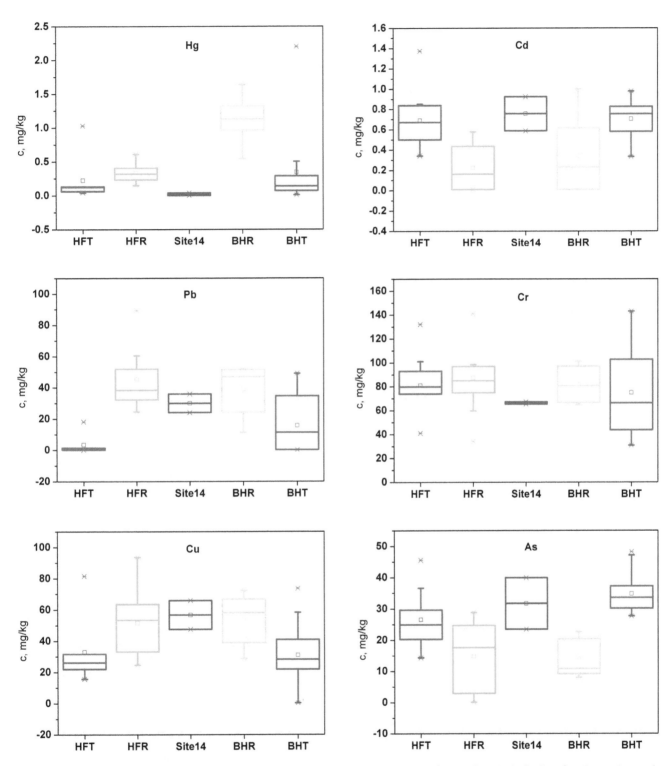

Figure 5. Comparison of heavy metal concentrations in sediments. (HFT: sites at inlets of main tributaries in the Hongfeng Reservoir, namely sites 1–5; HFR: representative sites within Hongfeng Reservoir, namely sites 6–13; BHT: sites at inlets of main tributaries in the Baihua Reservoir (except site 14), namely sites 15–22; BHR: representative sites within the Baihua Reservoir, namely sites 23–26).

upstream. On the other hand, the PCA failed to identify Pb sources in Baihua Reservoir, but only showed its strong negative associations with F1, indicating that there may be significant sources other than F1 and F2 for Pb. The spatial pattern of Pb in the Baihua Reservoir also showed much higher concentrations at sites within the reservoir than at sites in the tributaries (Fig. 3).

Although no main road crosses Baihua Reservoir directly, it is close to the urban district of Guiyang City, which has several roads and high traffic density (Fig. 1 only shows the main beltway and there are many other crisscrossed roads inside the beltway). In addition, Pb was added to gasoline in China until June 2000 [49]. Hence, Pb may originate from traffic pollutants deposited atmospherically. Moreover, highly positive correlations were found between Pb and Hg (p<0.01), implying that Hg might also come from atmospheric deposition (coal combustion) in addition to from the factors associated with F2.

Ecological Risk Assessment of Heavy Metals in Surface Sediments

Because there was no significant change in the mean heavy metal concentrations in sediment between the two field surveys, the average concentrations were used to study the ecological risk of single heavy metals and the combined ecological effects of six heavy metals for the two study reservoirs using both the SQG and Hakanson index methods (Table 6). The SQG method revealed that the metal concentrations were within the TEC and PEC ranges for Hg, Cd, Pb and Cu at 69.23%, 7.69%, 38.46% and 61.54% of the sites in the Hongfeng Reservoir, and for Cd, Pb and Cu at 7.69%, 38.46% and 46.15% of sites in the Baihua Reservoir. The heavy metals at the remaining sites in the corresponding reservoirs fell below the TEC. Cr and As exceeded the PEC at 15.38% of sites in the Hongfeng Reservoir and Hg, Cr and As exceeded the PEC at 30.77%, 7.69% and 38.46%, respectively, of sites in the Baihua Reservoir. Previous studies have shown that ecological risk assessments should consider the regional background values and that exceeding the SQG values does not always lead to adverse ecological effects [50]. Therefore, the As concentrations within background ranges in both reservoirs should be excluded. The Cr in the Hongfeng Reservoir and both Hg and Cr in the Baihua Reservoir may pose significant ecological risks for sediment-dwelling organisms, and they deserve special attention. Heavy metals within the TEC–PEC (viz., Hg in Hongfeng Reservoir and Cd, Pb and Cu in both reservoirs) are also a cause for concern because this uncertain area may be considered to be moderately polluted [10]. The toxicity, derived from mean PEC quotients, that results from the mixture of the six heavy metals at each sampling site in both reservoirs is shown in Figure 4a. Overall, mean PEC quotients for samples in the Baihua Reservoir (range 0.28–0.81) were slightly higher than those in the Hongfeng Reservoir (range 0.14–0.55). However, the mean PEC quotients for all of the samples in both reservoirs were well within the range of 0.1 to 1.0, indicating moderate toxicological risks for sediment-dwelling organisms, with a toxicity incidence of between 15 and 29% in the study areas (Table 2).

The Hakanson method expresses the threat to humans from fish consumption. The results from this index were quite different from those for the SQG method (Table 6). Both Hg and Cd posed high potential ecological risks at 7.69% and 15.38% of sites, considerable risks at 38.46% and 23.08% of sites and moderate risks at 23.08% and 53.85% of sites, respectively, in the Hongfeng Reservoir. The risks were higher in the Baihua Reservoir, in which there was a very high potential ecological risk from Hg at 23.08% of sites, high risks from Hg and Cd at 15.38% and 61.54% of sites, considerable risks from Hg and Cd at 7.69% and 15.38% of sites, and a moderate risk from Hg at 15.38% of sites. However, the other heavy metals (Pb, Cr, Cu and As) posed little potential ecological risks for all sites in both reservoirs, with E_r^i values lower than 40. The high concentrations and toxic-response factors of Hg and Cd in both reservoirs contribute to their posing higher ecological risks than the other metals we examined. RI illustrates the potential ecological risk from heavy metal mixtures, and RI at all sites in both reservoirs were higher than 95 (Fig. 4b). Site 1 showed the highest potential ecological risk ($RI = 484.25$) in the Hongfeng Reservoir, at a level that should cause concern because it poses a very high risk. Sites 2, 5 and 7–9 exhibited considerable ecological risks, while other sites showed moderate ecological risks

in the Hongfeng Reservoir. The combined ecological risk was more severe in the Baihua Reservoir. The RI at site 15 (Dongmenqiao tributary) and 23–25 (within the reservoir) were much higher than 380, which indicates a very high potential ecological risk. All of the other sites exhibited considerable ecological risks (except for the moderate ecological risks at site 17 and 18). Therefore, there are moderate to very high potential ecological risks from heavy metal mixtures in the sediments of both reservoirs. In addition, the contribution of the monomial potential ecological risk to RI for the six heavy metals in both reservoirs decreased in the following order: Hg ≈Cd > As > Cu >Pb > Cr, with the greatest ecological risk from Hg and Cd.

Overall, the ecological risks from either a single heavy metal or from mixed heavy metals were different for the two receptors (viz., sediment-dwelling organisms and human beings through fish consumption) in both prior contaminants and risk level. However, hot spots with higher ecological risks were similar even though two different methods were used, and they were mainly located in the North Lake and the Houwu area of the Hongfeng Reservoir and in the key tributaries and at all of the sites in the Baihua Reservoir. Therefore, the need for industrial wastewater and mining tailings treatment in upstream watersheds of both reservoirs should be highlighted, especially for the tributaries in the North Lake of the Hongfeng Reservoir and in the key tributaries of the Baihua Reservoir. Additionally, given that the lakes are sources for drinking water, continuous monitoring should be increasingly implemented in areas near their inflows. Finally, there is uncertainty in both the SQG and Hakanson index methods because the SQGs were developed in North America and the toxic-response factor in the Hakanson method is not very sophisticated. Therefore, further on-site or laboratory toxicological experiments should be carried out to ascertain the actual adverse effects on sediment-dwelling organisms and different fish species [10,50], as well as to determine the impacts on human health from consuming fish from the study area.

Influence of the Hongfeng Reservoir on the Baihua Reservoir

Reservoir construction generally leads to an increase in residence time and a decrease in suspended solids and turbidity. For an alkaline reservoir on the Yunnan-Guizhou Plateau such as the Hongfeng Reservoir, heavy metals tend to be adsorbed to suspended solids, and sediments then settle on the lake bed, resulting in fewer heavy metals in water. Hence, in heavily polluted areas, reservoirs may serve as a sink for pollutants and a buffer for downstream receiving areas. In this study, the catchment area upstream of the Hongfeng Reservoir occupies 84% of the Baihua Reservoir catchment area, and the heavy metal concentrations at site 14, the outlet of the Hongfeng Reservoir and inlet of the Baihua Reservoir, reflect the buffering effect of the Hongfeng Reservoir according to our two field surveys. The metals (Hg, Cd, Cr (VI) and As) in the surface water at site 14 had lower concentrations than they did at the inflow tributary sites of the Hongfeng Reservoir (results not shown). The sediment concentration of Hg was much lower at site 14 than at most sites at the tributaries and within the Hongfeng Reservoir (Fig. 5). He [48] also found that the Hongfeng Reservoir functioned as a net sink for Hg and that it intercepted a large amount of Hg before it was conveyed to the Baihua Reservoir. The sediment concentrations of Cd and As were lower at site 14 than the maximum concentration at the tributaries of Hongfeng Reservoir, although they were generally higher there than at sites within Hongfeng Reservoir (Fig. 5). Due to the different pollution sources (some indirect rather than through inflows), higher concentrations of Pb, Cr and Cu were found at the reservoir sites than at the tributary sites, and the sediment concentrations of Pb, Cr and Cu were generally lower at site 14 than at sites within the Hongfeng Reservoir (Fig. 5). On the other hand, the outflow of the Hongfeng Reservoir has accounted for an average of 70% of the total inflow of the Baihua Reservoir over the last 6 years (2005–2010, provided by the Administration of Hongfeng, Baihua and Aha Reservoirs), implying that the Hongfeng Reservoir may also serve as an important source for total metals in the Baihua Reservoir. For example, the concentrations of the metals (Hg, Cd, Cr (VI) and As) detected in water samples at site 14 fell in the mid-range of the concentrations in the other tributaries of the Baihua Reservoir (results not shown). The concentrations of all of the heavy metals in sediments at site 14 were generally within the concentration ranges of heavy metals in other tributaries of the Baihua Reservoir, with Hg at the low end, Cr in the medium range, and Cd, Pb, Cu and As at the high end (Fig. 5). However, the sediment concentrations of Cd and As at site 14 were generally higher than at sites within the Hongfeng Reservoir. This pattern was not found in the other heavy metals, resulting in uncertainty when considering whether the relatively high concentrations of Cd and As at site 14 came from the Hongfeng Reservoir or not (Fig. 5). Because Cd was mainly from industrial activities and no emission sources exist near site 14 (as determined by the field investigation), the high concentrations at site 14 might result from the pollution of site 1 (Fig. 3) because our sampling sites within the Hongfeng Reservoir did not cover the area near the outlet. In terms of the spatial distribution and the lithogenic source of As, the high concentrations at site 14 might be affected by other factors, such as soil type and land use of the nearby banks. Therefore, for heavy metal concentrations in sediment, the Hongfeng Reservoir might be an important source of Cd, Cu and Pb, and a moderately important source of Cr, but might not be an important source of Hg and As for the Baihua Reservoir. The results also indicate that the Hongfeng Reservoir is not always the most important source for total metals in the Baihua Reservoir, and other tributaries contribute large quantities of pollutants to the Baihua Reservoir, some of which even exceed the levels in the Hongfeng Reservoir. It should also be noted that our field surveys only indirectly reflect the potential long-term impacts from Hongfeng Reservoir on heavy metals in the sediment of the Baihua Reservoir, and continuous monitoring of inflows and outflows of both reservoirs are needed for the specific contribution of the upstream reservoir to the downstream reservoir in future studies.

Other factors contribute to the adverse effects of the Hongfeng Reservoir on the Baihua Reservoir. The Hongfeng Reservoir is a deep reservoir with thermal stratification from May to November [37], and its release water is mainly from the hypolimnion, which has a lower DO concentration, higher CO_2 concentration and lower pH than surface water during those months [21,37]. The water chemistry in the hypolimnion favors the release of heavy metals from the sediments and changes the speciation and toxicity of heavy metals. He et al. [37] found that the low DO and pH in hypolimnion accelerated Hg methylation at Houwu (near site 9) and enhanced the release of methylmercury from sediments at Daba (near the outlet) in the Hongfeng Reservoir in summer. He et al. [37] also concluded that the Hongfeng Reservoir was a net source of methylmercury for the Baihua Reservoir. In addition,

the release of hypolimnetic water has a cooling effect in the summer and a warming effect in winter, which may have a significant influence on the temperature downstream, and thus may indirectly influence the heavy metal distribution. Therefore, the outflow of the Hongfeng Reservoir may pose serious risks to ecosystems in the Baihua Reservoir. Further research is needed to help understand the influence of heavy metals and their chemical forms as they are transported downstream from reservoirs.

Conclusions

This study of heavy metal (Hg, Cd, Pb, Cr, Cu and As) concentrations in surface water and sediments from two adjacent drinking water reservoirs (the Hongfeng and Baihua Reservoirs) on the Yunnan-Guizhou Plateau, Southwest China, showed that surface water was polluted by Hg, and sediments were polluted by Hg, Cd, Pb, Cr and Cu. In both reservoirs, Cd and As mainly came from industrial activities and lithogenic source through inflows, respectively. The Pb, Cr and Cu in Hongfeng Reservoir may have arisen from a mixture of sources (traffic pollution and residual effect of former industrial influence), and they were present at higher concentrations at the junction of the North and South Lakes. Hg sources in the Hongfeng Reservoir might include the sources that contribute Hg through inflows, which were different for the North (industrial activities) and South Lakes (lithogenic origin), and atmospheric deposition resulting from coal combustion. For the Baihua Reservoir, Hg, Cr and Cu were primarily derived from upstream industrial activities, and the Pb originated from traffic pollution. Additionally, the Hg in Baihua Reservoir might have come from atmospheric deposition (coal combustion). Ecological risk was assessed using the SQGs and the Hakanson potential ecological risk index. There were moderate toxicological risks for sediment-dwelling organisms (with the main risks from Hg and Cr) and moderate to very high potential ecological risks for humans from fish consumption (with the main risk coming from Hg and Cd) in both reservoirs. Overall, the risks were higher in the Baihua Reservoir. Improved treatment of industrial wastewater and mining tailings in upstream watersheds would alleviate the pollution and ecological risk in both reservoirs, especially for tributaries of the North Lake of the Hongfeng Reservoir and the key tributaries of the Baihua Reservoir. Ecological restoration could be considered to counteract the residual effects from previous pollution; however, more research is needed in this area. In terms of heavy metal concentrations, the Hongfeng Reservoir acts as a buffer, but it is still an important source of Cd, Cu and Pb and a moderately important source of Cr for the Baihua Reservoir. The Hongfeng Reservoir also had adverse effects on the Baihua Reservoir and merits further research. These findings provide useful information about sediment quality in adjacent reservoirs on the Yunnan-Guizhou Plateau.

Author Contributions

Conceived and designed the experiments: GQW CML. Performed the experiments: BBW GQW JW QF. Analyzed the data: BBW JW. Contributed reagents/materials/analysis tools: BBW JW QF. Wrote the paper: BBW GQW.

References

1. Liu WX, Li XD, Shen ZG, Wang DC, Wai OWH, et al. (2003) Multivariate statistical study of heavy metal enrichment in sediments of the Pearl River Estuary. Environ Pollut 121: 377–388.

2. Chabukdhara M, Nema AK (2012) Assessment of heavy metal contamination in Hindon River sediments: A chemometric and geochemical approach. Chemosphere 87: 945–953.

3. Yi YJ, Yang ZF, Zhang SH (2011) Ecological risk assessment of heavy metals in sediment and human health risk assessment of heavy metals in fishes in the middle and lower reaches of the Yangtze River Basin. Environ Pollut 159: 2575–2585.

4. Uluturhan E, Kucuksezgin F (2007) Heavy metal contaminants in Red Pandora (Pagellus erythrinus) tissues from the Eastern Aegean Sea, Turkey. Water Res 41: 1185–1192.

5. Singh KP, Mohan D, Singh VK, Malik A (2005) Studies on distribution and fractionation of heavy metals in Gomti river sediments-a tributary of the Ganges. J Hydrol 312: 14–27.

6. Larsen B, Jensen A (1989) Evaluation of the sensitivity of sediment monitoring stationary in pollution monitoring. Mar Pollut Bull 20: 556–560.

7. Li XD, Wai OWH, Li YS, Coles BJ, Ramsey MH, et al. (2000) Heavy metal distribution in sediment profiles of the Pearl River estuary, South China. Appl Geochem 15: 567–581.

8. Agarwal A, Singh RD, Mishra SK, Bhunya PK (2005) ANN-based sediment yield river basin models for Vamsadhara (India). Water SA 31: 95–100.

9. Loska R, Wiechula D (2003) Application of principal component analysis for the estimation of source of heavy metal contamination in surface sediments from the Rybnik Reservoir. Chemoshere 51: 723–733.

10. Larrose A, Coynel A, Schafer J, Blanc G, Masse L, et al. (2010) Assessing the current state of the Gironde Estuary by mapping priority contaminant distribution and risk potential in surface sediment. Appl Geochem 25: 1912–1923.

11. Bai JH, Cui BS, Chen B, Zhang KJ, Deng W, et al. (2011) Spatial distribution and ecological risk assessment of heavy metals in surface sediments from a typical plateau lake wetland, China. Ecol Model 222: 301–306.

12. Persaud D, Jaagumagi R, Hayton A (1993) Guidelines for the protection and management of aquatic sediment quality in Ontario. Water Resources Branch. Ontario Ministry of the Environment, Toronto, 27.

13. Smith SL, MacDonald DD, Keenleyside KA, Ingersoll CG, Field J (1996) A preliminary evaluation of sediment quality assessment values for freshwater ecosystems. J Great Lakes Res 22: 624–638.

14. Ingersoll CG, Haverland PS, Brunson EL, Canfield TJ, Dwyer FJ, et al. (1996) Calculation and evaluation of sediment effect concentrations for the amphipod Hyalella azteca and the midge Chironomus riparius. J Great Lakes Res 22: 602–623.

15. Macdonald DD, Ingersoll CG, Berger TA (2000). Development and evaluation of consensus-based sediment quality guidelines for freshwater ecosystems. Arch Environ Contam Toxical 39: 20–31.

16. Word JQ, Albrecht BB, Anghera ML, Baudo R, Bay MS, et al. (2002). Predictive ability of sediment quality guidelines. In: Wenning RJ, Batley GE, Ingersoll CG, Moore DW, editors. Use of sediment quality guidelines and related tools for the assessment of contaminated sediments. Pensacola (FL): SETAC. p 121–162.

17. Bhavsar SP, Gewurtz SB, Helm PA, Labencki TL, Marvin CH, et al. (2010) Estimating sediment quality thresholds to prevent restrictions on fish consumption: Application to polychlorinated biphenyls and dioxins–furans in the Canadian Great Lakes. Integr Environ Assess Manag 6: 641–652.

18. Hakanson L (1980) An ecological risk index for aquatic pollution control: A sedimentological approach. Water Res 14: 975–1001.

19. Liu Y, Guo HC, Yu YJ, Huang K, Wang Z (2007) Sediment chemistry and the variation of three altiplano lakes to recent anthropogenic impacts in southwestern China. Water SA 33: 305–310.

20. Feng XB, Jiang HM, Qiu GL, Yan HY, Li GH, et al. (2009a) Mercury mass balance study in Wujiangdu and Dongfeng Reservoirs, Guizhou, China. Environ Pollut 157: 2594–2603.

21. Wang FS, Wang BL, Liu CQ, Wang YC, Guan J, et al. (2011a) Carbon dioxide emission from surface water in cascade reservoirs-river system on the Maotiao River, southwest of China. Atmos Environ 45: 3827–3834.

22. Wang FS, Liu CQ, Wang BL, Liu XL, Li GR, et al. (2011b) Disrupting the riverine DIC cycling by series hydropower exploitation in Karstic area. Appl Geochem 26: S375–S378.

23. Feng XB, Jiang HM, Qiu GL, Yan HY, Li GH, et al. (2009b) Geochemical processes of mercury in Wujiangdu and Dongfeng reservoirs, Guizhou, China. Environ Pollut 157: 2970–2984.

24. Han GL, Liu CQ (2004) Water geochemistry controlled by carbonate dissolution: a study of the river waters draining karst-dominated terrain, Guizhou Province, China. Chem Geo 204: 1–21.

25. Wang B, Liu CQ, Wu Y (2005) Effect of heavy metals on the activity of external carbonic anhydrase of microalga chlamydomonas reinhardtii and microalgae from karst lakes. Bull Environ Contam Toxicol 74: 227–233.

26. Lang YC, Liu CQ, Zhao ZQ, Li SL, Han GL (2006) Geochemistry of surface and ground water in Guiyang, China: Water/rock interaction and pollution in a karst hydrological system. Appl Geochem 21: 887–903.

27. He SL, Li CJ, Pan ZP, Luo MX, Meng W, et al. (2012) Geochemistry and environmental quality assessment of Hongfeng Lake sediments, Guiyang. Geophysical and Geochemical Exploration 36: 273–297 (in Chinese).

28. Wang Y, Luo TMZ (2001) Geostatistical and geochemical analysis of surface water leakage into groundwater on a regional scale: a case study in the Liulin karst system, northwestern China. J Hydrol 246: 223–234.

29. Sophocleous M (2002) Interactions between groundwater and surface water: the state of the science. Hydrogeol J 10: 52–67.

30. Feng XB, Foucher D, Hintelmann H, Yan HY, He TR, et al. (2010) Tracing mercury contamination sources in sediments using mercury isotope compositions. Environ Sci Technol 44: 3363–3368.

31. Yan HY, Feng XB, Shang LH, Qiu GL, Dai QJ, et al. (2008) The variations of mercury in sediment profiles from a historically mercury-contaminated reservoir, Guizhou province, China. Sci Total Environ 407: 497–506.

32. Kaiser HF (1960) The application of electronic computers to factor analysis. Educ Psychol Measure 20: 141–151.

33. Liu CW, Lin KH, Kuo YM (2003) Application of factor analysis in the assessment of groundwater quality in a Blackfoot disease area in Taiwan. Sci Total Environ 313: 77–89.

34. Macdonald DD, Carr RS, Calder FD, Long ER, Ingersoll CG (1996) Development and evaluation of sediment quality guidelines for Florida coastal waters. Ecotoxicology 5: 253–278.

35. Swartz RC (1999) Consensus sediment quality guidelines for PAH mixtures. Environ Toxicol Chem 18: 780–787.

36. Long ER, Ingersoll CG, Macdonald DD (2006) Calculation and uses of mean sediment quality guideline quotients: a critical review. Environ Sci Technol 40: 1726–1736.

37. He TR, Feng XB, Guo YN, Qiu GL, Li ZG, et al. (2008) The impact of eutrophication on the biogeochemical cycling of mercury species in a reservoir: A case study from Hongfeng Reservoir, Guizhou, China. Environ Pollut 154: 56–67.

38. Bai ZG, Wan GJ, Liu TS, Huang RG (2002) A comparative study on accumulation characteristics of ^{7}Be and ^{137}Cs in sediments of Lake Erhai and Lake Hongfeng, China, Geochimica 31: 113–118 (in Chinese).

39. Huang XF (2008) Studies on characteristics of pollution in sediments from Baihua Lake. Guiyang (in Chinese).

40. Huang XF, Qin FX, Hu JW, Li CX (2008) Characteristic and ecological risk of heavy metal polltuion in sediments from Hongfeng Lake. Res Environ Sci 21: 18–23 (in Chinese).

41. Zeng Y, Zhang W, Chen JA, Zhu ZJ (2010) Analysis of heavy metal pollution in the sediment of the inflow-lake rivers of the Hongfeng Lake, Earth Environ 38: 470–475 (in Chinese).

42. Liu F, Hu JW, Wu D, Qin FX, Li CX, et al. (2011) Speciation characteristics and risk assessment of heavy metals in sediments from Hongfeng Lake, Guizhou Province. Environmental Chemistry 30: 440–446 (in Chinese).

43. Tian LF, Hu JW, Luo GL, Ma JJ, Huang XF, et al. (2012) Ecological risk and stability of heavy metals in sediments from Lake Baihua in Guizhou Province. Acta Scientiae Circumstantiae 32: 885–894 (in Chinese).

44. NEPA: National Environmental Protection Agency (Presently known as MEP; Ministry of Environmental Protection) (1994) The Atlas of Soil Environmental Background Value in the People's Republic of China. China Environmental Science Press.

45. Hakanson L, Jasson M (1983) Principles of Lake Sedimentology. Springer Verlag, Berlin.

46. Bai JH, Cui BS, Wang QG, Gao HF, Ding QY (2009) Assessment of heavy metal contamination of roadside soils in Southwest China. Stoch Environ Res Risk Assess 23: 341–347.

47. Zhu ZM, Li ZG, Bi XY, Han ZX, Yu GH (2013) Response of magnetic properties to heavy metal pollution in dust from three industrial cities in China. J Hazard Mater 246–247: 189–198.

48. He TR (2007) Biogeochemical cycling of mercury in Hongfeng Reservior, Guizhou, China. Guiyang (in Chineses).

49. SEPA: State Environmental Protection Administration (Presently known as MEP; Ministry of Environmental Protection) (2000) Report on the State of the Environment in China. Available: http://english.mep.gov.cn/SOE/soechina2000/english/atmospheric/atmospheric_e.htm.

50. Farkas A, Claudio E, Vigano L (2007) Assessment of the environmental significance of heavy metal pollution in surficial sediments of the River Po. Chemosphere 68: 761–768.

2

What Magnitude Are Observed Non-Target Impacts from Weed Biocontrol?

David Maxwell Suckling[1,2]*, René François Henri Sforza[3]

1 Biosecurity Group, The New Zealand Institute of Plant and Food Research Ltd, Christchurch, New Zealand, 2 Better Border Biosecurity, Christchurch, New Zealand, 3 European Biological Control Laboratory, USDA-ARS, Campus International de Baillarguet, Montferrier-sur-Lez, France

Abstract

A systematic review focused by plant on non-target impacts from agents deliberately introduced for the biological control of weeds found significant non-target impacts to be rare. The magnitude of direct impact of 43 biocontrol agents on 140 non-target plants was retrospectively categorized using a risk management framework for ecological impacts of invasive species (minimal, minor, moderate, major, massive). The vast majority of agents introduced for classical biological control of weeds (>99% of 512 agents released) have had no known significant adverse effects on non-target plants thus far; major effects suppressing non-target plant populations could be expected to be detectable. Most direct non-target impacts on plants (91.6%) were categorized as minimal or minor in magnitude with no known adverse long-term impact on non-target plant populations, but a few cacti and thistles are affected at moderate (n = 3), major (n = 7) to massive (n = 1) scale. The largest direct impacts are from two agents (*Cactoblastis cactorum* on native cacti and *Rhinocyllus conicus* on native thistles), but these introductions would not be permitted today as more balanced attitudes exist to plant biodiversity, driven by both society and the scientific community. Our analysis shows (as far as is known), weed biological control agents have a biosafety track record of >99% of cases avoiding significant non-target impacts on plant populations. Some impacts could have been overlooked, but this seems unlikely to change the basic distribution of very limited adverse effects. Fewer non-target impacts can be expected in future because of improved science and incorporation of wider values. Failure to use biological control represents a significant opportunity cost from the certainty of ongoing adverse impacts from invasive weeds. It is recommended that a simple five-step scale be used to better communicate the risk of consequences from both action (classical biological control) and no action (ongoing impacts from invasive weeds).

Editor: Frederic Marion-Poll, AgroParisTech, France

Funding: This project was partially funded by the New Zealand Ministry of Business, Innovation and Employment (CO2X0501 Better Border Biosecurity, www. b3nz.org), and completed during a fellowship to DMS supported by the Organisation for Economic Cooperation and Development (Cooperative Research Programme: Biological Resource Management for Sustainable Agricultural Systems) at the European Biological Control Laboratory (USDA ARS), Montpellier. The funders had no role in study design, data collection and analysis, decision to publish, or preparation of the manuscript.

Competing Interests: The authors have declared that no competing interests exist.

* E-mail: Max.Suckling@plantandfood.co.nz

Introduction

Classical biological control of weeds involves the deliberate introduction of exotic organisms, or biological control agents, to manage weed problems in the invaded range. It offers an excellent and sustainable solution for invasive species [1,2]. Exotic weeds in natural and managed ecosystems have long been targeted, starting with the cases of prickly pear (*Opuntia* sp.) in India (1863) [3], then Sri Lanka (1865) [4], and Australia (1912) [3], and lantana (*Lantana camara* L.) in Hawai'i in 1902 [1]. After some assessment of cost-benefit ratio, the process involves collecting exotic natural enemies to control a target invasive weed, usually followed by importing, rearing, testing, and release from quarantine for establishment. Host specificity tests are conducted in artificial and field conditions, and increasingly combined with ecological and molecular evaluations [5]. Deliberate release of natural enemies is subject to official approvals.

Reported benefits in USA from the major weed biocontrol programs in the 20th century resulted in benefits (net of research costs) in excess of US$180M per annum [1], mainly from reduced ongoing costs of control using herbicides. Environmental benefits

of replacing pesticides can be considered to be proportional in magnitude to market economy benefits [6]. In South Africa, biocontrol of weeds contributes to prevention of substantial losses to the economy over the scale of decades, where it prevents the loss of ecosystem services that contribute to human well-being, including water [7]. Highly favorable results have emerged from similar analyses in Australia [8,9] and New Zealand [10–13]. Plant invasion continues to be a major concern nonetheless, with a lag phase of several decades, and new introductions further increasing net effects from the increase of global trade [14].

The increasing incidence and impact of invasive species is widely recognized as a major and increasing threat to food and fiber production, as well as ecosystem functioning [15], so it could be assumed that the need for classical biological control to mitigate costs is increasing. However, despite an increasing track record of success and specificity with improved scientific knowledge [16], classical biological control has been criticized in recent years, through emerging recognition of non-target impacts [17–20]. Solutions are clearly needed to better predict the risk of significant non-target impacts in order to gain societal, economic and environmental benefits, while mitigating risk. The obvious major risk is that of a host shift, or the preference for an adopted host (an indigenous species or a crop in the introduced environment), over the original host (the target). The threat is either to a native plant species at population level and to ecosystem function, or to a crop, by defoliation or seed predation resulting in a yield reduction [21,22]. The risks arise in this scenario because the newly-released organisms are self-perpetuating and self-dispersing, but these traits also offer the benefit of self-sustaining management [23]. That said, comparisons of realized host range with the predicted host range [24,25] can improve biosafety processes. However, a lack of agreement between retrospective laboratory tests and long-term field observations has led to the conclusion that very successful biological control agents without non-target impacts might never have been introduced because of overstated ecological risk in the laboratory [26]. Clearly both types of errors should be avoided from host range tests, where ecologically safe candidates are not released and the benefits of sustainable pest control are not realized, or, unsafe candidates are released and ecological damage results.

Until 2000, the frequency of cases of known non-target impacts from classical biological control of weeds was small, compared with the number of agents released [21,22]. Fowler et al. [22] reported that 12 biocontrol agents released against weeds had been recorded attacking non-target plants. Six of these cases (1.5% of agents released) were not anticipated. However, while this insect-centric perspective appears to offer some support for classical biological control, it has also been noted that release of thistle seed weevil *Rhinocyllus conicus* (Col. Curculionidae) in North America has led to about half of the (previously) recorded non-target impacts [27]. The release was undertaken during an era when rangeland management of economically important thistles was overriding. Crops were highly valued compared to natural values of indigenous North American thistles in the 1960s. As a consequence, incorporation of ecological considerations was limited [28]. Interestingly, there is no evidence of non-target impacts from plant pathogens thus far [29,30], but it may be too early to tell whether the organisms chosen are more host specific and therefore have a lower risk profile, since there have been far fewer introductions of plan pathogens so far. This field appears to offer good opportunity to avoid the mistakes of the past.

No overview of weed biological control studies has yet evaluated the of adverse non-target impact of all agents once released, separating effects reducing plant populations at ecological scale from effects which don't have such implications. In order to use an existing framework for such study, we followed Parker et al. [15] who suggested that the impact of an invader can be measured at five levels: (1) genetic effects, (2) effects on individuals (including demographic rates such as mortality and growth), (3) population dynamic effects (abundance, population growth), (4) community effects (species richness, diversity, trophic structure), and (5) effects on ecosystem processes (nutrient availability, primary productivity). The genetic effects are rather a special case, although the risk of hybridization with a native congener or other existing biological control agent can exist [31]. The remaining effects form a hierarchy of increasing impact from minimal to massive, detailing for each of the 5 descriptors, an impact or not, at every level from the individual plant to the ecosystem. Successful weed biological control can have indirect beneficial effects such as increased economic productivity, restored community or vegetation structure and ecosystem processes, and improved management effectiveness [32].

The risk assessment for weed biological control agents has seen standards rise over time, with increasing conservatism due to factors such as the Convention on Biodiversity [33]. Risk

assessment also varies between jurisdictions. One of the most highly-regarded regimes is that in New Zealand [33], under the Hazardous Substances and New Organisms Act (1996) [34]. The risk assessment for non-target impacts includes a consideration of beneficial and adverse effects. We have limited our consideration to adverse non-target impacts, which should logically take into account the impacts on individual plant taxa, irrespective of how many agents have been involved. We have reviewed the reported non-target impacts on plant species and assessed their magnitude of adverse impact on a five step scale that we have adapted from use with invasive species.

Methods

Updating the number of biological control agents released

In the 20th century, 1,120 releases of 365 species of biological control agents were made against 133 weeds in 75 countries [35], predominantly USA, Canada, Australia, South Africa and New Zealand. We reviewed the literature and contacted experts to identify a further 147 agents (Table 1), generating a new total of 512 organisms released for weed biological control, to May 2012.

Risk assessment scale

The Environmental Risk Management Authority of New Zealand (1996–2011) (and its successor, the Environmental Protection Authority, 2011-) uses a five step scale for risk assessment of new organisms such as weed biological control agents [36], for which we have proposed modifications to the accompanying text (Table 2), generated from known types of ecological consequences of invasive species [15]. Further, we propose that only items at moderate or above impact are to be considered "significant", since these definitions are based on plant populations declining, which we believe is a crucial point. Short-term impacts on individual plants with recovery should not receive the same weighting as impacts involving plant population decline, and this point is treated in detail here.

Retrospective application of the risk assessment scale

A systematic review was conducted of the specialized entomological literature sourced entries for a database of non-target impacts by plant taxon, agent and geographic location, from reviews [21,22,27,37,38] and primary peer-reviewed reports. In our search for literature to May 2012 we used as descriptors (weeds OR aquatic weeds OR weed control) AND (nontarget organisms OR nontarget effects OR host range OR host preferences OR host specificity OR risk) AND (biological control OR biological control agents) AND (insects). Filtering of results on the agents was recorded in a modified PRISMA chart (Preferred Reporting Items for Systematic Reviews and Meta-Analyses) (Fig. 1) [39], with results recorded by agent and plant taxa in a database (Table S1). Adverse effects were assessed where sufficient information was available (see citations) and each case was assigned a level of adverse impact within a five-step scale from "minimal to massive", based on Table 2. It was recognised that some minor effects might have been overlooked, but that non-target plant population suppression (moderate, major or massive impact) would probably be observed. Some plants or agents were included more than once as separate cases, with impacts at different magnitudes in different locations, due to parasitism or factors affecting non-target species abundance. Other attributes of each case were recorded, including year of introduction, evidence for the presence of a self-sustaining population, and type of plant (weed, native weed, crop, valued plant). Cases considered of

Table 1. Updated list of classical biological control agents released against weeds, since Julien and Griffiths [35].

Country	Insects	Mites	Pathogens	Nematodes	Total	Source
South Africa	32	1	3		36	[74]
Canada	11				11	[75]; R. Bourchier pers. com.
New Zealand	18				18	[34,49,76]
Australia	42	3	6		51	R. Winston, pers. com.
European Union	2				2	[77,78]
USA/Hawai'i	24	1	3	1	29	E. Coombs, pers. com.;[79]
Total	129	5	12	1	147	

negligible impact were not included (i.e. below minimal), such as the feeding on sunflower (reported in McFadyen [21] requiring cues from pollen [40], and treehoppers present but not feeding on many plants in Brisbane [41]. Self-sustaining populations did not include cases where herbivores established during an initial population explosion from nearby hosts but declined thereafter (e.g. the eriophyid mite *Aculus hyperici* on *Hypericum gramineum* [42]). These cases were considered to be minimal in magnitude, as a result of short-term non-target effects. We noted that some insects considered to be biological control agents were initially self-introduced, but were later deliberately distributed (e.g. *Larinus planus* (F.) Col. Cucurlionidae arrived from Europe by 1971, later released in Canada and the western US). Our analysis of the case of cactus moth *Cactoblastis cactorum* (Bergroth) includes the deliberate introduction to the Caribbean in the presence of native *Opuntia* species which were considered to be weeds in the 1950s, and where impacts have been recorded from its inadvertent and possibly natural spread to Florida [43].

Assessment of Indirect Effects

We also attempted a separate exercise on adverse indirect effects on ecosystems, but this was more difficult as there was less information, and most cases have only weak evidence, and may not be enduring enough to warrant inclusion (Table 2). It has been recognised for many years that removal of a weed through biological control may lead to either revegetation with native species or simply a change in weed species [44], and this effect was the most common source of indirect effects reported. For indirect effects [32], we have considered cases with an increased abundance of exotic species, only where weed problems were exacerbated. We again used Table 2; by our definition any indirect ecosystem effects start at a magnitude of moderate, with habitat modification.

Results

Effects on plants

Non-target effects on plants were recorded on a total of 193 cases affecting 152 plant taxa, of which 140 cases on 116 plant taxa were adequate for assessment of magnitude (Fig. 1), from 43 arthropods of 512 used as classical biological control agents of weeds. The details of each case that formed the basis of our assignments can be traced through Table S1, where details are limited here. Case studies to illustrate each magnitude are shown in Table 3, including potentially massive level adverse impacts judged to be underway on one *Opuntia* species in Florida, so far, from *Cactoblastis cactorum*. Major effects were assessed as underway on five other cactus species (*O. cubensis* Britton & Rose, *O. humifusa*

Table 2. Proposed scale for retrospectively assessing the magnitude of adverse environmental effects from biological control introductions.

Descriptor	Effects on individuals	Population dynamic effects	Community effects	Effects on ecosystem processes
Minimal	Feeding on non-target occasionally recorded, little successful development	-	-	-
Minor	Feeding damage	Seasonal feeding on non-target of <50% individuals, plant recovery	-	-
Moderate	Impact on fitness	Self sustaining population established on non-target, plant reproduction affected at population level	Minor detrimental habitat modification, or adverse effects on other biocontrols	-
Major	Plants killed and reduced reproduction	Impact on plant population readily detectable	Habitat modification detectable, impact on other organisms detectable	Minor effects on ecosystem processes
Massive	Plants killed before reproduction	Heavy impact and rapid population decline, species loss	Change in habitat structure of keystone species	Plant succession affected, changes to vegetation cover, loss of keystone species, ecosystem disruption

It is based on the system used in New Zealand under the HSNO Act (1996) for consideration of future risk following new organism introductions, redefined after Parker et al. [15].

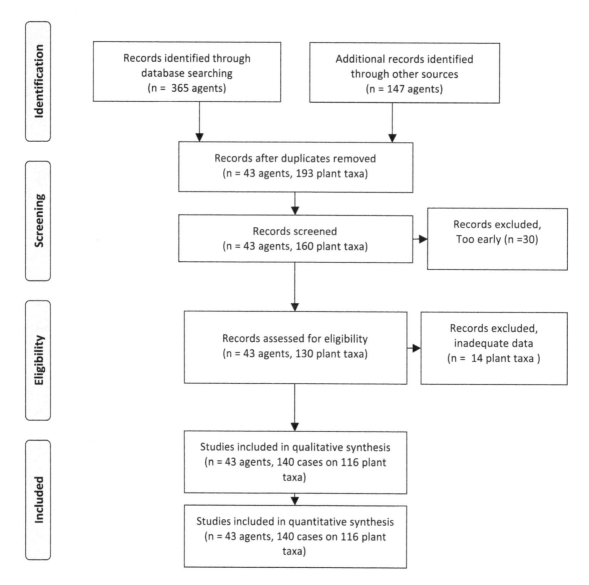

Figure 1. Modified PRISMA flow chart used in the systematic review process [39] for non-target impacts from classical biological control of weeds.

(Raf.) Raf., *O. stricta* (Haw.) Haw., *O. triacantha* (Willd.) Sweet, *O. cochenillifera* (L.) Mill.) in either Florida or Nevis and St Kitts [45], as well as on one thistle from thistle seed weevil (Table 3). Effects were assessed as moderate on two other thistles: *Cirsium undulatum* (Nutt.) Spreng. from *Larinus planus* and on *C. altissimum* L. Spreng. from *Trichosirocalus horridus* (Table 2), as well as effects from *C. cactorum* on *Opuntia triacantha* in Nevis and St Kitts [45]. All other impacts that could be assessed were judged as minor or minimal, with no enduring adverse effect on non-target plant populations (Table S1). About 8% of the non-target host plants with reported effects had above minimal-minor adverse impacts, which would be likely to affect plant reproduction (Fig. 2). Percentages in each magnitude are over the total number of non-target plant taxa (N = 140).

A few non-target plants were affected by more than one insect due to multiple introductions against related plants (e.g. *Rubus hawaiiensis* was judged to have had minimal impact from *Priophorus morio* and *Croesia zimmermani*, and minor impact from *Schreckensteinia festaliella* introduced to target other *Rubus*). It is clearly undesirable to have non-target herbivory effects accumulating.

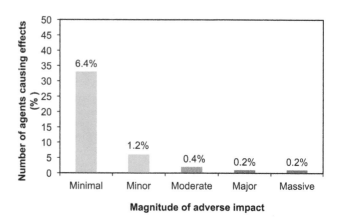

Figure 2. Number of biological control agents causing adverse impacts on non-target plant taxa, by magnitude.

Table 3. Examples of each magnitude of non-target impact on plants from weed classical biological control agents.

	Target species	Non-target species	Cause & predictability	Potential threat	References
Minimal	*Ulex europaeus* L.	*Genista monspessulana* (L.) L.A.S. Johnson	Deliberate release of two populations of *Cydia succedana* (Denis and Schiffermüller), one showed limited development on weeds including this one; Predictable	None, majority of examples	[24,61]
Minor	*Hypericum perforatum* L.	*Hypericum concinnum* Benth	Deliberate release of *Chrysolina quadrigemina* in California causing damage and varying impact on *H. concinnum*, it is still common; Predictable	None, some examples	[80]
Moderate	*Carduus nutans* L.	*Cirsium altissimum* L. Spreng	Deliberate release of *Trichosirocalus horridus* causing damage on non-target native thistles; Predictable	Too early to tell; impact is uncertain (moderate impact may be too high); rare.	[81,82]
Major	*Carduus nutans* L.	*Cirsium canescens*, Nutt.& *Carduus*	Deliberate release of *Rhinocyllus conicus* in the U.S. mainland following host range testing proving its safety for crops. Evidence of local population decline of *Cirsium canescens*; Predictable	High likelihood of some attacks on ~28 species of native thistles; otherwise rare.	[19,83]
Massive	*Opuntia lindheimeri*, Engelm., *O. stricta*, (Haw.) Haw., *O. triacantha* (Willdenow) Sweet	*Opuntia spinosissima* P.Mill.	Accidental release of *Cactoblactis cactorum* in the U.S. mainland following introduction against cacti in the Caribbean; severe feeding impact (threat of extinction without intervention?); Predictable	High likelihood of attacks on ~87 native cacti species (too early to tell for most); rare.	[27,45,46,70]

It was considered too early to tell the magnitude of impact for cactus moth attack on most *Opuntia* species in North America, although host range testing appears to place several species at major to massive risk (*Opuntia engelmannii* Salm-Dyck ex Engelm. var. engelmannii, *Opuntia engelmannii* Salm-Dyck ex Engelm. var. linguiformis (Griffiths) Parfitt & Pinkava, *O. ficus-indica* (L.) Mill., *O. stricta* (= *Opuntia dillenii*), *O. triacantha*), while several other species are likely to have plant resistance (*Consolea rubescens* (Salm-Dyck ex de Candolle), *Cylindropuntia acanthocarpa* (Engelmann and Bigelow) F.M. Knuth, *C. spinosior* (Engelmann), *O. leucotricha* de Candolle and *O. streptacantha* Lem.) [46].

Effects of insects

Of 512 agents released (Table 1 plus [35]), 91.6% of agents have had no known or recorded non-target impact (Fig. 3). Minimal impact (33 agents) or minor impact, with no reduction in plant population (6 agents) occurred with a further 7.6% of agents. Thus non-target plant populations were only adversely affected (moderate-massive range) from 0.8% of agents (n = 4 of 512). The four insect species accounted for all significant adverse effects on non-target plant populations (i.e. moderate-massive, Fig. 4); all were thistles or cacti and within the same genus as the target host plant. Of these, only two were deliberate introductions to the places where they have caused harm (*R. conicus* released in 1969 and thistle rosette weevil *Trichosirocalus horridus* released in 1974), with predictable outcomes that resulted from an earlier era of lower standards of biosafety than prevail today.

A total of 108 of 140 cases of recorded non-target impacts (77%) were in the same plant family as the target weed. About half (54%) were in the same genus as the target (Fig. 4). A few cases of minimal or minor non-target impacts occurred outside the host genus (stars, Fig. 4), or outside the host family (shaded bars, Fig. 4), but impacts in a different plant family are not known to cause plant populations to decline over time (i.e. have not been reported to our knowledge). All of the effects on crops and valued plants (such as ornamentals) were minimal in magnitude.

For deliberate releases, there were no massive effects determined. Major effects (n = 3 plant taxa) only occurred from deliberate releases in the 1950s–60s in Nevis and St Kitts (*C.*

cactorum on *O. stricta* (Haw.) Haw. and *O. cochenillifera* (L.) Mill., and *Rhinocyllus* on *C. canescens* Nutt.) (Fig. 5). Moderate effects occurred from three releases in the period 1958–1988 (n = 3 plant taxa, *C. cactorum* on *O. triacantha* (Willd.) Sweet, *Larinus* on *C. undulatum* (Nutt.) Spreng. and *Trichosirocalus* on *C. altissimum* L. Spreng. Minor effects (n = 39) occurred from releases in the period 1945–1992, while minimal effects occurred from releases in the period 1902–2001 (n = 71). In the case of both *Cactoblastis* and *Rhinocyllus*, a range from major to minimal impacts occurred on different plant taxa. Plant families varied in frequency of reported non-target impacts (Table 4). Table 4 gives a historical view of families with any negative impact from released biocontrol agents but i) does not reflect genetic linkages between plant taxa, and ii) does not rank risk of adverse impact between plant families. The potential obviously exists to further investigate the types of insects and plants showing any adverse effects, including minimal and minor effects, since these cases could be a harbinger of future problems.

Four cases of indirect non-target impacts were of moderate to major impact, although two of these we regarded as hypothetical or in progress, in our view awaiting better evidence at the time of

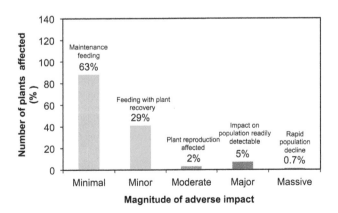

Figure 3. Number of plant taxa with non-target impacts from weed biological control agents, by magnitude.

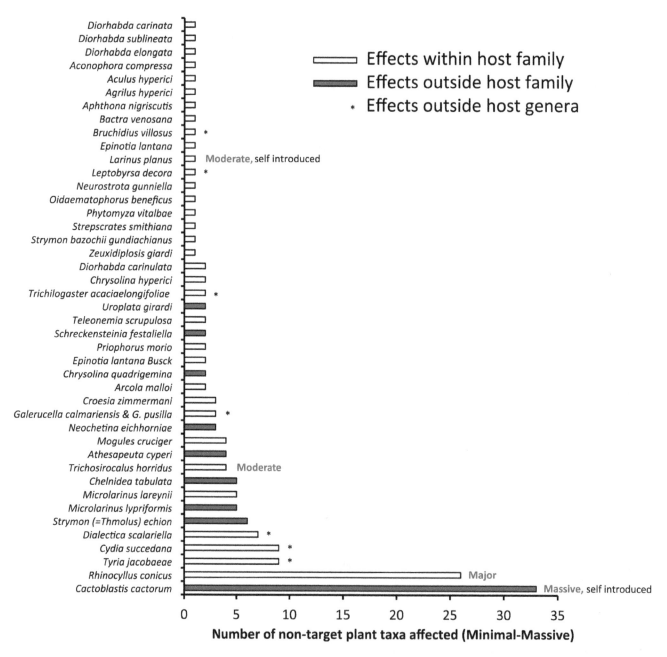

Figure 4. Number and phylogenetic proximity of non-target plant taxa known to be affected by weed biological control agents (minimal-massive).

writing (Table 5). Some cases involve the agent being effective as originally intended, and moderate effects have resulted from changes in plant cover. We have not included a somewhat similar case [47], as it was unclear whether the removal of ragwort exacerbated the thistles. One case (*Agapeta zoegana*) [48] did not meet our threshold for evidence of a real adverse ecological effect.

Assembling the meta-analysis

No known direct or indirect non-target impacts were found from 91.6% of agents released (Fig. 6). However, some of these agents have failed to establish, representing the risk of failure to achieve benefits. For example, about 36% of agents failed to establish in the history of weed biological control in New Zealand [13]. This risk of failure to get benefits may be declining [49]. Of

those that established and had non-target impacts, the majority of these were minimal or minor impacts that had no effect on plant population density. The majority of observed "effects" when considered by plant or by agent are actually in the no effect zone, when impact on non-target plant populations is considered. This leaves a few cases of impacts on plants from four introductions made some decades ago, where reasonably serious adverse non-target effects have been shown within the host genus. Two were deliberate introductions (and predictable) and two were not deliberate. In all of these cases (including those with minimal to minor impact), insect host range mainly spanned genera, although some had lower levels of non-target effects from limited host use outside host families. The benefit side of the equation has not been studied for the full range of agents globally, although two successes

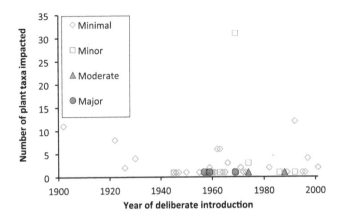

Figure 5. Year of deliberate introduction of arthropods used as weed biological control agents, sorted by magnitude of non-target impact on plant taxa.

Table 4. Plant taxa with any non-target impact recorded from classical biological control agents of weeds, sorted by family.

	Number of plant taxa
Asteraceae	47*
Cactaceae	31*
Fabaceae	12
Boraginaceae	7
Cyperaceae	5
Rosaceae	5
Clusiaceae	4
Zygopyhyllaceae	4
Lythraceae	3
Amaranthaceae	2
Hypericaceae	2
Verbenaceae	2
Euphorbiaceae	1
Myricacae	1
Pontederiaceae	1
Ranunculaceae	1

*significant impacts (moderate to massive) occurred within plant families.

with weed biological control were massive in beneficial effect in New Zealand, due to long term ecosystem removal of the target; 24% of New Zealand cases gave some clear benefit against target weeds (moderate-massive) [13].

Discussion

Effects on plants

Host range was found to be predictable for biological control agents released over the last century. The risk of host shifting has not been realised with any significant evidence. Population explosions have sometimes occurred at the initial phase of establishment, and occasionally crossed plant family boundaries. Examples of short-term spillovers include local feeding on melons and tomato by the prickly pear biological control agent *Cactoblastis cactorum*, after an initial population explosion and collapse of cacti [50], and *Teleonemia scrupulosa* (Hem.: Tingidae) on sesame when first introduced in East Africa against lantana [51]. For non-target plants receiving minimal or minor impacts, some species were reportedly capable of supporting self-sustaining populations (12/19), but this information was largely missing. This would seem to be a desirable standard for new host records. The effect of spillover onto other plants from initial large insect populations has generally not led to long term impacts on plant population levels, and was rated minimal or minor in magnitude in the vast majority of cases. The exceptions were cases involving four insects (*Cactoblastis*, *Rhinocyllus*, *Larinus* and *Trichosirocalus*). These cases illustrate the problem, but are not typical.

Effects of agents

We suggest that weed biological control has a 150 year historical biosafety track record of >99%, as far as is known. We acknowledge that unobserved impacts are possible because of the lack of post-release evaluation studies in most biocontrol programs [52], and we agree that greater efforts are needed to follow up previous introductions for improving assessment of both benefit and risk in future cases. However, the presence of completely unobserved moderate or greater long-term impacts seem unlikely in the majority of cases of releases. The importance of an atypical few (but very frequently cited) cases of major to massive potential impact is partly due to their large number of non-target hosts, as well as a general desire to avoid such non-target impacts. Of the four insects which have caused significant adverse impacts on plant populations, two were not even originally

deliberately introduced to the areas most affected. Impacts from the two deliberately introduced insects that have caused impacts (*Rhinocyllus* and *Trichosirocalus*) were foreseeable [27] and would not be permitted today. Not all cases with lower level impacts (minimal-minor) were necessarily within the same family and perhaps predictable, but of the cases with observed non-target feeding across plant families, all were minimal in magnitude. The original host range testing was, in a few cases, inadequate by modern standards, and later testing revealed that the non-target attack was predictable [53]. Retrospective analysis of predictability is not usually an easy task as laboratory host specificity tests are rarely published, but it has to be encouraged for improving risk assessment of potential new invaders [54,55].

Lawton [56] suggested that Diptera and Lepidoptera make worse biological control agents than Coleoptera and, possibly, Hemiptera. We did not find any evidence that the risk of non-target effects was greater with any particular insect order, as the data are limited to only three beetles and one moth causing significant adverse effects on non-target plant populations, as far as has been recorded.

The Rule of Tens

The "Rule of Tens" for biological invaders suggested that one tenth of organisms imported established self-sustaining populations in Britain, and of these one tenth became a pest [57]. They considered the case of biological control agents for weeds to be an exception, with greater probability of establishment (61%) and successful pest control (32%) partly due to deliberate release of large numbers. Our results suggest that another interpretation of the rule of tens may be valid for non-target impacts from weed biological control agents, because non-target impacts of any magnitude occurred with about 10% of introduced organisms (8.4%), and ~1% had an impact on non-target plant population dynamics (i.e. 0.8% became a pest at moderate to massive levels of adversity against the plant), according to our proposed scale.

Table 5. Assessment of potential magnitude of indirect adverse ecological effects from biological control agents (including target weed removal).

Agent	Target	Magnitude	Effect, comment	Reference
Urophora affinis and *U. quadri-fasciata*	*Centaurea maculosa*	Moderate	Elevating deer mouse populations, *Peromyscus maniculatus*, hypothetical	[84]
Diorhabda elongata	*Tamarix* spp.	Moderate	Loss of saltcedar vegetation[1] impacting bird nesting, mainly the southwestern willow flycatcher (*Empidonax traillii extimus*), hypothetical	[64]
Rhinocyllus conicus	*Carduus nutans*	Moderate	Declining populations of native picture-wing flies when seeds of their native thistle hosts were consumed by *R. conicus*	[85,86]
Chrysolina quadrigemina	*Hypericum perforatum*	Moderate to Major	Aggravating weeds, to *Bromus* spp., *Convovulus arvensis*, *Centaurea solstitialis*, *Taeniatherum caput-medusae*, common occurrence	[68]

[1]The absence of native trees is not the result of the biological control agent.

Distribution of impacts. Like weeds, which are sometimes seen as plants growing in the "wrong" place, biological control agents can become invasive. The scaling of magnitude of non-target impacts from biological control agents has not been attempted previously, although there has been much discussion and increasing assessment of the details of an apparently small number of such cases in recent decades [18–20]. The visibility and amplitude of the debate has raised the risk of reducing efforts on

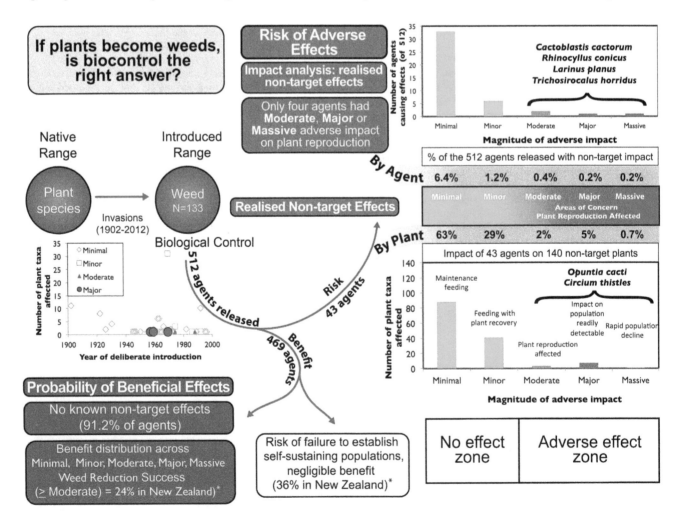

Figure 6. Impact analysis from biological control agents for weeds, with direct adverse impacts observed on non-target plants, and consideration of beneficial effects (*[13]).

biological control and achieving fewer net benefits in future. This is a risk because of the increased costs of providing better evidence of biological safety, with a demand for investigation of increasingly subtle effects, such as apparent competition, trophic cascades, and indirect mutualisms [20]. While these are putatively valid mechanisms for non-target impact, the available evidence for their importance as the source of indirect effects from insects introduced for weed biological control comes from increasingly complex experimental manipulations of multiple trophic levels [48,58]. While such complex interactive effects are of ecological and potentially management interest, the full effects and duration of impact remain unknown. As an example, the gall flies introduced against knapweed caused increased densities of native deer mice [20], which presumably could be beneficial for their predators, but logically this will last only until the knapweed declines or something else changes. Interactions between multiple biocontrol agents or trophic levels can sometimes produce negative management outcomes, although these types of interactions are dynamic [32]. Fowler and Withers [59] could find no indirect effects from weed biocontrol agents in New Zealand mediated via increased populations of natural enemies that exploit the introduced agent. The possibility of indirect competition may exist, but not be realised [60]. Such effects are unlikely to be large unless a "keystone species" is involved, but further analysis might indicate the frequency of this situation [16].

Predictability and systematic

Paynter et al. [53] identified several cases of inadequate procedures for host range testing which can explain some failures to predict field results. In the vast majority of cases, the laboratory-derived host range is wider than the observed host range in the field [61], and our results support the premise of predictability of insect host range in the field for the types of organisms considered for weed biological control. In some cases, beneficial collateral damage on weeds has occurred, as in the case of *Cydia succedana* (Lep.: Tortricidae), released into New Zealand after host range testing of one provenance. The situation, judged here to have minimal or minor non-target impacts (Table S1), is now understood as release of a mixture of two sources of the insect with different host ranges, highlighting the critical importance of adequate systematic support [24], as well as only releasing insects from sources which have been tested, as is now required in New Zealand.

After being famous for suppressing weedy cacti in Australia, cactus moth has become infamous as it heads to the *Opuntia*-rich regions of the southwestern USA and Mexico [62,63] and threatens one of the rarest plants in North America (Table 3). However, investigations on Nevis and St Kitts 50 years after the deliberate introduction (which found no extinctions) [45] and host range tests on a range of species [46] suggest that there will be significant differences between species and populations attacked by cactus moth.

The unpredictability of trophic cascades

Decisions to introduce new organisms can have adverse indirect consequences on weeds which have beneficial effects from supporting birds (Table 5). Many bird species, such as the endangered willow flycatcher, use saltcedar as breeding habitat [64]. Local reduction in saltcedar populations reduced nesting habitat [65], and this led the U.S. to adopt a moratorium on interstate movement of the agent [64]. However, the ecological indirect effects of the insects on saltcedar and bird populations are complex, with both positive effects (beetles used as prey), negative effects (loss of riparian habitat), and no change [65]. We have not

found any evidence reported of the effects of defoliation on nest failure. This example illustrates the complexity of indirect ecosystem effects in weed biocontrol where birds are considered as keystone species [66], benefiting to plants with a significant reduction observed in the level of leaf damage and plant mortality [67].

A weed after a weed

Non-target impacts from successful weed removal can include a shift from exotic species to native vegetation (obviously a desired outcome), or a result in a shift to other exotic weeds. If the new exotic weeds are worse, then biological control has had moderate or higher adverse indirect impact. If the weeds are equivalent, then there has been no obvious gain or loss, just an exchange of species [68]. The limited cases of weed succession listed in Table 5 reflect the scarcity of scientific data when ecological impacts occur at different levels of the trophic cascade. The shifting heterogeneity according to the geographic location described by Campbell and McCaffrey after the removal of Saint John's wort [68] shows that a multi-factorial approach (i.e. climatic, geological, edaphic, etc.) is necessary to understand subtle ecological processes. Observed ecological changes were not predicted before release of biological control agents, at least in a few cases.

To spread or not to spread?

Controlling the spread of a biological control agent becomes a double-edged sword because the potential benefits are reduced. Because of the magnitude of their impacts, *Cactoblastis*, *Rhinocyllus*, *Larinus* and *Trichosirocalus* raise wider questions for the feasibility of limiting redistribution of efficient biological control agents from one region to another. These cases posit the question: can we expect to limit the spread of any biological control agents to areas where their targets are weeds, not valued plants? These are cases of biological control where the servant has become a pest, and the search is underway for biocontrol agents of a biocontrol agent [69], the cactus moth. This possibility has been considered for some time [70], but will be complex because of native pyralid moths which could be placed at risk, as well as the risk of loss of weed biological control where it has been deemed desirable, such as Australia.

It is unclear how many of the thistles attacked are actually at risk of declining from the thistle seed weevil, but there is also a further risk of spread to the very rare Pitcher's thistle (*C. pitcheri*) should it disperse or be distributed into this rare plant's protected habitat [19,71]. Prudent conservation management suggests attempting to limit the further spread of such species which are capable of having significant non-target impacts, but this will limit benefits also. It seems probable that the existing worst four cases above will have extended host range utilization beyond the known (mostly congeneric) level under greater examination and geographic spread, as new similar hosts are encountered. Given scarcity of evidence for wide significant adverse impacts, it seems less likely that the number of significantly impacted hosts of other insects will expand rapidly upon greater scrutiny. New cases of non-target impact on plant populations can be expected to generally follow the observed distribution.

Our approach offers the benefit of providing a standardised framework for observing change in impact over time, since a number of effects are likely to be in flux, for example due to expanding geographic range. The application of the same five step scale used here to characterise benefits from weed biological control at the national level in New Zealand concluded that 24% of agents were successful at weed population suppression (the goal), and two cases were massive in benefit (long-term benefit at ecosystem scale) [13]. This goal needs to be balanced against the risks of non-target effects, which this study has examined and found to be low, if operated based on modern scientific methods. This is despite selected examples that indicate the problem (a tiny proportion of insects introduced for weed biological control are largely predictably adversely affecting two plant families).

Conclusions

It does appear that nearly all risk of significant non-target usage is borne by native plant species that are closely-related to target weeds, as suggested previously [27]. Non-target effects of any significance to plant populations arise only very rarely after more than a century of classical biological control of weeds. The risk of extinction of non-target cacti and thistles is an undesirable consequence of this human activity, but from a risk management perspective, classical biological control of weeds rates very low indeed compared with the environmental effects of invasive species from globalization, climate change, land use change and other human-induced factors which are rapidly accelerating the risks to rare and endangered species everywhere.

Unforeseen feeding outside the families of target host plants, although a recognised phenomenon in laboratory screening [72], has only proven to cause minimal or minor adverse impacts in the field, which are inconsequential to non-target plant population dynamics. The general lack of host shifts beyond the target plant family by weed biological control agents corroborates the proposition that most insects do not feed across more than one or two plant families [73], although polyphagy exists. Hence the choice of agents with a narrow host range and few or no native congeners to the target should mitigate the largest risks. This may lead weed classical biological control programs e.g. [74-79] to prioritise weed species with no direct congeners in the invasive range. Furthermore, choosing weed targets with few relatives anywhere would mitigate the risk of unforeseen movement. It seems likely that a review of the degree of genetic isolation in weed biocontrol targets from valued taxa would help to identify whether this is a valid approach to minimize non-target risks. In addition, consideration of the insect and plant families involved in non-target effects warrants further effort. Ecological cascades also require further investigation.

Acknowledgements

The authors thank E. Coombs (Oregon Dept. Agriculture, USA), R. Winston (Univ. of Idaho, USA) and R. Bourchier (Agric. Canada) for providing data and L. Smith (USDA-ARS, CA, USA), Stephen Hight (USDA-ARS, FLA, USA) and anonymous reviewers for comments. USDA is an equal opportunity provider and employer. DMS was a member of the board of the Environmental Risk Management Authority of New Zealand (2003–2011), and a signatory on binding decisions to approve or decline new organism introductions.

Author Contributions

Conceived and designed the experiments: DMS. Performed the experiments: DMS RFHS. Analyzed the data: DMS. Contributed reagents/materials/analysis tools: DMS. Wrote the paper: DMS RFHS.

References

1. Coulson JR, Vail PV, Dix ME, Nordlund DA, Kauffman WC (2000) 110 years of biological control research and development in the United States Department of Agriculture: 1883–1993. Washington D.C .: U.S. Dept. of Agriculture, Agricultural Research Service.

2. Clewley GD, Eschen R, Shaw RH, Wright DJ (2012) The effectiveness of classical biological control of invasive plants. J Appl Ecol 49: 1287–1295. doi:10.1111/j.1365-2664.2012.02209.x.

3. Julien M, McFadyen R, Cullen JM (2012) Biological control of weeds in Australia. MelbourneAustralia: CSIRO Publishing. 648 p.

4. Tryon H (1910) The "Wild Cochineal Insect", with reference to its injurious action on prickly pear (Opuntia spp.) in India, etc., and its availability for the subjugation of this plant in Queensland and elsewhere. Queensl Agric J 25: 188–197.

5. Gaskin JF, Bon M-C, Cock MJW, Cristofaro M, De Biase A, et al. (2011) Applying molecular-based approaches to classical biological control of weeds (Review). Biol Control 58: 1–21. doi:10.1016/j.biocontrol.2011.03.015.

6. Pimentel D, Acquay H, Biltonen M, Rice P, Silva M, et al. (1992) Environmental and economic costs of pesticide use. BioScience 42: 750–760.

7. van Wilgen BW, de Wit MP, Anderson HJ, Le Maitre DC, Kotze IM, et al. (2004) Costs and benefits of biological control of invasive alien plants: case studies from South Africa. South Afr J Sci 100: 113–122.

8. Page AR, Lacey KL (2006) Economic impact assessment of Australian weed biological control. Technical Series No.10 Adelaide, Australia: CRC for Australian Weed Management. 151 p.

9. McFadyen RC (2012) Benefits from biological control of weeds in Australia. Pak J Weed Sci Res 18: 333–340.

10. CRC (2001) Control of bitou bush: a benefit-cost analysis. In: The Co-operative Research Centre (CRC) for Weed Management Systems: An impact assessment. Adelaide, Australia: CRC for Weed Management Systems. pp. 27–34.

11. Nordblum T, Smyth M, Swirepik A, Sheppard A, Briese D (2001) Benefit-cost analysis for biological control of Echium weed species (Patterson's Curse/Salvation Jane). In: The Co-operative Research Centre (CRC) for Weed Management Systems: An impact assessment. Adelaide, Australia: CRC for Weed Management Systems. pp. 35–43.

12. Bourdôt GW, Fowler SV, Edwards GR, Kriticos DJ, Kean JM, et al. (2007) Pastoral weeds in New Zealand: status and potential solutions. NZ J Agric Res 50: 139–161. doi: 10.1080/00288230709510288.

13. Suckling DM (2013) Benefits from biological control of weeds in New Zealand range from minimal to massive: A retrospective analysis. Biol Control 66: 27–32. doi: 10.1016/j.biocontrol.2013.02.009.

14. McCullough DG, Work TT, Cavey JF, Liebhold AM, Marshall D (2006) Interceptions of nonindigenous plant pests at US ports of entry and border crossings over a 17-year period. Biol Invasions 8: 611–630. doi: 10.1007/s10530-005-1798-4.

15. Parker IM, Simberloff D, Lonsdale WM, Goodell K, Wonham M, et al. (1999) Impact: Toward a framework for understanding the ecological effects of invaders. Biol Invasions 1: 3–19. doi: 10.1023/A:1010034312781.

16. Fowler SV, Paynter Q, Dodd S, Groenteman R (2012) How can ecologists help practitioners minimize non-target effects in weed biocontrol? J Appl Ecol 49: 307–310. doi: 10.1111/j.1365-2664.2011.02106.x.

17. Turner CE, Pemberton RW, Rosenthal SS (1987) Host utilization of native Cirsium thistles (Asteraceae) by the introduced weevil Rhinocyllus conicus (Coleoptera: Curculionidae) in California. Environ Entomol 16: 111–115.

18. Simberloff D, Stiling P (1996) How risky is biological control? Ecology 77: 1965–1974. doi: 10.2307/2265693.

19. Louda SM, Rand TA, Russell FL, Arnett AE (2005) Assessment of ecological risks in weed biocontrol: input from retrospective ecological analyses. Biol Control 35: 253–264. doi:10.1016/j.biocontrol.2005.07.022.

20. Simberloff D (2011) Risks of biological control for conservation purposes. Biocontrol 57: 263–276. doi: 10.1007/s10526-011-9392-4.

21. McFadyen REC (1998) Biological control of weeds. Ann Rev Entomol 43: 369–393. doi: 10.1146/annurev.ento.43.1.369.

22. Fowler SV, Syrett P, Hill RL (2000) Success and safety in the biological control of environmental weeds in New Zealand. Austral Ecol 25: 553–562. doi: 10.1046/j.1442-9993.2000.01075.x.

23. van Lenteren JC, Babendreier D, Bigler F, Burgio G, Hokkanen HMT, et al. (2003) Environmental risk assessment of exotic natural enemies used in inundative biological control. Biocontrol Sci Technol 48: 3–38. doi: 10.1023/A:1021262931608.

24. Paynter Q, Gourlay AH, Oboyski PT, Fowler SV, Hill RL, et al. (2008) Why did specificity testing fail to predict the field host-range of the gorse pod moth in New Zealand? Biol Control 46: 453–462. doi: 10.1016/j.biocontrol.2008.05.004.

25. Pratt PD, Rayamajhi MB, Center TD, Tipping PW, Wheeler GS (2009) The ecological host range of an intentionally introduced herbivore: a comparison of

predicted versus actual host use. Biol Control 49: 146–153. doi: 10.1016/j.biocontrol.2009.01.014.

26. Groenteman R, Fowler SV, Sullivan JJ (2011) St. John's wort beetles would not have been introduced to New Zealand now: a retrospective host range test of New Zealand's most successful weed biocontrol agents. Biol Control 57: 50–58. doi: 10.1016/j.biocontrol.2011.01.005.

27. Pemberton RW (2000) Predictable risk to native plants in weed biological control. Oecologia 125: 489–494. doi: 10.1007/s004420000477.

28. Gassmann A, Louda SM (2000) *Rhinocyllus conicus*: initial evaluation and subsequent ecological impacts in North America. In: Wajnberg E, Scott JK, Quimby PC, editors. Evaluating indirect ecological effects of biological control. London, U.K.: CABI Publishing. pp. 147–183.

29. Barton J (2004) How good are we at predicting the field host-range of fungal pathogens used for classical biological control of weeds? Biol Control 31: 99–122. doi: 10.1016/j.biocontrol.2004.04.008.

30. Waipara NW, Barton J, Smith LA, Harman HM, Winks CJ, et al. (2009) Safety in New Zealand weed biocontrol: a nationwide pathogen survey for impacts on non-target plants. NZ Plant Prot 62: 41–49.

31. Gerard PJ, McNeill MR, Barratt BIP, Whiteman SA (2006) Rationale for release of the Irish strain of *Microctonus aethiopoides* for biocontrol of clover root weevil. NZ Plant Prot 59: 285–289.

32. Denslow JS, D'Antonio CM (2005) After biocontrol: Assessing indirect effects of insect releases. Biol Control 35: 307–318. doi: 10.1016/j.biocontrol.2005.02.008.

33. Sheppard AW, Hill R, DeClerck-Floate RA, McClay AS, Olckers T, et al. (2003) A global review of risk-benefit-cost analysis for the introduction of classical biological control agents against weeds: A crisis in the making? Biocontrol News Inf 24: 91N–108N.

34. EPA (2012) Biological control agents approved for use in New Zealand. Wellington, New Zealand: Environmental Protection Authority.

35. Julien MH, Griffiths MW (1998) Biological control of weeds: A world catalogue of agents and their target weeds. Wallingford, U.K.: CAB International.

36. EPA (2011) Decision making: A technical guide to identifying, assessing and evaluating risks, costs and benefits. Wellington, New Zealand: Environmental Protection Authority.

37. Willis AJ, Kilby MJ, McMaster K, Cullen JM, Groves RH (2003) Predictability and acceptability: potential for damage to nontarget native plant species by biological control agents for weeds. In: Spafford-Jacob H, Briese DT, editors. Improving the selection, testing and evaluation of weed biological control agents. Glen Osmond: South Australia CRC for Australian Weed Management. pp. 35–49.

38. Palmer WA, Heard TA, Sheppard AW (2010) A review of Australian classical biological control of weeds programs and research activities over the past 12 years. Biol Control 52: 271–287. doi: 10.1016/j.biocontrol.2009.07.011.

39. Moher D, Liberati A, Tetzlaff J, Altman DG (2009) The PRISMA Group (2009). Preferred reporting items for systematic reviews and meta-analyses: The PRISMA Statement. PLoS Med 6(6): e1000097. doi: 10.1371/journal.pmed.1000097.

40. Jayanth KP, Mohandas S, Asokan R, Visalakshy PNG (1993) Parthenium pollen induced feeding by *Zygogramma bicolorata* (Coleoptera: Chrysomelidae) on sunflower (*Helianthus annuus*) (Compositae). Bull Entomol Res 83: 595–598. doi: 10.1017/S0007485300040013.

41. Palmer WA, Day MD, Dhileepan K, Snow EL, Mackey AP (2004) Analysis of the non-target attack by the Lantana sap-sucking bug, *Aconophora compressa* and its implications for biological control in Australia. In: Sindel BM, Johnson SB, editors; Fourteeth Australian Weeds Conference "Weed Management - Balancing People, Planet, Profit", 6–9 September 2004, Sydney, Australia. pp. 341–344.

42. Willis AJ, Groves RH, Ash JE (1998) Interactions between plant competition and herbivory on the growth of *Hypericum* species: a comparison of glasshouse and field results. Aust J Bot 46: 707–721. doi: 10.1071/BT97025.

43. Habeck DH, Bennett FD (1990) *Cactoblastis cactorum* Berg (Lepidoptera: Pyralidae), a phycitine new to Florida. Florida Department of Agriculture & Consumer Services Entomology Circular 333: 1–4.

44. Cronk QCB, Fuller JL (1995) Plant invaders. London: Chapman and Hall. 241 p.

45. Pemberton RW, Liu H (2007) Control and persistence of native Opuntia on Nevis and St. Kitts 50 years after the introduction of *Cactoblastis cactorum*. Biol Control 41: 272–282. doi: 10.1016/j.biocontrol.2007.02.002.

46. Jezorek HA, Stiling PD, Carpenter JE (2010) Targets of an invasive species: Oviposition preference and larval performance of *Cactoblastis cactorum* (Lepidoptera: Pyralidae) on 14 North American opuntioid cacti. Environ Entomol 39: 1884–1892. doi: 10.1603/EN10022.

47. Pemberton RW, Turner CE (1990) Biological control of *Senecio jacobaea* in northern California, an enduring success. Entomophaga 35: 71–77. doi: 10.1007/BF02374303.

48. Callaway RM, DeLuca TH, Belliveau WM (1999) Biological control herbivores may increase competitive ability of the noxious weed *Centaurea maculosa*. Ecology 80: 1196–1201.

49. Fowler SV, Paynter Q, Hayes L, Dodd S, Groenteman R (2010) Biocontrol of weeds in New Zealand: an overview of nearly 85 years. In: Zydenbos SM, editor. Seventeenth Australasian Weeds Conference "New Frontiers in New Zealand: Together we can beat the weeds", 26–30 September 2010, Christchurch, New Zealand. New Zealand Plant Protection Society. pp. 211–214.

50. Dodd AP (1940) The biological campaign against prickly-pear. Brisbane: Commonwealth Prickly Pear Board (Australia). 77 p.

51. Greathead DJ (1968) Biological control of Lantana: a review and discussion of recent developments in East Africa. Int J Pest Manag Part C 14: 167–175. doi: 10.1080/05331856809432577.

52. Briese DT (2004) Weed biological control: applying science to solve seemingly intractable problems. Aust J Entomol 43: 304–317. doi: 10.1111/j.1326-6756.2004.00442.x.

53. Paynter QE, Fowler SV, Gourlay AH, Haines ML, Harman HM, et al. (2004) Safety in New Zealand weed biocontrol: a nationwide survey for impacts on non-target plants. NZ Plant Prot 57: 102–107.

54. Suckling DM, Charles JG, Kay MK, Kean JM, Burnip GM, et al. (2013) Host range testing for risk assessment of a polyphagous invader, painted apple moth. Agric For Entomol (Early view), October 25. doi: 10.1111/afe.12028.

55. Stephens AEA, Suckling DM, Burnip GM, Richmond J, Flynn A (2007) Field records of painted apple moth (*Teia anartoides* Walker: Lepidoptera: Lymantriidae) on plants and inanimate objects in Auckland, New Zealand. Aust J Entomol 46: 152–159. doi: 10.1111/j.1440–6055.2007.00571.x.

56. Lawton JH (1990) Biological control of plants: A review of generalisations, rules, and principles using insects as agents. For Res Inst Bull 155: 3–17.

57. Williamson M, Fitter A (1996) The varying success of invaders. Ecology 77: 1661–1666. doi: 10.2307/2265769.

58. Pearson DE, Callaway RM (2008) Weed-biocontrol insects reduce native-plant recruitment through second-order apparent competition. Ecol Appl 18: 1489-1500. doi: 10.1890/07-1789.1.

59. Fowler SV, Withers TM (2006) Biological control: Reducing the impact of invasive weeds and pests, or just another source of alien invaders? In: Allen RB, Lee WG, editors. Biological invasions in New Zealand. Berlin: Springer. pp. 355–369.

60. Paynter Q, Martin N, Berry J, Hona S, Peterson P, et al. (2008) Non-target impacts of *Phytomyza vitalbae* a biological control agent of the European weed *Clematis vitalba* in New Zealand. Biol Control 44: 248–258. doi: 10.1016/j.biocontrol.2007.08.003.

61. Withers TM, Hill RL, Paynter Q, Fowler SV, Gourlay A (2008) Post-release investigations into the field host range of the gorse pod moth *Cydia succedana* Denis & Schiffermüller (Lepidoptera: Tortricidae) in New Zealand. NZ Entomol 31: 67–76. doi: 10.1080/00779962.2008.9722168.

62. Johnson DM, Stiling PD (1998) Distribution and dispersal of *Cactoblastis cactorum* (Lepidoptera: Pyralidae), an exotic Opuntia-feeding moth, in Florida. Fla Entomol 81: 12–22.

63. Zimmermann HG, Bloem S, Klein H (2004) Biology, history, threat, surveillance and control of the cactus moth, *Cactoblastis cactorum*. Vienna, Austria: FAO/IAEA. 40 p.

64. Sogge M, Sferra S, Paxton E (2008) *Tamarix* as habitat for birds: implications to riparian restoration in the south- western United States. Restor Ecol 16: 146–154. doi: 10.1111/j.1526-100X.2008.00357.x.

65. Paxton EH, Theimer TC, Sogge MK (2011) Tamarisk biocontrol using tamarisk beetles: potential consequences for riparian birds in the Southwestern United States. The Condor 113: 255–265. doi: 10.1525/cond.2011.090226.

66. Corcket E, Gifford B, Sforza R (2013) Complexité des interactions au sein de la biocénose. In: Sauvion N, Calatayud PA, Thiery D, Marion-Poll F, editors. Interactions insects-plantes. Versaille Cedex: Editions Quae. pp. 443–460.

67. Mäntylä E, Klemola T, Laaksonen T (2011) Birds help plants: a meta-analysis of top-down trophic cascades caused by avian predators. Oecologia 165: 143–151. doi: 10.1007/s00442-010-1774-2.

68. Campbell CL, McCaffrey JP (1991) Population trends, seasonal phenology, and impact of *Chrysolina quadrigemina*, *C. hyperici* (Coleoptera: Chrysomelidae), and *Agrilus hyperici* (Coleoptera: Buprestidae) associated with *Hypericum perforatum* in Northern Idaho. Environ Entomol 20: 303–315.

69. Paraiso O, Hight SD, Kairo MTK, Bloem S (2011) Egg parasitoids attacking *Cactoblastis cactorum* (Lepidoptera: Pyralidae) in north Florida. Fla Entomol 94: 81–90.

70. Zimmermann HG, Moran VC, Hoffmann JH (2000) The renowned cactus moth, *Cactoblastis cactorum*: its natural history and threat to native *Opuntia* floras in Mexico and the United States of America. Divers Distrib 6: 259–269. doi: 10.1046/j.1472–4642.2000.00088.x.

71. Havens K, Jolls CL, Marik JE, Vitt P, McEachern AK, et al. (2012) Effects of a non-native biocontrol weevil, *Larinus planus*, and other emerging threats on populations of the federally threatened Pitcher's thistle, *Cirsium pitcheri*. Biol Conserv 155: 202–211. doi: 10.1016/j.biocon.2012.06.010.

72. Wapshere AJ (1974) A strategy for evaluating the safety of organisms for biological weed control. Ann Appl Biol 77: 201–211. doi: 10.1111/j.1744-7348.1974.tb06886.x.

73. Jaenike J (1990) Host specialization in phytophagous insects. Ann Rev Ecol Syst 21: 243–273. doi: 10.1146/annurev.es.21.110190.001331.

74. Klein H (2011) A catalogue of the insects, mites and pathogens that have been used or rejected, or are under consideration, for the biological control of invasive alien plants in South Africa. Afr Entomol 19: 515–549. doi: 10.4001/003.019.0214.

75. De Clerck-Floate R, Cárcamo H (2011) Biocontrol arthropods: new denizens of Canada's grassland agroecosystems. In: Floate KD, editor. Arthropods of Canadian grasslands: Inhabitants of a changing landscape. Ottawa, Canada: Biological Survey of Canada. pp. 291–321.

76. Ferguson C, Barratt AM, Hill R, Kean J (2007) BCANZ - Biological Control Agents introduced to New Zealand. http://b3.net.nz/bcanz, accessed October 31 2013..

77. Shaw RH, Tanner R, Djeddour D, Cortat G (2011) Classical biological control of *Fallopia japonica* in the United Kingdom – lessons for Europe. Weed Res 51: 552–558. doi: 10.1111/j.1365–3180.2011.00880.x.

78. Le Bourgeois T, Baret S, Desmier de Chenon R (2011) Biological control of *Rubus alceifolius* (Rosaceae) in La Réunion Island (Indian Ocean): from investigations on the plant to the release of the biocontrol agent *Cibdela janthina* (Argidae). Proceedings of the XIII International Symposium on the Biocontrol of Weeds, Hawai'i, USA, 11–16 September 2011.

79. APHIS (2012) Technical Advisory Group for Biological Control Agents of Weeds: TAG Petitions. APHIS Actions. March 2012 ed. Washington, D.C.: U.S. Department of Agriculture.

80. Andres LA (1985) Interaction of *Chrysolina quadrigemina* and *Hypericum* spp. in California. In: Delfosse ES, editor; Proceedings of the VI International Symposium on Biological Control of Weeds, Vancouver, Canada, 19–25 August, 1984. Agriculture Canada, Ottawa, Canada. pp. 235–239.

81. Takahashi M, Louda SM, Miller TEX, O'Brien CW (2009) Occurrence of *Trichosirocalus horridus* (Coleoptera: Curculionidae) on native *Cirsium altissimum* versus exotic *C. vulgare* in North American tallgrass prairie. Environ Entomol 38: 731–740. doi: 10.1603/022.038.0325.

82. Wiggins GJ, Grant JF, Lambdin PL, Ranney JW, Wilkerson JB (2009) First documentation of adult *Trichosirocalus horridus* on several non-target native *Cirsium* species in Tennessee. Biocontrol Sci Technol 19: 993–998. Doi: 10.1080/09583150903191343.

83. Dodge G, Louda SM, Inouye D (2005) Appendices to "Colonization of thistles by biocontrol agents". Digital Repository at the University of Maryland..

84. Pearson DE, Callaway RM (2003) Indirect effects of host-specifc biological control agents. Trends Ecol Evol 18: 456–461. doi: 10.1016/S0169-5347(03)00188-5.

85. Louda SM, Arnett AE, Rand A, Russell FL (2003) Invasiveness of some biological control insects and adequacy of their ecological risk assessment and regulation. Conserv Biol 17: 73–82. doi: 10.1046/j.1523–1739.2003.02020.x.

86. Louda SM, Kendall D, Connor J, Simberloff D (1997) Ecological effects of an insect introduced for the biological control of weeds. Science 277: 1088–1090. doi: 10.1126/science.277.5329.1088.

3

How Do the Chinese Perceive Ecological Risk in Freshwater Lakes?

Lei Huang[1], Yuting Han[1], Ying Zhou[2], Heinz Gutscher[3], Jun Bi[1]*

1 State Key Laboratory of Pollution Control & Resource Reuse, School of the Environment, Nanjing University, Nanjing, P.R. China, **2** Department of Environmental Health, Rollins School of Public Health, Emory University, Atlanta, Georgia, United States of America, **3** Department of Psychology, University of Zurich, Zurich, Switzerland

Abstract

In this study, we explore the potential contributions of a risk perception framework in understanding public perceptions of unstable ecosystems. In doing so, we characterize one type of common ecological risk– harmful algal blooms (HABs)–in four of the most seriously eutrophicated freshwater lakes in China. These lakes include Chaohu, Dianchi, Hongze, and Taihu, where a total of 2000 residents living near these sites were interviewed. Regional discrepancies existed in the pilot study regarding public perceptions of ecological changes and public concerns for ecological risk. Comparing HABs and other kinds of risks (earthquake, nuclear, and public traffic) through the psychometric paradigm method, *Knowledge, Effect,* and *Trust* were three key factors formulating the risk perception model. The results indicated that *Knowledge* and risk tolerance levels had significant negative correlations in the higher economic situation while correlations in the lower economic situation were significantly positive. *Effect* and risk tolerance levels had significant negative correlations in the high and middle education situation while correlations in the low education situation were close to zero or insignificant. For residents from Taihu with comparatively higher economic and educational levels, more investment in risk prevention measures and stronger policies are needed. And for residents from Hongze and Dianchi with comparatively low economic and educational levels, improvement of the government's credibility (*Trust*) was the most important factor of risk tolerance, so efforts to eliminate ecological problems with the stepwise development of economic and educational levels should be implemented and gradually strengthened. In turn, this could prevent public discontent and ensure support for ecological protection policies.

Editor: Maura Geraldine Chapman, University of Sydney, Australia

Funding: This research was supported by Chinese Natural Sciences Foundation (41271014 and 41171411), the Sino-America Collaborative Research Program (2010DFA91910) and 863 Program (2013AA06A309). The funders had no role in study design, data collection and analysis, decision to publish, or preparation of the manuscript.

Competing Interests: The authors have declared that no competing interests exist.

* E-mail: jbi@nju.edu.cn

Introduction

As a result of anthropogenic activities such as aquaculture, agriculture, waste discharges from industry, and human recreational activities [1], the rising ecological risk of harmful algal blooms (HABs) which may cause human disease [2] has become a worldwide concern [3], especially regarding China. HABs currently disturb four of China's largest fresh water lakes: Chaohu, Dianchi, Hongze, and Taihu, leading to different levels of water quality degradation, ecological damage, threats to human health, and socioeconomic losses [4,5]. An algal bloom event which contaminated 70% of the water plants in Wuxi attracted extensive national attention in June 2007; drinking water contamination prevented nearby 2 million citizens from obtaining potable water for several days, which resulted in excess demand for bottled water and consequently price inflation (nearly fivefold) [3]. Citizens may have varying levels of sensitivity to the ecological risks posed by frequent HABs, and there exists a diverse array of public risk perceptions regarding this kind of risk. Thus, risk perception research helps us to ascertain the attitudes of the public regarding ecological risk and to understand the factors that determine risk tolerance. Depending on career [6,7], education [8,9], gender [10,11], *etc.*, individuals hold different perceptions on how

dangerous an ecological risk could be. In addition to scientific and technical information [12], other factors (experience, familiarity, residency, etc.) influence individual decision-making abilities [2,6,13–15]. The psychometric paradigm method was first used in a risk analysis of a nuclear power plant [8], and it is currently the most influential method that sociologists apply in the field of risk analysis of public perception [16]. The "cognitive map" of hazards produced by the paradigm explains how people perceive the various risks they face [17], which uses several hazard characteristics (e.g., controllability, newness, dreadfulness, etc.) that hypothetically influence risk perception. Previous risk analysis studies [18] have surveyed a wide range of hazardous events which were divided into 4 main types [19]: technical (e.g., nuclear power plants), ecological (e.g., global warming), daily (e.g., public traffic), and natural (e.g., floods, landslides, earthquakes) hazards. These studies revealed the various factors that can significantly influence risk perception, such as how one's familiarity with public transportation impacts the degree of risk tolerance more significantly than one's familiarity with nuclear power. Furthermore, risk perception was influenced by individual-difference predictor variables, including demographics [10,11,20–23], expertise [7,8,24], the potential ecological impact [25]; social background (i.e. culture) [26], customs [27], economic development status [14],

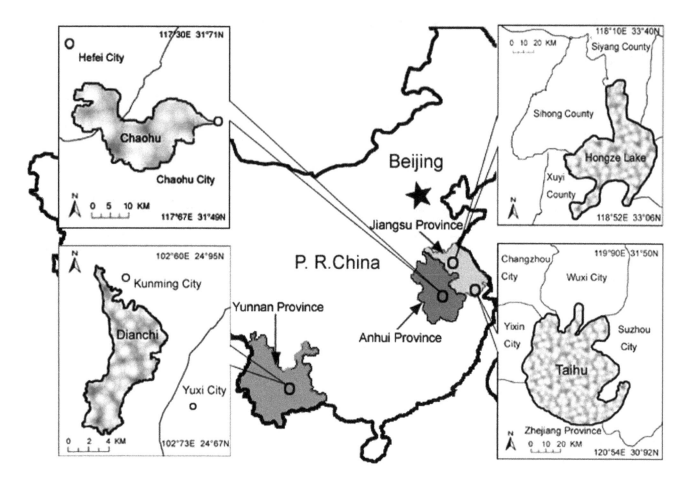

Figure 1. Location of investigating area and four large lakes in China.

and environmental conditions [28,29], all of which could also affect public actions and perceptions [30–32].

Currently, few studies have discussed the impact of lake ecological risks on public perception, and none have taken into account regional discrepancies. Regarding the public attitudes towards water quality, some considered clear water as the most important water quality characteristics followed by fewer HABs [33] and were willing to pay for a reduction in the health risks posed by HABs [34]. However, there are still some people who fail to consider the chronic effects, which result in their perceived risk different from the actual [35,36]. As public risk perception is likely to strongly influence behavior and result in different risk levels [35], improving risk perception through education for risk management and reduction seems very important. China has numerous lakes which are also vital drinking water resources, and the distribution of these lakes is widely dispersed. Understanding public risk perception can greatly inform policy officials on what constitutes publicly-acceptable water management strategies, while maintaining the ecological health of vital water bodies. Due to the 2007 algal-bloom outbreak in the Lake Taihu region, on September 7, 2011 the State Council promulgated the Regulation on the Administration of Taihu Lake Basin. In this way, the actions of the State Council underscored the dramatic importance of balancing ecological protection policies with local political governance [37]. Thus it is worth studying how the Chinese perceive ecological risks in these lake regions.

The present work aims to explore the potential contributions of a risk perception framework in understanding Chinese public perceptions of unstable ecosystems. In doing so, we use HABs as a primary example from which to characterize a lake ecological risk. This study aims to discover the following: 1) determining factors which influence public tolerance in the risk perception model by using a comparative analysis of four typical kinds of hazards; 2) perception factors which create discrepancies in local residents' risk tolerance levels of HABs; 3) and how demographic characteristics and social indicators influence the relationship between each perception factor and risk tolerance level.

Methods

Ethics Statement

This study was approved by the review board of Nanjing University. All participants were informed about the objectives and methods of the study before the investigation and written consent was obtained from all participants. The study did not involve participants from abroad.

Samples and Data Collection

In this study, we focused on the ecological risks posed by HABs that could cause human disease at four of the largest and most eutrophicated lakes in China, which include Chaohu, Dianchi, Hongze, and Taihu. The respondents were selected by a stratified random sampling of those living in cities or counties around these lakes. Each city or county was divided into several districts, with each district further divided into residential communities. House-

Table 1. The Background of Four Lakes [44,45,58].

Year	Chaohu	Dianchi	Hongze	Taihu
1983	22%II;78%III	100%V	18%II;50%III;32%IV	69%II;19%III;12%IV
1993	6.5%III;13.6%IV;79.9%V	100% worse than V	33.3%III;66.7%IV	16.3%II;75.5%III; 8.2%IV
2003	50%V;50%worse than V	100% worse than V	100%V	14.3%IV;14.3%V; 71.4%worse than V
Area	769 km^2	330 km^2	2069 km^2	3100 km^2
Surrounding cities and counties	Hefei, Chaohu	Kunming, Yuxi	Sihong, Xuyi, Siyang	Wuxi, Suzhou, Changzhou, Yixing

Note: Descriptions of II, III, IV, and V are shown in Table S3.

holds within these randomly selected communities were randomly selected. Fig. 1 shows the locations of these four sampled areas.

Questionnaire designs and surveys were conducted at all of the four lake areas in 2008. A total of 2,000 adults were surveyed based on sample proportions from large populations [38], as well as 120 adults in pilot study, and each area included 500 respondents. Because some respondents did not complete the survey or skipped questions, only 1,361 of the sampled respondents were qualified to be used in our analysis. The samples matched the respective populations well in terms of demographics, with 'education' as the exception. High level education (>12 years) was overrepresented in the survey compared to local demographics, which was coincident with many previous studies [14,29,39]. This phenomenon can be attributed to the fact that people with higher levels of education are more likely to participate in survey studies and complete questionnaires [40]. There is a similar overrepresentation of high education levels in each lake investigation; thus, the bias of the samples was likely of minor impact on the results of the comparison analysis among the four lakes. The survey for Lake Chaohu was conducted from July 2 to July 8, 2008, and 339 questionnaires qualified for analysis (a response rate of 68%). The survey for Lake Dianchi was

conducted from July 10 to July 17, and 267 qualified (a response rate of 53%). The survey for Lake Hongze was conducted from July 18 to July 25, and 315 completed questionnaires were filled out (a response rate of 62%). The survey for Lake Taihu was conducted from August 1 to August 9, creating a total of 440 completed questionnaires (a response rate of 88%). Please see Table S1 for detailed results of the surveys. The surveys were conducted at the respondent's workplace or home, and about 30 minutes were needed to finish all questions. Twenty senior students from Nanjing University's School of Environment conducted the survey. Before the start of the survey, they were trained in research methods and survey techniques. If the respondents were unclear about the questions, the students conducting the surveys could provide explanations but were forbidden from giving answers to the questions.

The survey collected information on four aspects: perceptions of ecological change, perceptions of risk characteristics, demographic variables, and social indicators. The first three were investigated by questionnaire survey. In this study, the questionnaire design was adapted from Slovic et al. [8], focusing on HABs and for reasons of comparison, three other kinds of hazards with typical risk characteristics: earthquakes (natural disaster), nuclear power

Table 2. Public Perception of Ecological Change of Four Lakes.

Survey questions	Chaohu (N = 339)		Dianchi (N = 267)		Hongze (N = 315)		Taihu (N = 440)	
	Average	SD	Average	SD	Average	SD	Average	SD
Water quality[a]	3.80	1.45	3.57	1.52	3.21	1.60	3.96	1.45
HABs[b]	3.81	1.44	3.53	1.48	3.15	1.44	4.12	1.17
Area[c]	3.57	1.25	3.61	1.32	2.97	1.38	3.40	1.40
Fish[d]	3.92	1.41	3.92	1.39	3.57	1.65	4.02	1.23
Birds[e]	3.87	1.22	3.55	1.41	3.93	1.06	4.09	1.08
Water plants[f]	3.66	1.36	3.56	1.44	3.17	1.57	3.46	1.51
Tourists[g]	2.37	0.97	1.79	0.81	1.93	0.85	1.95	1.02
Inhabitants[h]	2.14	1.31	2.28	1.01	3.05	1.42	1.77	0.71

[a]"Do you feel that the water quality of the lake has changed in the last decade?" Scale ranges from "much improved" (1) to "much worsened" (5);
[b]"Have you noticed that the lake's harmful algal blooms are occurring more frequently than a decade of ago?" Scale ranges from "strongly disagree" (1) to "strongly agree" (5);
[c]"Do you feel that the area of wetlands surrounding the lake has changed in the last decade?" Scale ranges from "much increased" (1) to "much decreased" (5);
[d]"Do you feel that the number species of fish in the lake has been changed in the last decade?" Scale ranges from "much increased" (1) to "much decreased" (5);
[e]"Do you feel that the number of species of birds inhabiting this lake has changed in the last decade?" Scale ranges from "much increased" (1) to "much decreased" (5);
[f]"Do you feel that the number of species of water plants in the lake has changed in the last decade?" Scale ranges from "much increased" (1) to "much decreased" (5);
[g]"Do you feel that the number of tourists to this lake has changed in the last decade?" Scale ranges from "much increased" (1) to "much decreased" (5);
[h]"Do you feel that the number of inhabitants living around the lake has changed in the last decade?" Scale ranges from "much increased" (1) to "much decreased" (5).

Table 3. Descriptive Statistics of the Risk Characteristic Variables of Four Kinds of Hazards.

Variables	HABs	Earthquake	Nuclear Power	Public Traffic	One-Way ANOVA (F value)
	Men(SD)	Mean(SD)	Mean(SD)	Mean(SD)	
Risk tolerance[a]	2.28(1.22)	1.94(1.37)	2.03(1.31)	2.85(1.39)	21.49***
Risk characteristic variables affecting risk perception					
1. Newness[b]	2.38(1.48)	4.12(1.39)	2.51(1.46)	2.24(1.37)	27.30***
2. Immediacy[c]	3.03(1.64)	2.78(1.67)	2.09(1.60)	3.15(1.55)	12.52***
3. Social risk[d]	2.99(1.49)	2.47(1.49)	3.61(1.37)	2.16(1.44)	15.21***
4. Personal effect[e]	3.10(1.05)	2.56(1.21)	3.38(1.19)	2.84(1.37)	17.64***
5. Knowledge[f]	2.76(1.41)	3.03(1.51)	2.22(1.33)	3.49(1.40)	19.38***
6. Benefit[g]	1.60(1.11)	1.64(1.21)	2.19(1.41)	3.02(1.56)	11.79***
7. Dread[h]	3.07(1.53)	3.19(1.73)	1.63(1.50)	3.20(1.45)	5.45*
8. Trust[i]	2.34(1.36)	2.80(1.55)	2.12(1.46)	2.72(1.37)	1.43

[a]"If your life or working surroundings contain this kind of risk, what is your tolerable degree of risk?" Scale ranges from "Not tolerable at all" (1) to "Very tolerable" (5);
[b]"Is the risk associated with each activity, substance, or technology new and non-familiar, or is it old and familiar? " Scale ranges from "Old" (1) to "New" (5);
[c]"Are the effects of the risk associated with each activity, substance, or technology immediate, or will they take place in the future?" Scale ranges from "Occurs immediately" (1) to "Occurs far in the future" (5);
[d]"How much risk is the national population subjected to as a product of each activity, substance, or technology?" Scale ranges from "No risk" (1) to "High risk" (5);
[e]"In what degree are you personally affected by the risk associated to each activity, substance, or technology?" Scale ranges from "Doesn't affect me" (1) to "Affects me" (5);
[f]"To what degree is the risk associated with each activity, substance, or technology known to you?" Scale ranges from "No knowledge" (1) to "High level of knowledge" (5);
[g]"How beneficial to you is the use, consumption, or accomplishment of each activity, substance or technology?" Scale ranges from "Low" (1) to "High" (5);
[h]"Is the risk associated with each activity, substance, or technology a common risk or a terrible risk?" Scale ranges from "Common" (1) to "Terrible" (5);
[i]"To what degree do you trust in the government or organizations?" Scale ranges from "No Trust at all" (1) to "Complete trust" (5).
*p<0.05;
***p<0.001.

(technical and fatal hazards), and public traffic (daily and familiar hazards). In addition, referring to the research of Sjöberg [29], possible risk characteristic variables that could be included in risk perception models were also selected (i.e., newness, immediacy, social risk, personal impact, knowledge, benefit, dread, and trust in government). On the basis of the pilot study, the questionnaire structure is proven valid and identical to Bartlett's Test of Sphericity. Then we modified vague expressions and created the formal questionnaire (Questionnaire S1). Perceptions of ecological change and risk characteristics were evaluated on a five-point Likert scale [8,11,16,29]. The demographic variables included sex, age, education, occupation, and annual income. Social indicators included socio-economic status, socio-educational level, and socio-environmental condition, most of which were derived from official statistics and fieldwork results [38,41,42]. Social indicators selected by extensive comparative and screening analysis with literature references [8,22,23,29,30,43,44] were used to analyze associations within tolerable risk levels and risk perceptions. The socio-economic status of each city or county in the researched lake areas could be expressed in three ways: GDP per capita, energy consumption per capita, and proportion of urban population. There were two variables in the socio-educational level: proportion of advanced education degrees and the quantity of media consumption (i.e., TV, radio, newspaper, etc.). Socio-environment conditions included the degree of lake service function changes (i.e., domestic water, industrial water, agricultural water, etc.) and a number of high-polluting enterprises. The data of variables depicted above come from local statistical yearbooks [38,41,42] except proportion of lake service function changes which is based on data from Lake (Chaohu, Dianchi, Hongze, and Taihu) Ecosystem Services Assessment Report (Table S2).

Data Analysis

Two statistical analyses were performed in this study. First, we analyzed the determining factors of ecological risk perception, comparing it to the three different kinds of hazards mentioned above (HABs compared to earthquakes, nuclear power, and public traffic). Comparisons between risk characteristics (*Risk tolerance, Newness, Immediacy, Social risk, Personal effect, Knowledge, Benefit, Dread, Trust*) of four kinds of hazards and four lakes were tested by the one-way ANOVA method. Following the traditional psychometric approach, a maximum likelihood factor analysis with oblique promax rotation was used to find the risk perception factors associated with each hazard. Three steps helped to identify the differences in perception of HABs in the four different lakes: factor analysis was conducted to find the risk perception factors, followed by a correlation analysis to test the validity of the factor analysis. The third step involved a multiple regression analysis to examine the determining factors of risk perception.

Second, a Kruskal-Wallis H test was conducted to examine regional demographic diversities of the researched areas, and the relationships between selected social indicators and risk tolerance were analyzed. In addition, factor space analysis [27], categorized by groups of people, was used to determine how demographic characteristics or social indicators influence the relationship between risk tolerance levels of HABs and each determining risk perception factor. We used these correlations to determine unit-length risk tolerance vectors for the participants in the given-samples factor space. The open symbols are projections of the endpoints of these vectors onto the three planes defined by pairs of oblique factors. Long projections (endpoints far from the origin) indicate that risk tolerance levels of HABs are highly correlated with the factors that define the plane, whereas short projections

Table 4. Factor Analysis Results of Risk Characteristics about Four Typical Kinds of Risks.

Variables	HABs			Earthquake		
	Total Variance Explained = 78%			Total Variance Explained = 76%		
	Factor1 Knowledge	Factor2 Effect	Factor3 Trust	Factor1 Knowledge	Factor2 Effect	Factor3 Trust
Newness	**0.720**	−0.158	0.147	**0.661**	0.254	0.208
Immediacy	0.079	**0.648**	−0.066	−0.139	**0.615**	0.226
Social risk	−0.010	**0.764**	0.007	0.046	**0.726**	0.312
Personal effect	0.052	**0.797**	0.201	−0.023	**0.812**	0.160
Knowledge	**0.643**	−0.101	−0.008	**0.579**	0.325	−0.008
Benefit	0.051	0.013	**0.619**	0.207	0.353	**0.768**
Dread	0.200	**0.672**	0.251	−0.080	**0.489**	*0.411*
Trust	−0.142	−0.067	**0.693**	*0.403*	0.121	**0.616**

Variables	Nuclear power			Public Traffic		
	Total Variance Explained = 75%			Total Variance Explained = 81%		
	Factor1 Knowledge	Factor2 Effect	Factor3 Trust	Factor1 Knowledge	Factor2 Effect	Factor3 Trust
Newness	**0.527**	0.341	0.082	**0.784**	0.208	−0.075
Immediacy	0.089	**0.603**	0.154	0.170	**0.637**	−0.037
Social risk	0.095	**0.653**	0.184	0.029	**0.793**	*0.439*
Personal effect	0.105	**0.722**	0.028	0.009	**0.805**	0.037
Knowledge	**0.643**	−0.077	0.217	**0.681**	−0.059	0.147
Benefit	0.125	0.146	**0.643**	−0.078	0.257	**0.628**
Dread	*0.413*	**0.474**	0.022	0.303	**0.635**	0.207
Trust	0.109	−0.139	**0.625**	0.048	0.059	**0.757**

Note: See Table 3 for variable description. |Factor pattern|>0.40 is in boldface type.

(endpoints close the origin) indicate that risk tolerance levels are not highly correlated with the same factors. The cosine of the angle between a pair of factors is equal to the correlation between those factors. Positive and negative correlations between vectors and factors can be judged relative to the dashed lines perpendicular to the factors. In this study, the length of the factor axes from the origin is 1.0. The symbols in different sizes and colors represent the risk tolerance level vectors for different groups of respondents for each demographic characteristic or social condition. The data analysis and reliability testing were performed by statistical package SPSS 17.0.

Results

Pilot Study on Public Perception of Ecological Changes

Table 1 provides background information on the changes that the four lakes have undergone in recent decades, showing deteriorating water quality and ecological changes occurring in each lake. There are both similarities and differences in perceived ecological changes in recent decades among residents living near the four lakes (see Table 2). The most noticeable phenomenon is that the residents from all lakes felt HABs occurred much more frequently, especially residents from Lake Taihu, who nearly unanimously agreed that HABs occur more frequently than the previous ten years (M = 4.12, SD = 1.17). As shown in Table 1, the water quality of Dianchi was the worst of all the lakes we investigated; however, Chaohu, Hongze and Taihu have greater deteriorated changes than Dianchi in the recent 10 years. In addition, some irrigation works brought great changes to Lake Hongze (e.g., lake volume shrank by a third) [44,45], but the residents did not perceive as strong a risk as those from Lake Taihu, while the actual changes of Lake Hongze was bigger than Lake Taihu in recent 10 years. Therefore, we must ascertain the key factors that impact public sensitivity and cause differences in individual perceptions of ecological changes.

What are the Determining Factors that Influence Public Tolerance in the Risk Perception Model?

Comparison between HABs and three other hazards. Earthquakes are natural disasters that have immediate

Table 5. Descriptive Statistics of the Risk Characteristic Variables of Four Large Lakes in China.

Variables	Chaohu (N = 339) Mean(SD)	Dianchi (N = 267) Mean(SD)	Hongze (N = 315) Mean(SD)	Taihu (N = 440) Mean(SD)	One-Way ANOVA (F value)
Risk tolerance	2.36(1.20)	2.09(1.26)	2.58(1.24)	1.97(1.19)	19.07***
Risk characteristic variables affecting risk perception					
1. Newness	3.74(1.54)	3.85(1.42)	3.10(1.53)	3.74(1.38)	18.99***
2. Immediacy	3.46(1.64)	2.93(1.68)	2.57(1.56)	3.16(1.60)	16.73***
3. Social risk	3.24(1.53)	3.00(1.54)	2.54(1.46)	3.17(1.40)	14.59***
4. Personal effect	3.12(1.27)	2.89(1.45)	2.70(1.32)	3.58(1.66)	19.42***
5. Knowledge	2.65(1.32)	2.65(1.41)	2.32(1.30)	3.21(1.41)	28.53***
6. Benefit	1.48(1.01)	1.69(1.21)	1.64(1.10)	1.58(1.10)	2.09
7. Dread	2.99(1.54)	2.92(1.50)	2.75(1.48)	3.46(1.51)	15.99***
8. Trust	2.13(1.29)	2.43(1.36)	2.50(1.40)	2.29(1.38)	4.34**

Note: See Table 3 for variable description.
**p<0.01;
***p<0.001.

Table 6. The Differences of Risk Characteristic Variables on a pairwise basis between Four Lakes.

Variables	Chaohu			Dianchi		Hongze	Taihu
	& Dianchi	**& Hongze**	**& Taihu**	**& Hongze**	**& Taihu**	**& Taihu**	**–**
Risk tolerance	−0.27**	0.15	−0.36***	−0.52***	−0.13*	0.61***	–
Risk characteristic variables affecting risk perception							
1. Newness	−0.11	−0.75***	−0.11	−0.64***	0.004	0.64***	–
2. Immediacy	−0.53***	0.36*	−0.23	0.89***	0.30	−0.59***	–
3. Social risk	−0.24	0.46***	−0.17	0.69***	0.07	−0.63***	–
4. Personal effect	−0.23	0.65*	0.49**	0.26	−0.82***	−0.88***	–
5. Knowledge	0.002	0.33*	−0.56***	0.33*	−0.56***	−0.89***	–
6. Benefit	0.22	0.05	0.11	−0.17	0.11	0.06	–
7. Dread	−0.06	0.17	−0.53***	0.23	−0.47***	−0.71***	–
8. Trust	0.29*	−0.08	0.14	−0.37**	−0.16	0.22	–

Note: See Table 3 for variable description. *$p<0.05$;
**$p<0.01$;
***$p<0.001$.

Table 7. Factor Analysis Results of Risk Characteristics about HABs in Four Lakes.

Variables	Chaohu			Dianchi		
	Total Variance Explained = 75%			**Total Variance Explained = 82%**		
	Factor1 _Knowledge_	**Factor2** _Effect_	**Factor3** _Trust_	**Factor1** _Knowledge_	**Factor2** _Effect_	**Factor3** _Trust_
Newness	**0.690**	0.065	−0.097	**0.801**	0.190	0.198
Immediacy	0.193	**0.621**	−0.103	0.104	**0.762**	−0.175
Social risk	−0.001	**0.794**	0.017	0.332	**0.693**	0.077
Personal effect	0.018	**0.753**	0.101	0.091	**0.583**	0.221
Knowledge	**0.671**	0.078	0.043	**0.604**	0.350	0.121
Benefit	−0.029	0.038	**0.980**	0.372	−0.009	**0.612**
Dread	0.292	**0.615**	0.142	0.084	**0.694**	0.139
Trust	0.072	−0.022	**0.712**	−0.096	0.043	**0.849**

Variables	Hongze			Taihu		
	Total Variance Explained = 76%			**Total Variance Explained = 79%**		
	Factor1 _Knowledge_	**Factor2** _Effect_	**Factor3** _Trust_	**Factor1** _Knowledge_	**Factor2** _Effect_	**Factor3** _Trust_
Newness	**0.778**	0.099	−0.018	**0.789**	−0.048	−0.150
Immediacy	0.090	**0.654**	−0.121	0.308	**0.626**	0.100
Social risk	0.217	**0.707**	−0.058	−0.115	**0.724**	0.057
Personal effect	0.007	**0.625**	0.251	−0.117	**0.819**	0.026
Knowledge	**0.598**	0.314	0.001	**0.640**	0.058	−0.113
Benefit	0.337	0.072	**0.693**	−0.097	−0.038	**0.877**
Dread	−0.129	**0.725**	0.135	0.155	**0.775**	−0.046
Trust	0.228	−0.109	**0.763**	0.501	0.028	**0.589**

Note: See Table 3 for variables description. |Factor pattern|>0.40 is in boldface type.

impact. Nuclear power is associated with involuntary, unknown, new, uncontrollable, and fatal risk characteristics. The risk of public traffic is a daily hazard with which people are familiar. Compared with these risks, HABs are hazardous ecological events that could afflict water bodies (eutrophic lake, coastal ecosystems, etc.) on an almost global scale and exert a long-lasting effect on the environment. The tolerable risk level of earthquakes was ranked last in the comparison analysis of risk tolerance, the frequency of which usually is regarded as a risk background value. However, such a low tolerable level of the respondents in this survey was thought mainly due to the Wenchuan Earthquake in China that occurred just two months prior to the date of the survey (May 12th, 2008). Table 3 shows that the four hazards had significantly diverse characteristics including _newness_ (F = 27.30, p<0.001), _immediacy_ (F = 12.52, p<0.001), _social risk_ (F = 15.21, p<0.001), _personal effect_ (F = 17.64, p<0.001), _knowledge_ (F = 19.38, p<0.001), _benefit_ (F = 11.79, p<0.001), and _dread_ (F = 5.45, p = 0.014). The results of a rotated factor pattern are shown in Table 4. All attributes related to _newness_ and _knowledge_ were loaded on Factor 1 (labeled as _Knowledge_). _Immediacy_, _social risk_, _personal effect_, and _dread_ were loaded on Factor 2 (labeled as _Effect_). Attributes related to _benefit_ and _trust_ were loaded heavily on Factor 3 (labeled as _Trust_). However, there were still a few minor differences among the hazards in the results of the factor analysis. For example, the characteristic variable _dread_ of the earthquake hazard was split between Factor 2 and Factor 3, which means that the degree of _dread_ of earthquakes was not only influenced by _Effect_ but also by _Trust_ in the government or organizations. The variable _social risk_ on public traffic was also split between Factor 2 and Factor 3, revealing that public traffic was not only influenced by _Effect_ but also by _Trust_ in the government or organizations. The variable of _dread_ in the hazard of nuclear power was split between Factor 1 and Factor 2, which implies that the degree of _dread_ the public holds for nuclear power plant is related to their _knowledge_ about the plant; however, factor analysis revealed that _Effect_ is still the primary factor.

Although the types of hazards were significantly different from each other, we found that a three-factor solution (_Knowledge_, _Effect_ and _Trust_) explained 75%–81% of the total variance in each hazard's characteristics. Therefore, the selected risk factors

Table 8. Using Factor Scores about HABs to Explain Mean Risk Tolerance of Four Lakes.

Factors	Chaohu		Dianchi		Hongze		Taihu	
	β	SE	β	SE	β	SE	β	SE
Factor1:*Knowledge*	0.226*	0.105	0.248**	0.040	0.245*	0.115	−**0.156***	0.057
Factor2:*Effect*	−**0.224***	0.097	−0.061	0.019	−0.013	0.012	−**0.187***	0.082
Factor3:*Trust*	0.234*	0.113	0.348***	0.056	0.264**	0.098	0.237*	0.114
Adjusted R^2	0.485		0.522		0.380		0.401	
F-value	25.94***		27.86***		15.32***		23.51***	

*p<0.05,
**p<0.01,
***p<0.001.
Note: The dependent variables are mean tolerance judgment for the HABs, calculated by averaging judgments over participants from different lakes. Unstandardized regression coefficients are from regression models with factor scores from the factor analysis as independent variables.

appropriately described the hazards, but their impacts on public perception and risk tolerance of each hazard were diverse.

Correlation analysis of characteristic variables for HABs. The first item in Table 5 shows the respondents' tolerance levels of HABs, implying that the residents around Lake Hongze and Lake Chaohu were comparatively willing to accept HABs risk (M = 2.58, SD = 1.24; M = 2.36, SD = 1.20, respectively), while the residents around Lake Taihu were the most reluctant to accept the risk (M = 1.97, SD = 1.19). It should be noted that

risk tolerance levels of HABs were significantly different among the four lakes (F = 19.07, p<0.001), as were all the risk characteristic variables of HABs, except benefit (F = 2.09, p = 0.182). Differences on a pairwise basis between four lakes can be found in Table 6.

Many risk characteristic variables were correlated with each other [6,15,18,19]. For example, *knowledge* correlated with both *social risk* (a = 0.190, p<0.01) and *newness* (a = −0.315, p<0.01), and *social risk* correlated with *immediacy* (a = 0.336, p<0.01). In particular, *dread* correlated with most variables, including *newness*

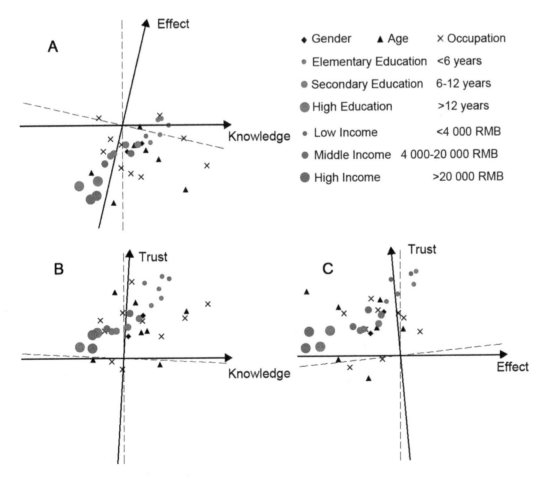

Figure 2. Individual risk perception factors for different demographic characteristics plotted in the factor space.

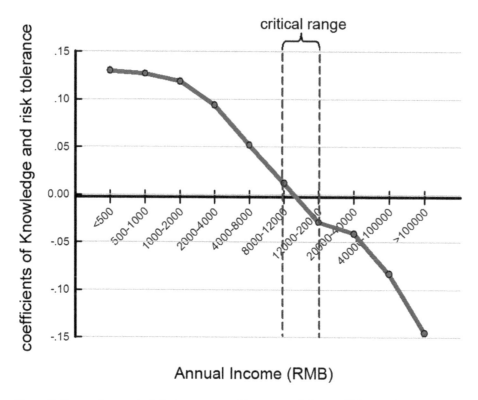

Figure 3. Regression analysis between annual income and the coefficients of Knowledge and risk tolerance.

(a = 0.234, p<0.01), *immediacy* (a = 0.240, p<0.01), *social risk* (a = 0. 367, p<0.001), *personal effect* (a = 0.322, p<0.001), and *benefit* (a = −0.144, p<0.05), so factor analysis was necessary to eliminate inter-correlation.

What are the Perception Factors Creating Discrepancies in Risk Tolerance?

A separate factor analysis was conducted to determine the relationships within risk characteristic variables in different regions. The results (Table 7) were generally similar to the factor analysis results for all participants in hazard comparison analysis (Table 4).

Risk perception models were established to obtain the regression coefficients for respondents' risk tolerance levels of HABs, using the three risk perception factors listed in Table 7 as independent variables (see Table 8). The results from the respondents around Lake Hongze and Lake Dianchi had weaker (i.e., less negative) correlations between *Effect* (Factor 2) and risk tolerance of HABs. *Knowledge* about hazards (Factor 1) and *Trust* in government or organizations (Factor 3) had positive coefficients, which indicated that tolerable levels of HABs increased as *Knowledge* and *Trust* increased, except in the Taihu area where there existed a significant negative association between *Knowledge* and risk tolerance. As a result, the impacts of these three factors on risk tolerance were not always of the same directions, especially the impacts of *Knowledge* (factor 1) and *Effect* (Factor 2). Overall, four perception models can explain 38.0%–52.2% of the variance in tolerable levels, and the independent variables considered in the risk perception model all played an important role in the decision process.

How do Demographic Characteristics and Social Indicators Influence the Relationship between each Perception Factor and Risk Tolerance?

Impact of demographic variables on individual risk perception of HABs. Table S1 shows the results of the regional demographics of the researched areas derived from the Kruskal-Wallis H (x^2) test. To visualize the discrepancy in risk perception among different demographic characteristics, factors representing individual judgments of risk perception were plotted using factor space analysis. Fig. 2 reveals that more knowledge about a hazard is significantly associated with a higher risk tolerance level in low-income groups. However, it is worth noting that these correlations are significantly negative in high-income groups, indicating that among wealthy people, risk tolerance levels decrease as they learn more about a hazard. However, the factor *Knowledge* is not necessarily accurate: it only measures respondents' perceptions of their levels of understanding of risks. Respondents may misgauge their own knowledge levels due to lack of access to relevant information.

Regression analysis was used to test the impact of an individual income on risk perception. The coefficients of *Knowledge* and risk tolerance- defined as the dependent variables- were categorized by annual income level, whose range was defined as the independent variable. The analysis revealed that annual income has a positive or negative correlation between *Knowledge* and risk tolerance within a critical range (8,000–20,000 RMB, or $1,250–$3,125 in U.S. dollars) (see Fig. 3).

Impact of social variables on public risk perception of HABs. As the results shown in Fig. 4, significant relationships are existed between risk tolerance level of HABs and some social variables, especially energy consumption per capita $(R^2 = 0.91)$, and proportion of urban population $(R^2 = 0.91)$, and proportion of advanced education degrees $(R^2 = 0.94)$, which indicate that

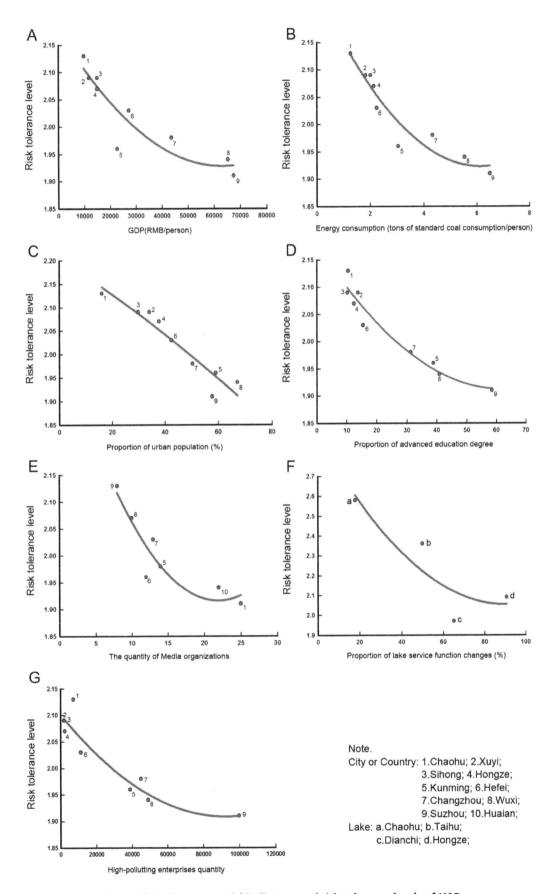

Figure 4. Regression analysis between social indicators and risk tolerance levels of HABs.

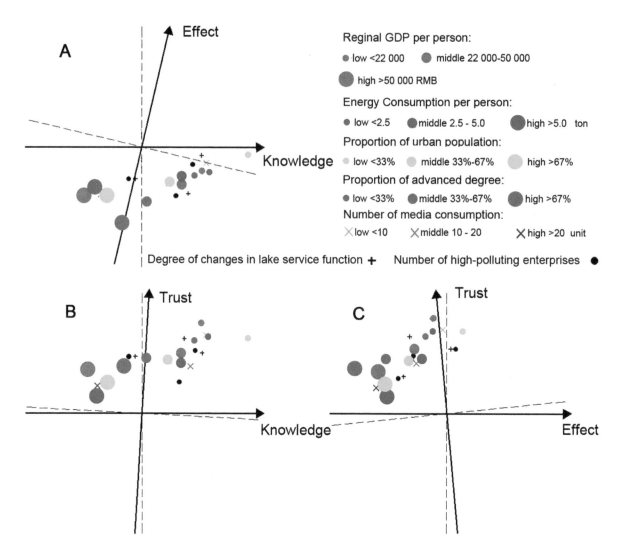

Figure 5. Public risk tolerance for different social conditions plotted in the factor space.

different socio-economic status or socio-educational level may correlated with different risk tolerance levels of local residents.

To visualize the diversity of risk perception in different social conditions, three risk perception factors were plotted with social variables in the factor space (Fig. 5). Clearly, in the regions with a low urban population or low GDP per capita, more Knowledge about the hazard was positively associated with higher risk tolerance levels, which is consistent with the previous analysis of individual income levels; the factor *Effect* and risk tolerance levels showed no significant correlations. In the regions with a high proportion of advanced education degrees, we found that the factor *Effect* was significantly correlated with higher risk tolerance levels, which is consistent with the previous analysis of individuals' education levels. However, the factor Trust had a more significant impact on risk tolerance levels in the regions with few media organizations, while the factor *Effect* had a more significant impact on tolerable risk levels in the regions with many media organizations. More information about regional variance of risk tolerance according to different perception factors can be found in Fig. 5.

Discussion

Currently, public risk tolerance levels of HABs in the four researched areas were all comparatively low (M = 2.28, SD = 1.22). On one hand, risk eliminating measures should be strengthened to reach publicly tolerable levels; on the other hand, the government should help the public to make rational judgments regarding perceived risks, which could help prevent public unease and ensure their support for ecological protection policies. Thus, understanding the factors that determine risk tolerance is very important for protecting both the environment and public safety. This study of risk perception models applied to HABs found that Chinese residents' risk tolerance levels depend primarily on three factors: *Knowledge* (i.e., how much does one know about this kind of risk?), *Effect* (i.e., how would this kind of risk affect humans or society?), and *Trust* (i.e., how much can residents trust the government or organizations?).

Furthermore, the impact of these three perception factors varied among the respondents from the four large lakes. We found that regional economic development could explain most of this variance. For example, a negative relationship existed between *Knowledge* and risk tolerance for the respondents from Lake Taihu, while the correlation was significant but in a positive relationship for respondents from the other three lakes. In the demographic analysis, the proportion of high-income groups in the region of Taihu (53%) is much higher than that in the other three regions (Chaohu, 33%; Dianchi, 24%; Hongze, 22%) (See Table S1). This explains why the regression coefficient for respondents from Lake Taihu is opposite to that of respondents from the other lakes. This phenomenon indicates that people whose annual income is lower than a critical range of 8,000–20,000 RMB ($1,250–$3,125 U.S. dollars) care more about their own livelihoods and less about the ecological system, and even ignored a familiar risk in HABs. Although they understand the HABs phenomenon, they do not feel it intolerable. But public attitudes can be changed with an improved economic status. Three socio-economic indicators revealed this tendency. The coefficients of the social variables were all negative (see Fig. 4-A, B, C), indicating that risk tolerance levels decreased as socio-economic levels increased. These results support the findings from previous studies by Bi et al. and Lind [46,47], and can explain regional differences in public sensitivity to ecological changes in the pilot study (Table 2).

Educational level also had a significant impact on the sensitivity of risk perception, especially for *Effect*. It was found that *Effect* and risk tolerance levels had significant negative correlations in the groups of high and middle education levels while correlations in low education level groups were close to zero or insignificant (Fig. 2). In demographic analyses, the proportions of groups with higher education in the regions of Taihu (47%) and Chaohu (42%) were higher than those of Dianchi (29%) and Hongze (26%), which could explain the higher coefficients of the factor *Effect* in the perception models of the former two regions. At the same time, the impacts of socio-educational conditions, namely the proportion of advanced education degrees and quantity of media consumption, displayed similar trends (Fig. 5).

Economic development in the Lake Taihu area has reached the level of moderately developed countries, with an average annual income of about 20,000 RMB ($3,125 U.S. dollars) per person [38], which was above the critical range of annual income. In addition, the proportion of urban population and GDP per capita was at a relatively higher level. This confirmed that public risk perception and socio-economic factors hold a strong correlation. The results showed that the residents from Lake Taihu considered their living environment a high priority and paid more attention to health risks, which prompted them to have the lowest risk tolerance levels for HABs. Furthermore, higher socio-educational levels could promote more concerns about the consequences of ecological risks. After a theoretical disaster turned into a reality, risk tolerance levels sharply decreased, which may eventually lead to social disorder or public panic. Therefore, more investment in risk prevention measures and stronger policies than those contained in the government's existing restoration plan of Lake Taihu are needed.

With lower levels of economic development in the Hongze and Dianchi areas yielding an average annual income of about 8,000 RMB ($1,250 U.S. dollars) per person [37,41], the top priority of the public in those areas is rapid economic development, and more economic burdens from lake restoration investments could lead to public discontent. Respondents from these areas weakly considered the *Effect* of HABs and were insensitive to the negative consequences of environment problems and ecological risks. Their perceived risk tolerance was evidently correlated with *Trust* in government or organizations. As in previous studies about nuclear power risk perception in Japan [48] and China [49], insensitive people who had higher trust in government would judge the risk depending on heuristic processing [50,51] which is generally irrational. So if the government lost its credibility, these groups of

people, who tend not to rely on rational judgment, would sharply decrease their risk tolerance levels and may unnecessarily become alarmed. Sensitivity to risk will affect individual behavior [35,52]. For example, Cabrera et al. [53] found that the sensitivity to health risk of pesticide will affect the farm workers' behavior. In many exposure assessment studies [54,55], researchers have found that individuals' behaviors would result in the difference of actual risk. Huang et al. [56] and Marchwinska et al. [57] believed improving access to education and better risk communication to increase sensitivity will reduce high-risk behaviors and strengthen local governments' credibility for risk reduction, control and management. So, the findings of this study suggested that existing political structures played an important role in risk tolerance [6]. In order to strengthen local governments' credibility for risk control and management, it's necessary to improve access to education and better risk communication for those biased people, offering accurate information and guiding them on adjusting behaviors to avoid high risks. According to the stepwise development of economy and education, gradual efforts should be made to eliminate ecological problems such as HABs in lakes.

These results provided some insight into how policymakers can target certain groups for effective risk communication and public education programs. For example, public communicators may effectively achieve better aims by targeting different groups of people divided by income levels. Thus, risk communication can not only prevent inattentive or unwitting individuals from ignoring the genuine level of risk, but it can also raise public awareness of environmental protection and attract more attention to ecological changes, thereby reducing individual activities (such as tourism, farming, etc.) that degrade the ecological environment. In addition, the results may also suggest possible social discontent occurring in different sites according to the relationships between risk tolerance and social characteristics.

This study explores Chinese public perception of a specific eutrophication problem occurring in lake ecosystems, a phenomenon which has rarely concerned researchers but is gradually becoming one of the most serious ecological risks in China. The importance of this study lies in the finding that Chinese residents hailing from different regions may have different cognitions and perceptions about ecological risks, which suggests that regional characteristics influence public attitudes towards ecological problems. Due to complex social conditions and the sensitive conformist nature of the Chinese public, it is crucial to detect target groups in different regions, which helps to promote effective risk communication in improving ecosystem health and maintaining social stability.

Nevertheless, some important factors related to the risk perception of HABs still need to be addressed with further research, including emotion, action tendency, and willingness-to-pay. The role of many social factors and individual demographic variables also has not been sufficiently studied. Information about the participant's factual knowledge of HAB risk which will be used to examine the factual correctness of their knowledge has not been collected. In addition, the Chinese public's perception of different kinds of hazards may change significantly over time. All of these issues should be considered and discussed in future studies.

Supporting Information

Table S1 Demographic Data of Study Participants in the Four Lakes.

Table S2 Proportion of lake service function changes.

Acknowledgments

We would like to thank a visiting researcher Van der K.jan from Harvard University for the English revisions.

References

1. McFarlane BL, Witson DO (2008) Perceptions of ecological risk associated with mountain pine beetle (Dendroctonus ponderosae) infestations in Banff and Kootenay National Parks of Canada. Risk Anal 28: 203–212.
2. Kuhar SE, Nierenberg K, Kirkpatrick B, Tobin GA (2009) Public perceptions of Florida red tide risks. Risk Anal 29: 963–969.
3. Guo L (2007) Doing battle with the green monster of Taihu Lake. Science 317: 1166–1166.
4. Backer LC, McNeel SV, Barber T, Kirkpatrick B, Williams C, et al. (2010) Recreational exposure to microcystins during algal blooms in two California lakes. Toxicon 55: 909–921.
5. Chen J, Zhang D, Xie P, Wang Q, Ma Z (2009) Simultaneous determination of microcystin contaminations in various vertebrates (fish, turtle, duck and water bird) from a large eutrophic Chinese lake, Lake Taihu, with toxic Microcystis blooms. Sci Total Environ 407: 3317–3322.
6. Huang L, Duan B, Bi J, Yuan Z, Ban J (2010) Analysis of Determining Factors of the Public's Risk Acceptance Level in China. Hum Ecol Risk Assess 16: 365–379.
7. Lazo JK, Kinnell JC, Fisher A (2002) Expert and layperson perceptions of ecosystem risk. Risk Anal 20: 179–194.
8. Slovic P (1987) Perception of risk. Science: 280–285.
9. Slimak MW, Dietz T (2006) Personal values, beliefs, and ecological risk perception. Risk Anal 26: 1689–1705.
10. Flynn J, Slovic P, Mertz CK (1994) Gender, race, and perception of environmental health risks. Risk Anal 14: 1101–1108.
11. Slovic P (1989) Trust, emotion, sex, politics and science: surveying the risk-assessment battlefield. In: Bazerman M MD, Tenbrunsel A, Wade-Benzoni K, editor. San Francisco: The New Lexington Press. 277–313.
12. Arvai J, Gregory R (2003) Testing alternative decision approaches for identifying cleanup priorities at contaminated sites. Environ Sci Technol 37: 1469–1476.
13. Salihoglu G, Karaer F (2004) Ecological risk assessment and problem formulation for Lake Uluabat, a Ramsar State in Turkey. Environ Manage 33: 899–910.
14. Bronfman NC, Cifuentes LA (2003) Risk perception in a developing country: the case of Chile. Risk Analy 23: 1271–1285.
15. Hohl K, Gaskell G (2008) European Public Perceptions of Food Risk: Cross-National and Methodological Comparisons. Risk Analy 28: 311–324.
16. Slovic PFB, Lichtenstein S (1980) Facts and fears: understanding perceived risk. In: Schwing RC AW, editor. New York: Springer Press. 181–216.
17. Siegrist M, Keller C, Kiers HA (2005) A new look at the psychometric paradigm of perception of hazards. Risk Anal 25: 211–222.
18. McDaniels T, Axelrod LJ, Slovic P (2006) Characterizing perception of ecological risk. Risk Anal 15: 575–588.
19. Ho MC, Shaw D, Lin S, Chiu YC (2008) How do disaster characteristics influence risk perception? Risk Anal 28: 635–643.
20. Rogers GO, Reduction H, Center R (1985) On determining public acceptability of risk. In: Whipple C CV, editor. New York: Plenum Press. 483–504.
21. Hammitt JK, Liu J-T (2004) Effects of disease type and latency on the value of mortality risk. J Risk Uncertain 28: 73–95.
22. Mrozek JR, Taylor LO (2002) What determines the value of life? A meta-analysis. J Policy Anal Manage 21: 253–270.
23. Viscusi WK, Aldy JE (2003) The value of a statistical life: a critical review of market estimates throughout the world. J Risk Uncertain 27: 5–76.
24. Chung JB, Kim HK, Rho SK (2008) Analysis of local acceptance of a radioactive waste disposal facility. Risk Anal 28: 1021–1032.
25. Peters E, Slovic P (1996) The role of affect and worldviews as orienting dispositions in the perception and acceptance of nuclear Power1. J Appl Soc Psychol 26: 1427–1453.
26. Bronfman NC, Cifuentes LA, deKay ML, Willis HH (2007) Accounting for variation in the explanatory power of the psychometric paradigm: The effects of aggregation and focus. J Risk Res 10: 527–554.
27. Willis HH, DeKay ML (2007) The roles of group membership, beliefs, and norms in ecological risk perception. Risk Anal 27: 1365–1380.
28. Rizak SN, Hrudey SE (2006) Misinterpretation of drinking water quality monitoring data with implications for risk management. Environ Sci Technol 40: 5244–5250.
29. Sjöberg L (2004) Local acceptance of a high-level nuclear waste repository. Risk Anal 24: 737–749.
30. Grothmann T, Reusswig F (2006) People at risk of flooding: Why some residents take precautionary action while others do not. Nat Hazards 38: 101–120.

Author Contributions

Conceived and designed the experiments: LH YH JB. Performed the experiments: LH YH. Analyzed the data: LH YH. Contributed reagents/materials/analysis tools: LH YZ HG. Wrote the paper: LH JB.

31. Wallquist L, Visschers VH, Siegrist M (2010) Impact of knowledge and misconceptions on benefit and risk perception of CCS. Environ Sci Technol 44: 6557–6562.
32. Cervantes O, Espejel I, Arellano E, Delhumeau S (2008) Users' perception as a tool to improve urban beach planning and management. Environ Manage 42: 249–264.
33. Kosenius A-K (2010) Heterogeneous preferences for water quality attributes: The Case of eutrophication in the Gulf of Finland, the Baltic Sea. Ecol Econ 69: 528–538.
34. Hunter PD, Hanley N, Czajkowski M, Mearns K, Tyler AN, et al. (2012) The effect of risk perception on public preferences and willingness to pay for reductions in the health risks posed by toxic cyanobacterial blooms. Sci Total Environ 426: 32–44.
35. May H, Burger J (2006) Fishing in a polluted estuary: fishing behavior, fish consumption, and potential risk. Risk Anal 16: 459–471.
36. Sukharomana R, Supalla R (1998) Effect of risk perception on willingness to pay for improved water quality. Presentations, Working Papers, and Gray Literature: Agric Econ: 46.
37. You M (2012) Annual review of Chinese environmental law developments: 2011. Environ Law Rep 42: 10482–10488.
38. Jiangsu Statical Bureau (2010) Jiangsu Statistical Yearbook, Beijing, P.R.C.: China Statistics Press.
39. De Groot JI, Steg L (2010) Morality and nuclear energy: perceptions of risks and benefits, personal norms, and willingness to take action related to nuclear energy. Risk Anal 30: 1363–1373.
40. Purvis-Roberts KL, Werner CA, Frank I (2007) Perceived risks from radiation and nuclear testing near Semipalatinsk, Kazakhstan: A comparison between physicians, scientists, and the public. Risk Anal 27: 291–302.
41. Anhui Statistical Bureau (2010) Anhui Statistical Yearbook. Beijing, P.R.C.: China Statistics Press.
42. Yunnan Statistical Bureau (2010) Yunnan Statistical Yearbook. Beijing, P.R.C.: China Statistics Press.
43. McCall CH (1982) Sampling and Statistics: Handbook for Research. Iowa, U.S.A.: Iowa State University Press.
44. The state of enviroment in China in 2003 (2004): Ministry of Environmental Protection of the People's Republic of China. Available: http://jcs.mep.gov.cn/hjzl/zkgb/2003/. Accessed 2011 Mar 1.
45. XC J (1995) The environment of Chinese lakes. Beijing, P.R.C.: Ocean Press.
46. Bi J, Yu C (1996) New water quality indices for the era of sustainable development in China. GeoJournal 40: 9–15.
47. Lind N (2002) Social and economic criteria of acceptable risk. Reliab Eng Syst Saf 78: 21–25.
48. Katsuya T (2002) Difference in the formation of attitude toward nuclear power. Polit Psychol 23: 191–203.
49. Huang L, Bi J, Zhang B, Li F, Qu C (2010) Perception of people for the risk of Tianwan nuclear power plant. Front Environ Sci Eng in China 4: 73–81.
50. Parker GR (1989) The role of constituent trust in congressional elections. Public Opin Q 53: 175–196.
51. Petty RE, Cacioppo JT (1981) Attitudes and Persuasion: Classic and Contemporary Approaches, Dubuque, IA: WC Brown Co. Press.
52. Brewer NT, Weinstein ND, Cuite CL, Herrington JE (2004) Risk perceptions and their relation to risk behavior. Ann Behav Med 27: 125–130.
53. Cabrera NL, Leckie JO (2009) Pesticide risk communication, risk perception, and self-protective behaviors among farmworkers in California's Salinas Valley. Hisp J Behav Sci 31: 258–272.
54. Qu CS, Ma ZW, Yang J, Liu Y, Bi J, et al. (2012) Human exposure pathways of heavy metals in a lead-zinc mining area, Jiangsu Province, China. PLoS One 7: 1–11.
55. Samuel D, Preethi P, Meredith W, Daniel OH, Li H, et al. (2012) Health Risks of Limited-Contact Water Recreation. Environ Health Perspect 120: 192–197.
56. Huang L, Sun K, Ban J, Bi J (2010) Public Perception of Blue-Algae Bloom Risk in Hongze Lake of China. Environ Manage 45:1065–1075.
57. Marchwinska-Wyrwal E, Teaf CM, Dziubanek G, Hajok I (2012) Risk assessment and risk communication in environmental health in Poland. Eur J Public Health 22: 742–744.
58. Environmental quality standard for surface water (2002) Ministry of Environmental Protection of the People's Republic of China. Available: http://www.tba.gov.cn:89/eword/uploadfile/20080327103039192.pdf. Accessed 2011 Mar 1.

4

Managing for Interactions between Local and Global Stressors of Ecosystems

Christopher J. Brown[1]*, **Megan I. Saunders**[1], **Hugh P. Possingham**[2], **Anthony J. Richardson**[3,4]

1 The Global Change Institute and the School of Biological Sciences, The University of Queensland, St Lucia, Queensland, Australia, **2** Australian Research Council, Centre for Excellence for Environmental Decisions, School of Biological Sciences, The University of Queensland, St Lucia, Queensland, Australia, **3** Climate Adaptation Flagship, Commonwealth Scientific and Industrial Research Organisation, Marine and Atmospheric Research, Dutton Park, Queensland, Australia, **4** Centre for Applications in Natural Resource Mathematics, School of Mathematics and Physics, The University of Queensland, St Lucia, Queensland, Australia

Abstract

Global stressors, including climate change, are a major threat to ecosystems, but they cannot be halted by local actions. Ecosystem management is thus attempting to compensate for the impacts of global stressors by reducing local stressors, such as overfishing. This approach assumes that stressors interact additively or synergistically, whereby the combined effect of two stressors is at least the sum of their isolated effects. It is not clear, however, how management should proceed for antagonistic interactions among stressors, where multiple stressors do not have an additive or greater impact. Research to date has focussed on identifying synergisms among stressors, but antagonisms may be just as common. We examined the effectiveness of management when faced with different types of interactions in two systems – seagrass and fish communities – where the global stressor was climate change but the local stressors were different. When there were synergisms, mitigating local stressors delivered greater gains, whereas when there were antagonisms, management of local stressors was ineffective or even degraded ecosystems. These results suggest that reducing a local stressor can compensate for climate change impacts if there is a synergistic interaction. Conversely, if there is an antagonistic interaction, management of local stressors will have the greatest benefits in areas of refuge from climate change. A balanced research agenda, investigating both antagonistic and synergistic interaction types, is needed to inform management priorities.

Editor: Sam Dupont, University of Gothenburg, Sweden

Funding: CJB's contribution was supported by the Australian Research Council (ARC) Centre of Excellence for Environmental Decisions. MIS was supported by an ARC Super Science Postdoctoral Fellowship. AJR was supported by an ARC Future Fellowship FT0991722. This work forms part of the ARC Discovery Grant DP0879365. ARC: http://www.arc.gov.au/. The funders had no role in study design, data collection and analysis, decision to publish, or preparation of the manuscript.

Competing Interests: The authors have declared that no competing interests exist.

* E-mail: c.brown5@uq.edu.au

Introduction

Ensuring the persistence of critical habitats, dependent communities and ecological processes requires simultaneous management of multiple local and global stressors caused by human activities [1,2]. A stressor is an environmental variable that negatively affects individual physiology or population performance when it is beyond its normal range of variation [3]. Generally, local stressors can be manipulated directly by management. Examples include improving water quality to halt declines of seagrass, coral reef and near-shore communities [4], maintaining riparian forest to buffer streams from run-off [5], and creating reserves to slow deforestation of tropical forests [6]. Increasingly, stressors with global causes are major drivers of ecosystem change. In particular, global climate change threatens habitats, ecological communities and ecological processes [7,8]. For instance, extreme temperature events threaten the persistence of seagrass beds in the Mediterranean [9], drought and fire threaten fragmented forests [10], heat waves and ocean acidification threaten coral reef habitat and dependent fish communities [11,12], and warming threatens numerous species with extinction [13]. Reducing global stressors requires collaboration among countries or regional management bodies, so they are not amenable to manipulation directly by

management at a local scale. Therefore, management at a local scale can only act on impacts of global stressors indirectly, by reducing local stressors.

Stressors can have interactive effects on populations and ecosystems. If there is no interaction, the combined effect of two stresses is said to be additive, which is the sum of their effects in isolation. Interactions between stresses can be synergistic, where the combined effect of two stresses is greater than the additive expectation. Interactions may also be antagonistic, where the combined effect is less than the additive expectation. Stressors and their interactions can act at different levels of ecological organisation. A species may be subject to a synergistic interaction when the presence of one stressor reduces the physiological tolerance of individuals to additional stresses. For example, some corals may be hyper-sensitive to thermal bleaching if they are already physiologically stressed by poor water quality [14]. At a population level, if individuals tolerant of one stressor are sensitive to another, multiple stressors will tend to have a synergistic effect on mortality [3]. Antagonisms may occur when a population is made up of individuals that are either tolerant or sensitive to stress (co-variability), regardless of the stressor's identity. For instance, if one stressor removes the most sensitive individuals, the remaining population will be tolerant of additional stressors [3]. More rarely, antagonisms can have mitigative effects on individuals. High sediment levels can reduce survival of coral colonies [15], but, sediments can be beneficial for corals at risk of bleaching, because reduced water clarity may reduce physiological light stress on corals [15]. At the community level, co-variability in stressor tolerance among species can also lead to antagonisms or synergisms, in a similar way that variability among individuals in a population can lead to interactive effects on populations [3].

We propose there are three prevailing views about interactions in the management of global stressors of ecosystems. The first is that synergisms are prevalent [16]. Synergisms are of concern, because future rates of ecosystem decline predicted on the basis of individual stressor effects will be underestimated if there are synergistic interactions between stressors [16,17]. Synergisms will also cause more rapid declines in ecosystems than additive or antagonistic interactions. This view implies that management of local stressors can benefit ecosystems impacted by global stressors. A second view is that multiple stressors have cumulative impacts on ecosystems (e.g. [1,18]). While useful for identifying the large-scale impacts of humans on ecosystems, such studies assume additive interactions and imply that management that addresses the largest stressor will have the greatest benefit [18,19]. The final view is managing for ecological resilience [20]. This generally entails managing a local stressor to reduce the likelihood of ecological transitions to alternative degraded states, such as coral reefs to macro-algal dominated reefs, or desertification of grasslands [21,22]. This may include reducing a local stressor, such as fishing, to improve recovery rates from, or resistance to, uncontrollable disturbances, such as hurricanes and climate change [23,24].

None of these views addresses the prevalence of antagonistic interactions between stressors; they all assume that managing a local stressor improves the ecosystem. Antagonisms imply that local management actions cannot compensate for global stressors such as climate change impacts. Recent meta-analyses of experimental studies from marine, freshwater and terrestrial systems indicate that antagonisms are just as common as synergisms. At both population and community levels; antagonistic and synergistic interactions each made up approximately a third of all interactions [16,17]. There is also evidence that individuals and species with lower metabolic rates tend to be more tolerant to multiple different kinds of physiological stress [25]. This implies co-tolerance in sensitivity to multiple stressors and suggests that antagonisms will be prevalent. There is concern that the focus on synergisms over antagonisms will result in ineffective management actions and wasted management effort [20,26]. While this concern follows intuitively from empirical studies, there is a dearth of modelling studies for understanding the effectiveness of local management actions on outcomes for populations and communities.

Our approach is to build simple models to illustrate how interactions between climate and local stressors matter for the management of populations and communities. We use two case study models to illustrate interactions at both population and community levels. First, we use a population model of a seagrass bed to examine the expected number of years before a seagrass bed is degraded beyond recovery. The gain local management can make is calculated by comparing the outcomes when a local stressor, poor water quality, is or is not improved. We compare scenarios where poor water quality interacts additively, antagonistically or synergistically with ocean warming. There are multiple physiological mechanisms by which heat stress and poor water quality affect seagrass mortality [27–30]. Warming may physiologically stress seagrass by increasing respiration rates more rapidly than photosynthetic rates. Algal epiphytes may also outcompete seagrass at warmer temperatures [31]. Increased water column nitrate, as a consequence of terrestrial run-off from fertilizer use, may also promote phytoplankton and epiphyte growth and reduce light to seagrass. Eutrophication may also directly stress seagrass physiology [27].

Second, we use a community model to show that interactive effects of local and global stressors on a community of species affects species richness when the local stressor is remediated [3]. Each of these models is explained in more detail below. These examples show synergisms can accelerate declines in ecosystems, but also provide the greatest opportunity for management to benefit ecosystems. Our simple models also highlight that antagonisms can be more challenging to manage. We suggest that identifying the type and strength of interactions has not received the necessary research focus needed to support management of climate change impacts through local actions and for setting achievable management targets.

Methods

Population Response to Stressor Interactions: Model of Seagrass

We examined how management outcomes depended on interaction types using a model of Mediterranean seagrass density [9]. The original model was used to predict the year a seagrass bed reached a critically low density, assuming additive effects of global warming and poor water quality stress on mortality rate. We adapted this model by incorporating interactions between these stressors. It is currently not known whether synergisms or antagonisms are more likely for warming and water quality impacts on seagrass [9]. Warming and nutrient inputs may worsen declines of seagrass by simultaneously increasing growth of phytoplankton and epiphytes (multiplicative synergism) [32]. Alternatively, there may be an antagonism if nutrient inputs dominate the stress response of seagrass growth (stressor dominance, antagonism) [27]. Finally, while we are not aware of a demonstration of mitigative antagonisms for seagrass, for other aquatic primary producers poor water quality can mitigate heat stress, by reducing light stress [15].

In analyses we considered both antagonistic and synergistic interactions to demonstrate their importance for management and we hope this will stimulate further studies that will quantify the type and strength of the interaction.

We modelled declines in seagrass density using an exponential model [9]:

$$N_t = N_0 \exp(t(R - M_t)) \qquad (1)$$

where N_t was seagrass density at time t, R was the recruitment rate and M_t was the seagrass mortality rate at time t. We assumed the recruitment rate was constant over time, whereas mortality varied with temperature and a local stressor. We modified the original model for mortality rate [9] to include interactions between poor water quality (local stressor) and warming stress (global stressor):

$$M_t = a_1 \, Warming + a_2 \, Local + a_3 \, Warming \, Local + K \qquad (2)$$

This linear additive effects model is commonly used for estimating effect sizes (a_i) of stresses and their interaction on ecological properties from empirical data in linear statistical models (Folt et al. 1999).

Adding an interaction term to Jorda *et. al*'s [9] equation increases the overall mortality rate for a synergism, or decreases the mortality rate for an antagonism. This would be unrealistic, because the overall mortality rate was estimated from field studies. Hence, for each value of a_3 (the interaction) we recalculated the intercept, K, so that the mortality rate in a base scenario was constant and consistent with field measurements (Table 1). This ensures that our scenarios of decline were consistent with those originally presented in Jorda *et al.* [9]. Including an additional interaction term in this way was not unrealistic, because the additive effects of warming and water quality stressors on mortality were estimated separately, so interactions were not considered. This was also consistent with how effect sizes are often estimated from field studies. We also explored the outcome when the interaction term was an additional source of mortality; thus, the overall mortality is higher with a synergism and lower with an antagonism. This situation would occur if the mortality rate due to individual stressors is estimated in experimental studies, but the overall mortality rate is not estimated in the field.

Mortality due to the local stressor was present in scenarios without management and absent in scenarios with management, and we used the same value as Jorda *et al.* [9] (Table 1). We increased temperature linearly at a rate of 0.38°C per decade. Jorda *et al.* [9] considered inter-annual variability in temperature and variability among climate model predictions and thus predicted mean declines with confidence intervals for seagrass. Our intent was to show the effect of interactions, rather than to quantify likely rates of seagrass decline, so we used a linear rate of temperature change to avoid unnecessary complexity. Regardless, the long-term rate of temperature change gave a similar result to the model mean in Jorda *et al.* [9]. We thus compared times to reach the 10% density threshold under different interaction types with and without management. The 10% density threshold was chosen because seagrass recovery is unlikely beyond this point [9]. We considered three classes of interaction: additive $(a_3 = 0)$, synergistic $(a_3 > 0)$ and antagonistic $(a_3 < 0)$.

There is danger in extrapolating linear models so we kept our interaction strengths within bounds that gave overall mortality rates no greater than those in Jorda *et al.* [9]. We did not vary interaction strengths to be greater than 10% of the effect of temperature. This value is plausible, given interaction strengths estimated in experiments of nutrient and warming stress on seagrass [27].

The study that this model was based on has a number of inherent assumptions and limitations. In particular the methodology used to obtain the estimates of seagrass mortality rates in Jorda *et al.* [9] has been criticised [33], but see [34]. As such, the base mortality rate and the temperature effect on mortality used by

Jorda *et al.* [9] may have been under or over-estimated and subsequently, the predicted year in when seagrass reaches a critically low density may be inaccurate. Further, Jorda *et al.* [9] were not able to estimate the impact of local stressors on seagrass directly using empirical measurements. To account for these criticisms, additional analyses were conducted where the base mortality rate, the temperature dependent mortality rate, the recruitment rate, and the mortality caused by the local stressor were varied. Our intent was to show qualitatively how interactions affect the outcome of management, rather than provide quantitative estimates of the year seagrass reached a critically low density. Thus, when interpreting the results from the additional analyses, we focussed on whether our qualitative results are changed, rather than quantitative differences in the year seagrass reached a critically low density.

Community Response to Stressor Interactions: Modelling Co-tolerance

First we explain a conceptual model for community responses to multiple stressors and then below we explain how to modify the model to examine the impacts of interactions on the outcomes of management. We used the species co-tolerance model (Fig. 1, [3]). In this model, the interactive effect of two stressors on an aggregate community property, such as species richness, depends on the co-tolerance of each species to the stressors. Stressors can be pulse disturbances operating over a short time, in which case the model predicts the short-term community response. Stressors can also be press disturbances operating over a long time, or a series of pulse disturbances, in which case the model predicts the long-term community response. A stressor of a certain magnitude is assumed to remove all sensitive species. In the co-tolerance model, additional stresses only affect the community aggregate property if they affect species not sensitive to the first stress. For instance, consider an assemblage of coral reef fish species. Global warming, coral bleaching and subsequent loss of habitat over the long-term may cause local extinction of coral dependent fish species and a lower fish species richness. In this conceptual model, an additional stress, fishing, will only further reduce species richness if the remaining species are sensitive to fishing [12].

An additive effect occurs if individual species tolerances to two stressors are randomly distributed with respect to each other (Fig. 1). In this instance, habitat loss from global warming that affects 50% of the fish species will reduce species richness by 50%. Fishing pressure that also affects 50% of fish species will consequently reduce species richness by only 25%, because the 25% of species that are highly sensitive to both stressors have already been affected (a stressor dominance effect). In communi-

Table 1. Parameter values for the seagrass model scenarios where growth was modelled $R - M_t$ and $M_t = a_1$ *Warming* + a_2 *Local* + a_3 *Warming Local* + K.

Interaction scenario	Recruitment rate	Climate effect (a_1)	Local effect (a_2)	Interaction (a_3)	Intercept (K)	Mortality rate in 2010*
Additive	0.05	0.021	0.02	0	−0.471	0.096
Synergistic	0.05	0.021	0.02	0.001	−0.497	0.096
Antagonistic – dominance	0.05	0.021	0.02	−0.0005	−0.458	0.096
Antagonistic – mitigative	0.05	0.021	0.02	−0.001	−0.445	0.096

*with both stresses.
Interaction values were chosen so that overall mortality rates were never greater than those in the model of [9]. The interaction strength (a_3) is weaker for the dominance antagonism than the mitigative antagonism, to reflect the different processes that cause these types of interactions.

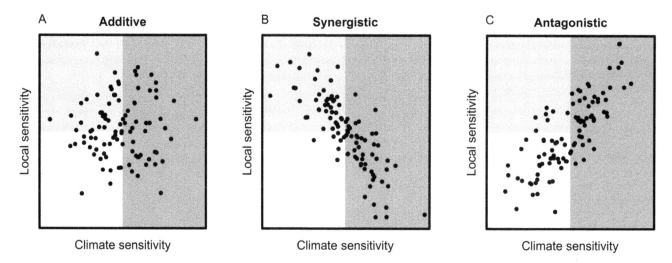

Figure 1. Co-tolerance of species to both climate and local stressors for three types of interactions. Species tolerances were generated randomly (nominal scales) for an additive interaction (random co-tolerance, $\rho = 0$), a synergistic interaction (negative co-tolerance, $\rho = -0.8$), and an antagonistic interaction (positive co-tolerance, $\rho = 0.8$). Each point represents the tolerances of a single species to the two stressors. Species in the dark grey region will be threatened by climate change stress, the local stressor will additionally affect species in the light grey region. Species in the white region will be unaffected by either stressor. The most species will be lost with a synergism and the least with an antagonism [3].

ties, synergistic and antagonistic interactions occur if co-tolerances are negatively or positively correlated respectively (Fig. 1). If there is strong negative co-tolerance, two stresses that each affect 50% of species will together affect close to all species, because species not sensitive to habitat loss will likely be sensitive to fishing. Whereas, if there is positive co-tolerance, species that are sensitive to habitat loss are also sensitive to fishing, so cumulative stressors will have only small further effects on the community.

We modelled community responses to multiple stressors by representing species' co-tolerances as a multivariate normal distribution. Thus, species' marginal tolerances are normally distributed with respect to each stressor and their co-tolerance was described by the correlation between stressor tolerances, ρ. For an extreme synergism, $\rho = -1$, whereas $\rho = 0$ for an additive interaction and $\rho = 1$ for an extreme antagonism. Increases in species richness from management were calculated by comparing the number of species affected by either or both of the local and global stressors, compared to the number of species affected by only the global stressor (see Appendix S1). Using this analysis, for the coral reef fish assemblage, we might ask how much greater species richness will be on reefs threatened by bleaching if management protects reefs from fishing by placing them in marine reserves?

Other distribution forms could also be used, but we chose the multivariate normal because it is a realistic way to simulate species tolerances to environmental factors. We further assumed that species tolerances to stressors are fixed. Stress tolerance could also be dynamic [3], however, we do not include this additional complexity, because we have illustrated dynamic responses in the seagrass case study.

In reality, co-tolerance patterns may deviate from a linear bivariate relationship. A negative convex co-tolerance curve between a local stressor (fishing) and climate change was recently described for a coral reef fish assemblage [12]. In this study, Graham *et al.* [12] assessed fish species vulnerability to population declines caused by climate and fishing. They used scientific theory and empirical assessments to assign vulnerability scores for each stressor to each species and tested these scores against independent empirical data. We used Graham *et al.*'s [12] vulnerability scores

for fish species to calculate management gains from creating marine reserves (thereby reducing the fishing stressor to zero), in the presence of the climate stressor. Species tolerances are log-normally distributed in these data, however, the results are similar to those assuming a normal distribution.

Results

Population Response to Stressor Interactions: Seagrass Population Mortality

We first used the additive effects equation with interactions (equation 2) to predict how local and global stressors affected seagrass mortality rate for different interaction types. Reducing the local stressor decreased the mortality rate when there was no interaction or a synergistic interaction (Fig. 2). Improvements were greater when the synergism was stronger (not shown). For a dominance antagonistic interaction, mortality rate decreased if management reduced the local stressor, but by a smaller amount than for an additive or synergistic interaction. By contrast, for a mitigative antagonistic interaction, mortality rate increased when the local stressor was reduced. This counter-intuitive result occurs because the antagonistic effect benefitted the ecosystem by a greater amount than the direct additive impacts.

Next we simulated seagrass density using the various scenarios for mortality rate. Declines in seagrass density with poor water quality were almost identical for three scenarios of interaction types without management, because we adjusted the base rate mortality to compensate for interaction effects (Fig. 3a). As in Jorda *et al.* [9], the 10% density threshold was reached in the year 2049, except for stronger antagonisms where the decline was slightly slower and the threshold is reached in 2050. When the interaction was additive, improving water quality yielded a gain of 13 years (2049 to 2062) and when the interaction was synergistic, management yielded a gain of 38 years (2049 to 2087). When the interaction was a mitigative antagonism, the density threshold was reached four years earlier if the local stressor is reduced (2050 to 2046). When the interaction was a weaker dominance antagonism, the density threshold was reached only four years later when the local stressor was reduced (2049 to 2053).

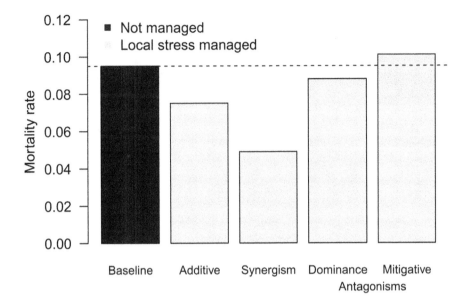

Figure 2. Mortality rate of seagrass for different interaction types. Mortality rate is high with both warming and the local stressor (water quality, dark grey bar). If the local stressor is improved (light grey bars), mortality rate is reduced by 0.02 per year for an additive interaction. Management with a synergistic interaction between warming and local stressor gains a greater reduction in mortality rate, whereas the reduction is small with a dominance antagonism. If there is a mitigative antagonism, mortality rate increases if the local stressor is improved.

There was an exponential relationship between year of loss and interaction strength when the local stressor was improved. Stronger synergistic interactions (positive values) exponentially increased the year the density threshold was reached (Fig. 3b). Antagonistic interactions that were >~3% of the temperature effect size were mitigative, because improving water quality at this point meant the density threshold was reached earlier than without management.

The general result, that synergistic interactions gave greater gains from managing the local stressor, was robust to alternative parameter combinations. Larger temperature effect sizes reduced the benefits of improving the local stressor (Fig. 4a). Increasing the recruitment rate led to greater gains for management when there

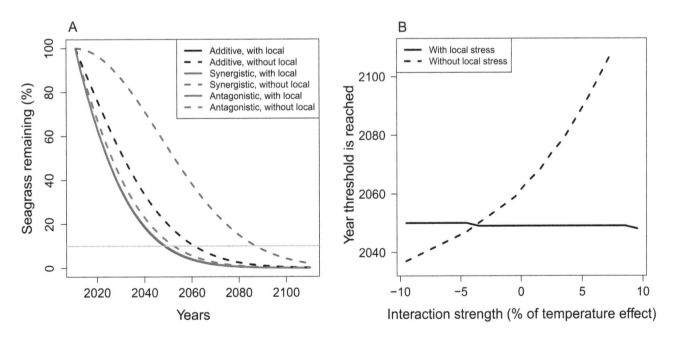

Figure 3. Seagrass density for different interactions with and without management of the local stressor. (A) Decline in seagrass density with and without improving the local stressor (water quality) for additive, synergistic (5% of temperature effect size), and antagonistic interactions (−2.5% of temperature effect size). The grey line represents the 10% seagrass density threshold where seagrass loss is believed to be irreversible [9]. The interaction scenarios with the local stressor almost perfectly overlay each other. (B) Year of seagrass loss for a range of interaction strengths (positive is synergistic, negative is antagonistic) when water quality is not managed (solid line) and when water quality is improved (dashed line).

was a synergism and greater losses when there was a mitigative antagonism (Fig. 4b). Decreasing the initial mortality rate had a similar effect to increasing the recruitment rate (Fig. 4c). Increasing the mortality caused by the local stressor meant the density threshold was reached earlier (Fig. 4d). This also improved the gain from management, because by removing the local stressor management affected a greater fraction of the overall mortality rate.

We next considered a case where the interaction is an additional source of mortality, so that overall mortality was higher with a synergism and lower with an antagonism. In this case, the rate of decline with both stressors was greater with a synergism and slower with an antagonism (Fig. 5a). Whereas, if management reduces the local stressor, the outcome was the same regardless of the interaction. The year the threshold was reached was constant amongst interaction types when the local stressor was mitigated (Fig. 5b). Whereas the year of loss occurs exponentially earlier for increasingly synergistic interactions when the local stressor was present. The general conclusion that more synergistic interactions provide greater gains still held.

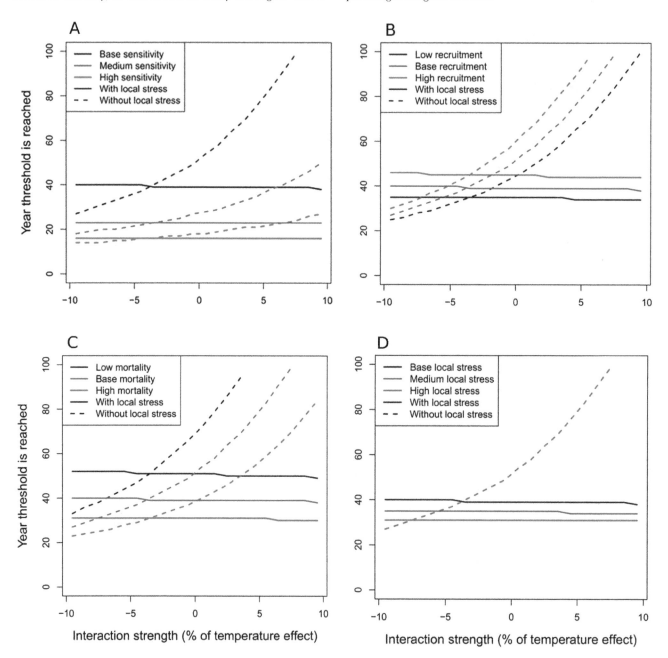

Figure 4. Year of seagrass loss for a range of interaction strengths, when the seagrass model parameters are varied. For each scenario, one parameter was varied while other parameters were held constant. Model scenarios when the local stressor is not managed are indicated with solid lines and scenarios when the local stressor is improved are indicated with dashed lines. Each colour indicates a different parameter value. (A) Varying warming effect sizes (parameter a_1, black $a_1 = 0.021$, blue $a_1 = 0.023$, red $a_1 = 0.025$). (B) Varying recruitment rates (parameter R, black $R = 0.04$, blue $R = 0.05$, red $R = 0.06$). (C) Varying base mortality rates (Mortality in year 2010, M_{2010}, black $M_{2010} = 0.076$, blue $M_{2010} = 0.096$, red $M_{2010} = 0.116$). (D) Varying the effect of the local stressor on mortality rate (parameter a_2, black $a_2 = 0.02$, blue $a_2 = 0.03$, red $a_1 = 0.04$). In (D), the simulations without the local stressor overlay each other.

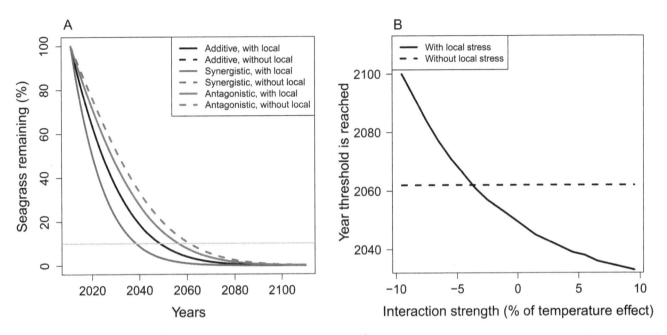

Figure 5. Seagrass density declines for different interaction types when mortality from the interaction term is additional to the base rate mortality. (A) Seagrass density decline with and without the local stressor for additive, synergistic (5% of temperature effect size) and antagonistic interactions (−2.5% of temperature effect size). The grey line represents the 10% seagrass density threshold where seagrass loss is believed to be irreversible. The interaction scenarios with the local stressor almost perfectly overlay each other. (B) Year of seagrass loss for different interaction strengths (positive is synergistic, negative is antagonistic) when the local stressor is not managed (solid line) and when the local stressor is removed (dashed line).

Community Response to Stressor Interactions: Stressor Co-tolerance of Coral Reef Fish

We first considered the number of species remaining when subject to both a global stressor (e.g. habitat loss by coral bleaching) and a local stressor (e.g. fishing pressure) for different species co-tolerance patterns. Regardless of the co-tolerance type, the number of species conserved by removing the local stressor was the greatest when the magnitude of the climate stressor was the smallest (Fig. 6). If the number of species lost in response to a global stressor was a large proportion of the total number of species, then removing the local stressor had little benefit. The type of co-tolerance had the greatest effect on the number of species conserved when the global stressor was of intermediate magnitude (Fig. 6). The most species were benefitted by local management if there is a negative co-tolerance relationship (synergism), whereas the fewest species benefit if there was a positive co-tolerance relationship (antagonism). Hence, while antagonisms (positive co-tolerance) resulted in loss of fewer species, management that assumed an additive or synergistic interaction would not be effective.

In reality, coral reef fish may show negative co-tolerance to climate and fishing, with convex rather than linear co-tolerance. Modelling the response of coral reef species richness to reductions in fishing indicated that large gains in species richness were made for intermediate climate impacts (Fig. 7). Thus, this empirical example with a convex rather than linear relationship, and with species distributed log-normally on the stressor axes, was consistent with the theoretical model of species co-tolerance (Fig. 7).

Discussion

We explored the impact of managing local stressors when there are interactions with global stressors using two kinds of ecosystem model. Both models demonstrated that management to reduce a local stressor has the largest benefits for ecosystems when there are synergistic interactions. By contrast, reducing the local stressor gave smaller benefits when there was an antagonism, or could even worsen stressor impacts, if there was a mitigative antagonism. Therefore, knowing the type of interaction is important to determine the expected benefits of reducing a local stressor.

There was little that local management could do to counter severe impacts of climate change on populations and communities, even with synergistic interactions or negative co-tolerance relationships. In these cases, reduction of global greenhouse gas emissions is necessary to slow degradation of ecosystems. Without mitigation of global warming, seagrass populations in the Mediterranean Sea will likely fall below the 10% density threshold within the next 100 years regardless of local management interventions [9] (but see [33]) or the type of interaction between local and global stressors. However, management of local stresses did buy more time when there was a synergistic interaction. This added time may provide an opportunity for evolutionary adaptation, development of alternative local management actions, and for mitigation of global warming. Further, climate impacts are spatially variable so there may be opportunities to identify refuges where local management can have the greatest benefits [26].

The co-tolerance model implies that greater numbers of species will be benefitted by local actions when there are negative co-tolerance relationships (synergisms). Whether these species are important for community assembly and ecological functions is an important next question. For instance, different functional groups of coral reef fish species may fall at different places on the co-tolerance curve [12]. This has important implications for ecosystem function because different groups of coral reef fish perform important functional roles on reefs, such as herbivory [22,35]. For coral reef fish, functionally important species tend to have greater sensitivity to fishing than to climate, so management of fishing can contribute to conserving these functions [12]. It may

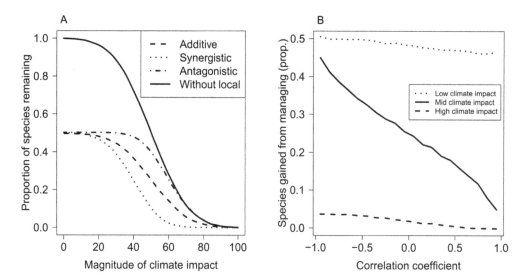

Figure 6. Predicting species loss with co-tolerance relationships. (A) The proportion of species remaining out of 10 000 for different magnitudes of warming temperature. Dashed lines show species remaining with local and global stressors for different interactions. The local stressor was assumed to affect half the species in the absence of the climate stressor. Increasing magnitudes of the climate stressor reduce the proportion of species remaining. The solid line shows the species remaining without the local stress (same for all interaction types). (B) Species gained by reducing the local stressor for different cotolerance strengths (x-axis is the correlation coefficient for stressor responses, negative is synergistic and positive is antagonistic). Management will have the greatest benefit at low climate impact sites (dotted line) and little benefit at high climate impact sites (dashed line), regardless of the interaction type. At moderate impact sites however, there are greatest management gains when there is negative co-tolerance.

be the case in other systems that functionally important species are more sensitive to climate than the local stressor. In these systems, conservation should focus on reducing stressors in refuges from climate change impacts.

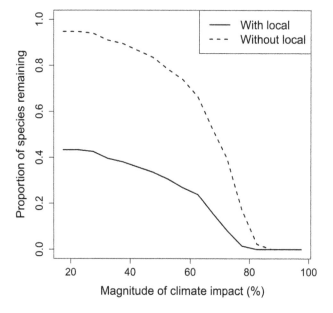

Figure 7. Empirical example of management effectiveness for negative co-tolerance. Proportion of coral reef fish species remaining out of the 134 observed for different magnitudes of climate change impacts (data from [12]). The solid line is for a fishing stressor affecting 50% of species and the dashed line without the fishing impact.

Measuring Interactions

Estimates of interaction effect sizes and types between local and global stressors are helpful for identifying appropriate management actions to global impacts on ecosystems. A major caveat to our models was that appropriate estimates of interaction effect sizes were not readily available. Manipulative studies in the field and laboratory can empirically estimate interaction effect sizes. Concurrent time-series data of ecosystem indicators and stressor values can also be used to estimate interactions from monitoring data. To be most informative for management, studies of multiple stressors could consider interactions between local and global stressors, rather than solely between global stressors [36,37].

While it has not been widely applied to date, the co-tolerance model provides a complementary method for predicting interaction effects on community traits. For instance, Graham *et al.* [12] characterised the climate and fishing sensitivity of a coral reef fish community as a negative co-tolerance (synergistic) relationship. This conceptual model is still in early stages of development and field tests of how well it predicts community responses to multiple stressors are needed (e.g. [12]). Importantly, immediate stress responses and recovery dynamics of species may lead to very different communities. Species' environmental responses may better be characterised on three or more axes, which encompass groups of species that are numerically dominant in stable productive environments, stressful environments and post-disturbance [38,39]. The community response to multiple stressors will therefore depend on both the intensity and frequency of disturbances and the variability in recovery rates of species.

A challenge for predicting responses to management of a local stressor is that stressors indirectly affect state variables of management interest, through their direct effects on physiological and population processes, such as mortality rates. For the seagrass population model, warming and water quality stressors had linear effects on mortality rate, but because mortality is cumulative, the stresses had non-linear effects on seagrass density. Making empirical measurements of processes is more time consuming than measurements of state and often not practical in experimental settings. For instance, estimating mortality rate requires at least two counts of population size for each stressor's level [29], and it is difficult to obtain unbiased estimates of mortality in many aquatic organisms, such as seagrass [33,40]. It is important to estimate mortality rate accurately for predicting declines and recovery in populations. However, our qualitative findings from the population model were robust to higher and lower mortality rate estimates.

Stress responses of populations and communities should not always be expected to be as straightforward as in the simple models used here. Studies of seagrass responses to warming and eutrophication indicate that stress responses can involve multiple traits, including changes to photosynthesis and respiration rates and interactive effects can vary for these different physiological traits [27]. Further, the physiological response to an environmental stressor can be non-linear, so extrapolation from linear models may fail to predict the ecological response to management. If the physiological response is a monotonic function of the stresses, our general result, that managing under a synergism gives greater benefits than under an antagonism, will hold for population density. However, physiological responses to stressors can also be non-monotonic. For instance, growth, mortality and reproductive rates for a species increase from cool to moderate temperatures, but rapidly decline beyond an optimal temperature [41]. Whether seagrass responds to changes in an environmental variable as a stress or a benefit may also vary seasonally [30]. For community traits, species may shift their tolerance of one stressor in response to presence of another stressor, meaning that co-tolerance curves can be dynamic [3]. For instance, fished species may be more sensitive to climate change impacts than unfished species [42].

Investigating non-linear and multiple trait stress responses in manipulative studies requires large amounts of replication and often logistically unfeasible experimental designs. For practical reasons, most experimental studies of interactions use only two levels for each stressor [16,17,37]. Further, there are often scale disparities in stressor impacts that make empirical estimation of interactions challenging. For instance, the effects of ocean acidification on physiology of marine organisms are amenable to laboratory experiments, whereas the effects of fishing on food webs are not. Conducting adequately replicated manipulative experiments in field sites may be prohibitively expensive and often, politically unacceptable [43].

Models to Inform Management

There are several models that could be used to inform appropriate management actions when there are interactions between local and global stressors. Stressor co-tolerance relationships can be built on the basis of literature reviews and may be a rapid way of estimating interactions. Analyses such as Graham et al.'s [12] can inform management directly, such as in predicting how placing marine reserves will affect reef communities impacted by warming. Measurements of stressor interactions on multiple physiological traits, such as seagrass photosynthesis and respiration, can be integrated into process-based models to extrapolate outcomes for ecological states under different management responses (e.g. [44]). Process-based models can also integrate stressor impacts from experimental studies and larger-scale field studies, such as effects of fishing and ocean acidification on marine food webs [45,46]. Considering interactions between species may be particularly important, because species interactions may often be a mechanism for mitigative effects. This is a caveat to the co-

tolerance model, which currently does not consider dynamic interactions.

Implications and Conclusions

Antagonisms are often perceived as less of a concern than synergisms, because impacts of multiple stressors on ecosystems will be smaller, so management in these circumstances may be viewed as less urgent. This view has contributed to the bias in the present literature towards analyses of synergisms [16]. Our analysis demonstrates a need for a more balanced research agenda that identifies both synergisms and antagonisms. Reducing stressors that interact antagonistically may be a lower priority for management, but if an antagonism is assumed to be additive or synergistic, attempts at improving ecosystems may be foiled. Acting when there are antagonisms is unlikely to have negative effects on the ecosystem, but it does waste effort and resources that could be used to have greater benefits elsewhere. For instance, global warming and fishing may have an antagonistic effect on coral reefs, so marine reserves will be of greatest benefit to corals if they are placed in refuges from warming [26].

Resources to determine the type of interaction may not be available when a decision regarding management of a local stressor must be made, so management must assume that a particular type of interaction is prevalent. Previous research has suggested it is conservative to assume additive interactions (e.g. [1,47]). Assuming that additive interactions are common is an appropriate middle ground if the aim is to estimate the potential impact of stressors on an ecosystem and it is not known whether synergistic or antagonistic interactions are more likely. Our research indicates that for making management decisions, it can be more conservative to assume either antagonistic or synergistic interactions, depending on the goals of management [23] and the action being taken to address a local stressor. For instance, the choice may be the location at which to act on a local stressor (such as where to place marine reserves to reduce fishing pressure) and the goal to ensure some sites maintain healthy ecosystems. In this instance, the most conservative strategy is to assume an antagonistic interaction, and undertake remediation of local stressors in climate refuges. Reducing local stressors in sites impacted strongly by climate change will provide no benefit if there is an antagonism. Alternatively, the choice may be to act, or not act, to reduce a local stressor in one location. In this case, assuming an additive or synergistic interaction is more conservative from an environmental perspective, because it implies taking action rather than no action. Further research, using risk analysis techniques, is needed to elucidate what the most conservative assumption is for different management contexts when interactions cannot be estimated.

Mitigative antagonisms may be particularly challenging for management, because outcomes may ultimately be worse than under no management. In particular, interactions need to be considered across food webs, rather than just individual species, because mitigative antagonisms commonly show up in food web models [45]. Future syntheses are needed that determine the prevalence of mitigative antagonisms, so the magnitude of this management risk can be identified.

The challenges antagonisms pose for ecosystem management have previously been pointed out by other authors, where it has been argued that management of ecosystems with antagonisms requires action on both global and local stresses [48]. With delays on mitigation of global stressors, management requires alternative approaches that can work at local scales. This study suggests management priorities can be adapted to accommodate interactions with climate change, provided climate impacts are not severe. For instance, local stresses that interact synergistically, rather than antagonistically, with climate change should be a priority for management action. Climate change impacts are also spatially variable, so management faced with antagonisms could identify refuges from climate change where management of local stresses will have greatest benefits (e.g. [23]). Incorporating these interactions into schemes for prioritising management action (e.g. [18,19,49]) is an important next step.

Acknowledgments

We thank Nicholas Graham for providing the coral reef fish data and for comments that improved the manuscript and Gary Griffith for discussions. We are also thankful for comments from two anonymous reviewers that improved this manuscript.

Author Contributions

Conceived and designed the experiments: CJB MIS HPP AJR. Performed the experiments: CJB. Analyzed the data: CJB MIS HPP AJR. Contributed reagents/materials/analysis tools: CJB MIS. Wrote the paper: CJB MIS HPP AJR.

References

1. Halpern BS, Walbridge S, Selkoe KA, Kappel CV, Micheli F, et al. (2008) A global map of human impact on marine ecosystems. Science 319: 948–952.

2. Hof C, Araujo MB, Jetz W, Rahbek C (2011) Additional threats from pathogens, climate and land-use change for global amphibian diversity. Nature 480: 516–519.

3. Vinebrooke RD, Cottingham KL, Norberg J, Scheffer M, Dodson SI, et al. (2004) Impacts of multiple stressors on biodiversity and ecosystem functioning: the role of species co-tolerance. Oikos 104: 451–457.

4. Morgan CL (2011) Limits to Adaptation: A Review of Limitation Relevant to the Project "Building Resilience to Climate Change - Coastal Southeast Asia". Gland, Switzerland: IUCN.

5. Saunders DL, Meeuwig JJ, Vincent ACJ (2002) Freshwater protected areas: Strategies for conservation. Conserv Biol 16: 30–41.

6. Wright SJ (2005) Tropical forests in a changing environment. Trends in Ecology & Evolution 20: 553–560.

7. Parmesan C, Yohe G (2003) A globally coherent fingerprint of climate change impacts across natural systems. Nature 421: 37–42.

8. Harley CDG, Randall Hughes A, Hultgren KM, Miner BG, Sorte CJB, et al. (2006) The impacts of climate change in coastal marine systems. Ecology Letters 9: 228–241.

9. Jorda G, Marba N, Duarte CM (2012) Mediterranean seagrass vulnerable to regional climate warming. Nature Climate Change 2: 821–824.

10. Mantyka-Pringle CS, Martin TG, Rhodes JR (2012) Interactions between climate and habitat loss effects on biodiversity: a systematic review and meta-analysis. Global Change Biology 18: 1239–1252.

11. Hoegh-Guldberg O, Mumby PJ, Hooten AJ, Steneck RS, Greenfield P, et al. (2007) Coral reefs under rapid climate change and ocean acidification. Science 318: 1737–1742.

12. Graham NAJ, Chabanet P, Evans RD, Jennings S, Letourneur Y, et al. (2011) Extinction vulnerability of coral reef fishes. Ecology Letters 14: 341–348.

13. Thomas CD, Cameron A, Green RE, Bakkenes M, Beaumont LJ, et al. (2004) Extinction risk from climate change. Nature 427: 145–148.

14. Carilli JE, Norris RD, Black BA, Walsh SM, McField M (2009) Local stressors reduce coral resilience to bleaching. Plos One 4: e6324.

15. Anthony KRN, Connolly SR, Hoegh-Guldberg O (2007) Bleaching, Energetics, and Coral Mortality Risk: Effects of Temperature, Light, and Sediment Regime. Limnology and Oceanography 52: 716–726.

16. Darling ES, Cote IM (2008) Quantifying the evidence for ecological synergies. Ecology Letters 11: 1278–1286.

17. Crain CM, Kroeker K, Halpern BS (2008) Interactive and cumulative effects of multiple human stressors in marine systems. Ecology Letters 11: 1304–1315.

18. Klein CJ, Ban NC, Halpern BS, Beger M, Game ET, et al. (2010) Prioritizing Land and Sea Conservation Investments to Protect Coral Reefs. Plos One 5: e12431.

19. Halpern BS, Lester SE, McLeod KL (2010) Placing marine protected areas onto the ecosystem-based management seascape. Proceedings of the National Academy of Sciences 107: 18312–18317.

20. Cote IM, Darling ES (2010) Rethinking ecosystem resilience in the face of climate change. PloS Biology 8: e1000438.

21. Rietkerk M, van de Koppel J (1997) Alternate stable states and threshold effects in semi-arid grazing systems. Oikos 79: 69–76.

22. Mumby PJ, Hastings A, Edwards HJ (2007) Thresholds and the resilience of Caribbean coral reefs. Nature 450: 98–101.

23. Game ET, McDonald-Madden E, Puotinen ML, Possingham HP (2008) Should We Protect the Strong or the Weak? Risk, Resilience, and the Selection of Marine Protected Areas. Conserv Biol 22: 1619–1629.

24. Ling SD, Johnson CR, Frusher SD, Ridgway KR (2009) Overfishing reduces resilience of kelp beds to climate-driven catastrophic phase shift. Proceedings of the National Academy of Sciences of the United States of America 106: 22341–22345.

25. Parsons PA (1991) Evolutionary Rates: Stress and Species Boundaries. Annual Review of Ecology and Systematics 22: 1–18.

26. Darling ES, McClanahan TR, Côté IM (2010) Combined effects of two stressors on Kenyan coral reefs are additive or antagonistic, not synergistic. Conservation Letters 3: 122–130.

27. Touchette BW, Burkholder JM, Glasgow HB (2003) Variations in eelgrass (Zostera marina L.) morphology and internal nutrient composition as influenced by increased temperature and water column nitrate. Estuaries 26: 142–155.

28. Burkholder J, Tomasko D, Touchette B (2007) Seagrasses and eutrophication. Journal of Experimental Marine Biology and Ecology 350: 46–72.

29. Marba N, Duarte CM (2010) Mediterranean warming triggers seagrass (Posidonia oceanica) shoot mortality. Global Change Biology 16: 2366–2375.

30. Rasheed MA, Unsworth RKF (2011) Long-term climate-associated dynamics of a tropical seagrass meadow: implications for the future. Marine Ecology-Progress Series 422: 93–103.

31. Short FT, Neckles HA (1999) The effects of global climate change on seagrasses. Aquatic Botany 63: 169–196.

32. Orth R, Marion S, Moore K, Wilcox D (2010) Eelgrass (Zostera marina L.) in the Chesapeake Bay region of mid-Atlantic coast of the USA: Challenges in conservation and restoration. Estuaries and Coasts 33: 139–150.

33. Altaba CR (2013) Climate warming and Mediterranean seagrass. Nature Climate Change 3: 2–3.

34. Jorda G, Marba N, Duarte CM (2013) Climate warming and Mediterranean seagrass. Nature Climate Change 3: 3–4.

35. Bellwood DR, Hughes TP, Nystrom M (2004) Confronting the coral reef crisis. Nature 429: 827–833.

36. Brown CJ, Schoeman DS, Sydeman WJ, Brander K, Buckley LB, et al. (2011) Quantitative approaches in climate change ecology. Global Change Biology 17: 3697–3713.

37. Wernberg T, Smale DA, Thomsen MS (2012) A decade of climate change experiments on marine organisms: procedures, patterns and problems. Global Change Biology 18: 1491–1498.

38. Darling ES, Alvarez-Filip L, Oliver TA, McClanahan TR, Côté IM (2012) Evaluating life-history strategies of reef corals from species traits. Ecology Letters 15: 1378–1386.

39. Grime JP (1977) Evidence for the existence of three primary strategies in plants and its relevance to ecological and evolutionary theory. American Naturalist 111: 1169–1194.

40. Ebert TA, Williams SL, Ewanchuk PJ (2002) Mortality estimates from age distributions: Critique of a method used to study seagrass dynamics. Limnology and Oceanography: 600–603.

41. Portner HO, Farrell AP (2008) Physiology and Climate Change. Science 322: 690–692.

42. Planque B, Fromentin JM, Cury P, Drinkwater KF, Jennings S, et al. (2010) How does fishing alter marine populations and ecosystems sensitivity to climate? Journal of Marine Systems 79: 403–417.

43. Grantham HS, Bode M, McDonald-Madden E, Game ET, Knight AT, et al. (2010) Effective conservation planning requires learning and adaptation. Frontiers in Ecology and the Environment 8: 431–437.

44. Carr JA, D'Odorico P, McGlathery KJ, Wiberg PL (2012) Modeling the effects of climate change on eelgrass stability and resilience: future scenarios and leading indicators of collapse. Marine Ecology Progress Series 448: 289–301.

45. Griffith GP, Fulton EA, Richardson AJ (2011) Effects of fishing and acidification-related benthic mortality on the southeast Australian marine ecosystem. Global Change Biology 17: 3058–3074.

46. Griffith GP, Fulton EA, Gorton R, Richardson AJ (2012) Predicting interactions among fishing, ocean warming, and ocean acidification in a marine system with whole-ecosystem models. Conserv Biol 26: 1145–1152.

47. Ban NC, Alidina HM, Ardron JA (2010) Cumulative impact mapping: Advances, relevance and limitations to marine management and conservation, using Canada's Pacific waters as a case study. Marine Policy 34: 876–886.

48. Selig ER, Casey KS, Bruno JF (2012) Temperature-driven coral decline: the role of marine protected areas. Global Change Biology 18: 1561–1570.

49. Mumby PJ, Elliott IA, Eakin CM, Skirving W, Paris CB, et al. (2011) Reserve design for uncertain responses of coral reefs to climate change. Ecology Letters 14: 132–140.

5

Identification of Land-Cover Characteristics Using MODIS Time Series Data: An Application in the Yangtze River Estuary

Mo-Qian Zhang, Hai-Qiang Guo, Xiao Xie, Ting-Ting Zhang, Zu-Tao Ouyang, Bin Zhao*

Coastal Ecosystems Research Station of the Yangtze River Estuary, Ministry of Education Key Laboratory for Biodiversity Science and Ecological Engineering, Institute of Biodiversity Science, Fudan University, Shanghai, P.R. China

Abstract

Land-cover characteristics have been considered in many ecological studies. Methods to identify these characteristics by using remotely sensed time series data have previously been proposed. However, these methods often have a mathematical basis, and more effort is required to better illustrate the ecological meanings of land-cover characteristics. In this study, a method for identifying these characteristics was proposed from the ecological perspective of sustained vegetation growth trend. Improvement was also made in parameter extraction, inspired by a method used for determining the hyperspectral red edge position. Five land-cover types were chosen to represent various ecosystem growth patterns and MODIS time series data were adopted for analysis. The results show that the extracted parameters can reflect ecosystem growth patterns and portray ecosystem traits such as vegetation growth strategy and ecosystem growth situations.

Editor: Gil Bohrer, The Ohio State University, United States of America

Funding: This research was financially supported by the National Basic Research Program of China (No. 2013CB430404), the Natural Science Foundation of China (grant No 31170450), and the National Key Technology R&D Program (No. 2010BAK69B15). The funders had no role in study design, data collection and analysis, decision to publish, or preparation of the manuscript.

Competing Interests: The authors have declared that no competing interests exist.

* E-mail: zhaobin@fudan.edu.cn

Introduction

Land-cover characteristics and their dynamics have captured much attention in the field of ecology, since land-cover exerts a huge influence over ecosystem biodiversity, water budget [1], energy flow [2], and carbon cycling [3]. Remotely sensed time series data provide an opportunity to identify land-cover characteristics at the temporal scale, which often reflect the features of ecosystem growth patterns. Ecosystem growth patterns can be categorized into four types (adapted from [4]): (i) undisturbed ecosystems; (ii) ecosystems that have suffered coverage damage that either lasted the whole growing season or followed by vegetation restoration in the growing season; (iii) ecosystems that have suffered a phenology change that is expressed as either a shift in the growing season or a shortened growing season; and (iv) ecosystems that underwent changes in both coverage and phenology. However, it is challenging to extract desired land-cover characteristics while remaining independent of inter-annual and inter-class variations [1]. Therefore, proper land-cover characteristic identification methods are needed.

Methods that take into account the temporal features of time series data to identify land-cover characteristics have been developed in recent decades; such methods can be roughly classified into two types. The first type is based on signals observed at different temporal scales: vegetation information is often present at seasonal and inter-annual scales, while noise typically has a higher frequency. By decomposing data into different temporal frequencies, noises can be excluded and parameters can be obtained to reflect long-term trends or seasonal patterns. Research

based on this kind of method includes land-cover classification [5] and long-term vegetation dynamic study [6]. However, the ecological meaning of parameters obtained by this kind of method is often limited, and the relations between parameters and land-cover dynamics need further investigation. The second type of methods is based on land surface phenological stages. The phenological stages recognized by time series data include: (i) constant low/no leaf period in winter when the vegetation is dormant, (ii) rapid vegetation growth period in spring, (iii) a period with relatively stable high aboveground biomass in summer, and (iv) rapid senescence period in autumn [7]. Research based on such methods can provide more detailed ecological information (Table 1) that can be applied to study land surface phenology [8], vegetation response to changing climate [9], zoology [10], and so on.

Though methods based on phenological stages have been widely used in ecological studies, phenological stages are often detected based on mathematical criteria such as choosing a certain threshold or detecting curve changes [8,11]. However, it is difficult to choose a mathematically ideal technique [11], and different analysis methods sometimes provide conflicting results on the same research topic (such as the long-term greenup trend in North America [8]). In this study, we propose a method to identify land-cover characteristics from the ecological perspective of sustained vegetation growth. During the analysis, phenological growth stages were first identified based on sustained vegetation growth trends, and parameters designed to reflect land-cover characteristics were extracted accordingly. Improvement was also made in parameter

Table 1. Summary of vegetation metrics used in time series analysis.

Vegetation metric	Interpretation	References
Greenup	Time represents the start of growing season when plant grows and photosynthesis begins	[10,21,22]
Maturity	Time when green leaf area stabilizes with high photosynthesis activity	[22]
Senescence	Time when plant begins senescence either expressed by green biomass decrease or reduced photosynthesis	[22]
Dormancy	Time represents the end of growing season when photosynthesis reaches its minimum and plants become dominant	[10,21,22]
Length of growing season	Time span between greenup and dormancy which represents the duration of photosynthetic activity	[10,21]
Maximum VI	Highest VIs level in growing season	[21]
Timing of maximum VI	Time when VIs reaches its maximum	[10,21]
Seasonal amplitude	VIs value difference between vegetation dormancy and have the highest aboveground biomass	[10,21]
Annual integration	Sum of VIs values in growing season	[21]
Greenup rate	Growth rate during the period between greenup and mature	[10,21]
Senescence rate	Senescence rate during the period between senescence and dormancy	[10,21]

extraction, which was inspired by a technique used for extracting the hyperspectral red edge position.

Materials and Methods

Ethics Statement

As a field survey conducted for remote sensing research, we did not conduct any activities concern field samplings of soil, plants, or animals in the work. All lands where we conducted the survey are non-fenced public areas and accessed to everyone, thus we do not need to ask for any official permission.

Site Description

This study was conducted on the Chongming Island and the Changxing Island, two alluvial islands in the mouth of the Yangtze River, China (121°10′49″ –121°59′10″E, 31°17′4″ –31°54′20″N, Fig. 1). The area is subject to the northern subtropical monsoon climate, with an average annual temperature of 15.3°C and a total annual precipitation around of 1000 mm. Several large land reclamations have taken place since 1960s, the reclaimed areas are much larger than ordinary farmland and neighboring areas are often under the same land management schemes. Diverse land use and a relatively large reclamation area make the study area suitable for identifying land-cover characteristics with remote sensing data.

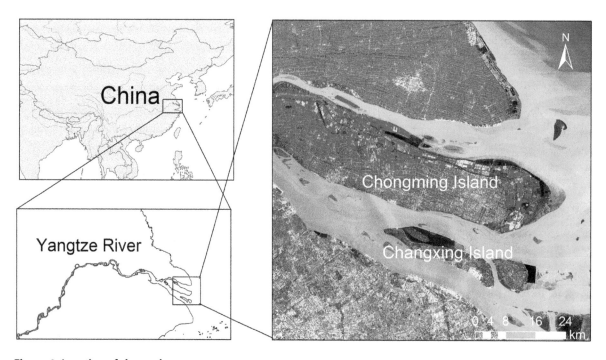

Figure 1. Location of the study area.

Table 2. Descriptions of different land-cover types in study area.

Land-cover types	Description	Vegetation coverage	Disturbance pattern
Urban	Urban area	Low to medium	No
Orchard	Orange tree plantation area	Low to medium	No
Fallow	Farmland where no farming activities conducted, usually covered by natural herbaceous plants such as weed and common reed	Medium to high	No
Cropland-1	Single-cropping farmland with only rice planted from late May to October	High	Yes; Happened early in the year
Cropland-2	Double-cropping farmland with winter wheat planted from late last November to early May and rice planted the same time as cropland1	High	Yes; Happened in the mid year

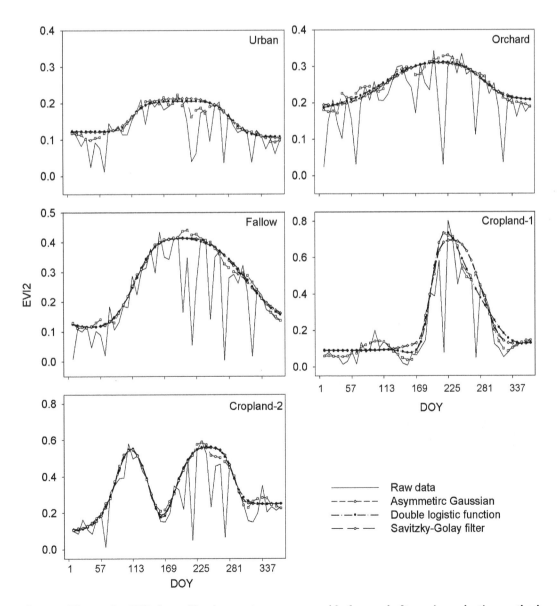

Figure 2. Time series EVI2 data of land cover types processed before and after noise reduction methods.

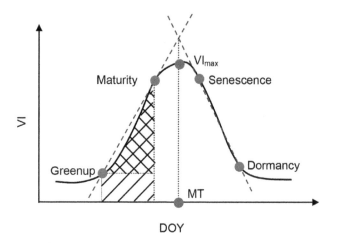

Figure 3. Diagram of parameter extraction from time series vegetation index (VI) data. Blue points represent time points separating different vegetation growth stages, while red points are parts of the extracted parameters.

Analysis Preparation

Remote sensing data. A 250 m 8-day composite surface reflectance data set (MOD09Q1) was used for this study. Satellite quality assurance (QA) data were obtained for further noise reduction, and selected data were derived from MOD09A1 because QA data from MOD09Q1 are insufficient to deliver the actual condition. Remote sensing data for the year 2009, when several field surveys were conducted, were used for analysis. All the remote sensing data used were downloaded from NASA (LP DAAC).

Vegetation indices (VIs) are specially designed indicators that reflect certain properties, such as vegetation coverage (*e.g.*, NDVI, EVI & MSAVI) and land surface water content (*e.g.*, LSWI). A two-band EVI (EVI2) was selected in this study for its superiority over the widely used NDVI [12]. EVI2 is calculated as follows:

$$EVI2 = 2.5 \times (N - R)/(N + 2.4 \times R + 1) \qquad (1)$$

where N and R are reflectance in the near-infrared (NIR) and red bands of MODIS data, respectively.

Field survey. In order to acquire the actual land-cover conditions in different seasons, we conducted three field surveys across the year of 2009. Before the first field survey, historical TM and airborne imageries were studied in the laboratory to identify the relatively homogenous regions for field surveys. During the field surveys, a portable Global Position System (GPS) was used to localize the target ground objects such as cultivated lands, fallow lands, orchards, and buildings. To aid this task, color maps of TM and airborne images were printed beforehand and taken with the investigators for field checks. The field notes were also made and taken to the laboratory for further analysis, such as location check, classification and accuracy assessment.

Land-cover selection. The studied land-cover types were chosen based on ecosystem growth patterns, and five land-cover types (urban, orchard, fallow, and two types of croplands) were chosen for further analysis (Table 2). Among them, urban, orchard, and fallow were used to represent ecosystems that experience a loss in coverage throughout the growing season; cropland-2 was used to represent ecosystems with short-term coverage loss; and cropland-1 was used to represent ecosystems under a growing season shift. Since in the study area no land-cover

type showed the characteristics of ecosystems under a shortened growing season, this ecosystem growth pattern was not included in the present analysis. To better analyze land-cover characteristics, remote sensing pixels that represent only one land-cover type were used in the analysis.

Analysis Techniques

Noise reduction. The asymmetric Gaussian method [13] and double logistic function [14] were chosen for noise reduction in this study, since their ability to maintain the integrity of signals is proven [15]. The Savitzky-Golay filter was also chosen since it could capture detailed variations in time series data and has shown good performance when applied in study related to China [16]. Noise reduction was achieved by using TIMESAT [13,17,18]. Ancillary weights of each data were set according to the QA data. Weights were set at high values for best-quality data (described as *clear* in QA data), at moderate values when data were acquired under less ideal conditions (*cloud shadow* or *mixed*), and at low values when data represent cloudy pixels. Fig. 2 shows the data of different land-cover types represented by EVI2 before and after noise reduction.

Phenological stages discrimination. Though rates of changes in vegetation coverage may vary, the vegetation growth trend inherited in each phenology stage (sustained increase/decrease, or consistency) remained constant for a certain time; therefore, we propose to discriminate phenological stages based on the sustained vegetation growth trend. The sustained trend was recognized by the following procedure: if the increment/decrement between neighboring data was larger than a certain numerical value (the theoretical increase/decrease threshold), we defined it as an increase or a decrease; and if the increment/decrement remained constant for some time (for instance more than one month), the period would be identified as showing a sustained increase/decrease trend. Time points (greenup, maturity, senescence, and dormancy; see Table 1) that separate these phenological stages were identified accordingly. Greenup and maturity were identified as the beginning and ending of the period when vegetation showed a sustained increase trend, respectively; the beginning and ending of the sustained decrease trend were termed as senescence and dormancy, respectively.

The theoretical increase/decrease threshold was calculated as:

$$Threshold = (EVI2_{max} - EVI2_{min})/n \qquad (2)$$

where $EVI2_{max}$ represents the maximum value of each time series data. Because the aboveground biomass of evergreen vegetation may vary in winter, when calculating the theoretical increase/decrease threshold, $EVI2_{min}$ used the minimum value in the first/second half of the year, respectively. The variable n represents the period when vegetation biomass increases/decreases. The length of this period can be determined from long-term field observations. As the theoretical increase/decrease threshold is not supposed to give a quantitative value, the time period used can be longer than actual value. In this study, we simply assumed that the growing season spans the whole year, with vegetation biomass increase and decrease period accounting for half a year each. Further, the corresponding number of MODIS data was used to represent this period. If there were more than one sustained increase periods, the first period was used to identify greenup and maturity. Senescence and dormancy were identified in a similar way, except that the last sustained decrease period was used for the identification when more than one sustained decrease period existed.

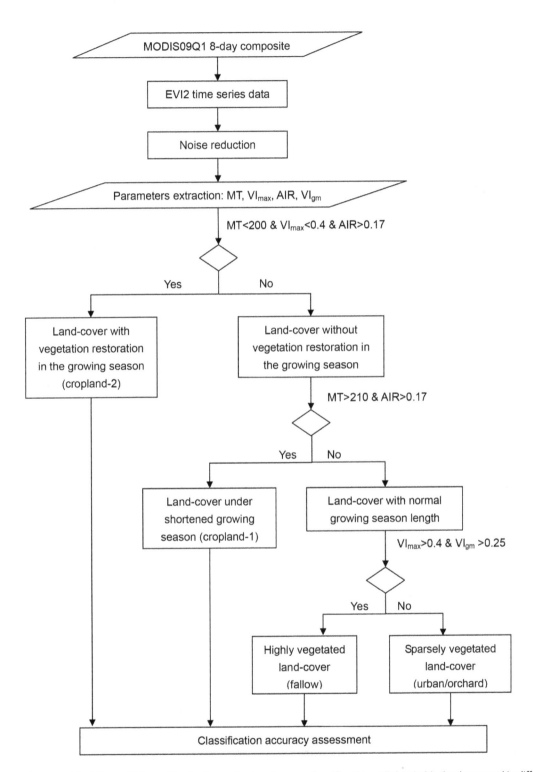

Figure 4. Workflow of hierarchical scheme for land-cover classification. All thresholds that been used in different noise reduction methods were same.

Parameters extraction. The time at which aboveground biomass reaches its maximum (MT, a date) was first identified. MT was extracted by extrapolating two straight lines across the time points that discriminate phenological stages (Fig. 3). This process was inspired by a technique used in hyperspectral analysis, which stabilizes the red edge position when there are multiple peaks in the first derivative curve of hyperspectral data [19]. The

EVI2 extracted on day MT was used to represent the maximum vegetation coverage (VI_{max}, dimensionless). If MOD09Q1 data were missing for that day, VI_{max} was linearly interpolated between the previous and following data.

Two other parameters were further extracted to reflect the vegetation growth status. The average increase rate (AIR, dimensionless) between greenup and maturity was calculated to

Table 3. The mean value and standard deviation (SD) of parameters extracted from time series vegetation index (VI) data with asymmetric Gaussian method (A), double logistic function (B), and Savitzky-Golay filter (C).

A	MT		VI_{max}		AIR		VI_{gm}	
	mean	SD	mean	SD	mean	SD	mean	SD
Urban	200	12.493	0.198	0.035	0.043	0.006	0.149	0.006
Orchard	205	31.842	0.370	0.047	0.059	0.021	0.308	0.041
Fallow	194	10.799	0.496	0.052	0.167	0.025	0.318	0.033
Cropland-1	228	6.799	0.618	0.082	0.250	0.049	0.391	0.035
Cropland-2	166	12.356	0.200	0.070	0.223	0.031	0.327	0.028

B	MT		VI_{max}		AIR		VI_{gm}	
	mean	SD	mean	SD	mean	SD	mean	SD
Urban	204	13.345	0.197	0.037	0.038	0.006	0.149	0.031
Orchard	216	26.139	0.372	0.047	0.053	0.019	0.310	0.043
Fallow	191	10.796	0.504	0.056	0.177	0.029	0.313	0.033
Cropland-1	229	10.878	0.625	0.085	0.252	0.067	0.390	0.040
Cropland-2	168	11.090	0.188	0.081	0.216	0.032	0.327	0.029

C	MT		VI_{max}		AIR		VI_{gm}	
	mean	SD	mean	SD	mean	SD	mean	SD
Urban	189	20.587	0.188	0.081	0.034	0.015	0.140	0.037
Orchard	185	40.105	0.350	0.052	0.063	0.025	0.290	0.055
Fallow	192	10.531	0.495	0.069	0.154	0.041	0.288	0.055
Cropland-1	228	9.770	0.628	0.092	0.285	0.077	0.349	0.047
Cropland-2	169	12.665	0.199	0.076	0.229	0.032	0.331	0.035

reflect how vegetation grows from the minimum vegetation coverage to a relatively stable status. The mean EVI2 between greenup and maturity (VI_{gm}, dimensionless) focuses on the average status in the sustained growing period

$$AIR = \sum_{a}^{b} \Delta VI_i \Big/ (T_{growth} - 1) \qquad (3)$$

$$VI_{gm} = \sum_{a}^{b} VI_i \Big/ T_{growth} \qquad (4)$$

where a and b represent greenup and maturity respectively; $\sum_{a}^{b} \Delta VI_i$ represents the accumulated increments of EVI2 in the sustained growing period (backslash region in Fig. 3); $\sum_{a}^{b} VI_i$ is the EVI2 accumulation in the same period (slash region in Fig. 3); T_{growth} represents the time span between greenup and maturity (Fig. 3), and we used the number of MODIS data to represent this period.

Different ecosystem growth patterns can be expressed by parameter differentiations. Coverage differentiation would be most evident for ecosystems that have suffered vegetation coverage loss lasted the growing season, and the maximum vegetation coverage (hence the values of VI_{max}) would then reduce accordingly. Since total vegetation coverage increase/decrease status is related to the maximum coverage, the values of AIR and VI_{gm} will also decrease. In ecosystems that are under short-term vegetation loss, VI_{max} will more or less represent the coverage during the period of vegetation damage and not the maximum coverage in the growing season; therefore, VI_{max} value will decrease. However, the changes in phenology or total vegetation coverage (and hence parameters of MT, AIR, and VI_{gm}) depend on the severity and duration of the damage. Phenology differentiation is the most obvious characteristic of an ecosystem undergoing a growing season shift, and the value of MT would change accordingly. Ecosystems with shortened growing seasons exhibit slight shifts of phenology and accelerated vegetation coverage increase/decrease rates. All parameters would change in ecosystems that undergo both coverage and phenological changes.

Land-cover classification. In order to test whether the extracted parameters could be used for actual land-cover change detection, a hierarchical classification scheme was adopted to classify the studied land-cover types (Fig. 4). The parameters used in each classification level were chosen based on the aim of the classification, with each parameter aimed to discriminate only one aspect of the land-cover (sparsely/densely planted, with/without phenological shift, or high/low growth rate). For example, in this study, both coverage and phenology in cropland-2 are distinct from other land-cover types, hence three parameters, MT, VI_{max}, and AIR were chosen in the first classification level. Land-cover types were not sub-classified artificially from each other if no apparent differences in ecosystem features were detected.

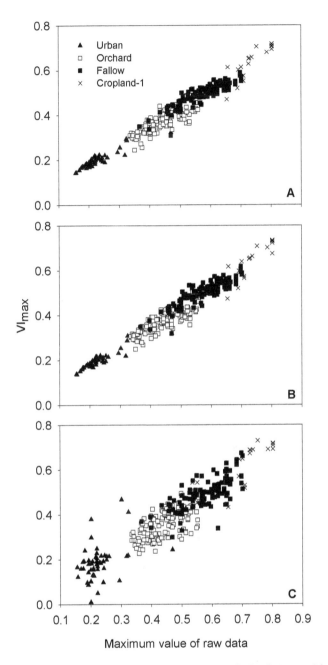

Figure 5. Results evaluation by comparing derived VI_{max} with corresponding maximum values of raw data. Frames showed results that obtained with asymmetric Gaussian method (A) ($R^2 = 0.944$), double logistic function (B) ($R^2 = 0.947$), and Savitzky-Golay filter (C) ($R^2 = 0.837$) respectively.

The thresholds used for classification were set roughly according to predefined criteria rather than on the basis of training data, and hence all original data were used for validation. The thresholds were defined by the following criteria. The threshold of VI_{max}/VI_{gm} was set as the arithmetic mean value of soil background and the highest/mean VI value of pixels with the highest vegetation coverage. MT, being representative of phenological information, would change in ecosystems under phenological changes. The threshold of AIR was set as the arithmetic mean value of the observed highest values and the lowest ones. All thresholds used in classification are the same for data processed by different noise

reduction methods. Confusion matrixes were used to evaluate classification accuracies.

Results

Basic Characteristics of Land-cover Types

The extracted parameters can reflect the basic characteristics of different land-cover types (Table 3). MT value changes reflect changes in vegetation phenology. In cropland-1, MT values changes as the growing season has shifted, and these values are the largest among all land-cover types. In cropland-2, since human interference has actually altered vegetation phenology, the MT values have also shifted and are the smallest of all land-cover types. Urban, orchard, and fallow have intermediate MT values, which reflects the fact that the vegetation phenology has not changed here.

VI_{max} reflect the changes in ecosystem coverage. Since cropland-1 did not undergo vegetation coverage loss, this land-cover type has the highest VI_{max} values (Table 3). VI_{max} values of fallow, orchard, and urban decrease with reduced vegetation coverage. Because the tree density in orchard areas is not high, vegetation coverage of orchards is no larger than that of fallow areas (as indicated in Fig. 5); hence, it is understandable that the average VI_{max} values of orchard are lower than those of fallow. In cropland-2, as MT occurs during the time right after rice transplantation at when the land is barely covered, the values of VI_{max} are not high.

AIR and VI_{gm} are parameters that reflect vegetation growth status. On the whole, the change patterns of AIR and VI_{gm} values are similar to those of VI_{max}, with cropland-1 having the highest values, followed by fallow and orchard, and urban areas having the smallest values. However, in cropland-2, as VI_{max} values do not reflect the maximum vegetation coverage in the growing season, AIR and VI_{gm} values do not follow the trend exhibited in VI_{max}.

MT Results Evaluation

An evaluation was performed on MT results to illustrate the variation in the values (Table 3), because although MT values are quite similar for data processed by different noise reduction methods in fallow, cropland-1, and cropland-2, the results of urban and orchard varied with methods and much lower MT values were obtained when using the Savitzky-Golay filter. As there is no readily available evaluation method, we chose an indirect means of assessment. VI_{max} is based on the position of MT, and a departure of MT from the time of the highest aboveground biomass would result in a decrease in the VI_{max} value. Thus, the VI_{max} value can serve as an indicator for MT evaluation, and comparisons of VI_{max} with the corresponding maximum values of raw data are shown in Fig. 5. The asymmetric Gaussian method and double logistic function provided satisfying results; however, the coefficient of the relationship (R^2) is much lower when using the Savitzky-Golay filter, which indicates greater errors in data processed by this method. Therefore, the observed MT variations in Table 3 should be the result of unstable performance of the Savitzky-Golay filter when it is applied for areas with low vegetation coverage, as vegetation signals are weak and more sensitive to noise in such areas. Cropland-2 was excluded from this evaluation because its VI_{max} values do not reflect the maximum vegetation coverage in the growing season.

Ecosystem Traits Detection

Because the species composition varied among ecosystems, the vegetation growth condition expressed at ecosystem scale differed,

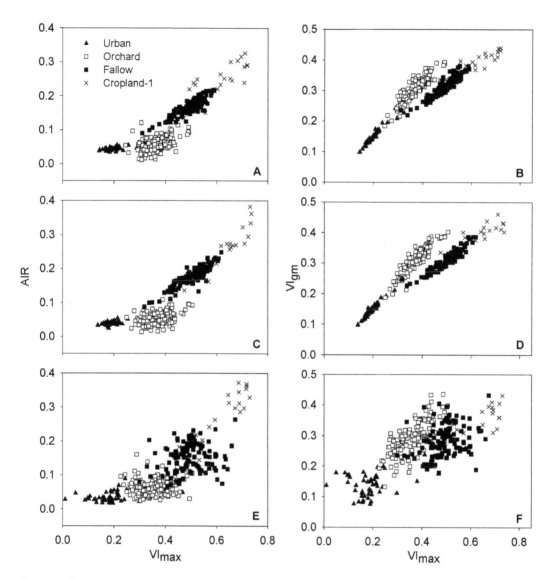

Figure 6. The cross-comparisons among parameters extracted. The left column represent the comparisons between AIR and VI_{max} obtained via asymmetric Gaussian method (A), double logistic function (C), and Savitzky-Golay filter (E), respectively. The right column represent the comparisons of VI_{gm} with VI_{max} by using asymmetric Gaussian method (B), double logistic function (D), and Savitzky-Golay filter (F).

and this trait is inherent in ecosystems. However, AIR and VI_{gm} cannot be used for direct detection of such differences, because the influence of coverage would hide them. By using VI_{max}, the coverage differentiation can be partially minimized (Fig. 6), and vegetation growth traits can be conveyed through slope changes.

The differences between vegetation growth rates are shown in Fig. 6A, C, and E. In Fig. 6A and C, slopes of cropland-1 and fallow are larger than those of urban and orchard, indicating that under same coverage, cropland-1 and fallow grow faster. Although intra-class variations in data processed by Savitzky-Golay filter are larger than those in the data processed by the other two methods, a similar pattern could also be observed (Fig. 6E). Since orchard areas comprising woody plants and trees are the dominate urban vegetation in the study area, the slope differences seen in the figures indicate the differences between herbaceous vegetation and woody plants, as herbaceous plants grow faster that woody plants under suitable conditions.

The differences between ecosystem average growing conditions are shown in Fig. 6 B, D, and F. In these figures, the slopes of

orchard are higher than those of cropland-1 and fallow, indicating that the average coverage of orchard is larger at ecosystem level. Since the woody plants in orchard are evergreen and herbaceous plants senescence every year, the aboveground biomasses are different when spring comes, therefore, the coverage of woody plants increased faster. Because some of the trees are deciduous in urban area, the slopes of urban are slightly lower than those of orchard. Cropland-2 was excluded in this part of analysis since its VI_{max} values do not reflect the maximum vegetation coverage in the growing season.

Classification Accuracy Assessment

Confusion matrixes were used to evaluate classification accuracy, (Table 4) and were made for data processed by each different noise reduction method. All noise reduction methods achieve relatively high overall classification accuracy. The user's accuracy of fallow is the lowest in all methods, and errors mainly arise from misclassification of urban/orchard. In our study, the chosen land-cover types are only based on actual land surface situations and we

Table 4. Land classification accuracy assessments of data processed by different smoothing methods (A) asymmetric Gaussian method, (B) double logistic function, and (C) Savitzky–Golay filter.

A	Reference data					
Classification	**Urban/orchard**	**Fallow**	**Cropland-1**	**Cropland-2**	**Total**	**User's accuracy (%)**
Urban/orchard	144	5	0	29	178	80.90
Fallow	36	115	2	9	162	70.99
Cropland-1	0	2	19	0	21	90.48
Cropland-2	0	0	0	439	439	100
Total	180	122	21	477		
Producer's accuracy (%)	80	94.26	90.48	92.03		
Overall accuracy: 89.63%			**Kappa: 0.825**			
B	Reference data					
Classification	**Urban/orchard**	**Fallow**	**Cropland-1**	**Cropland-2**	**Total**	**User's accuracy (%)**
Urban/orchard	147	5	0	34	186	79.03
Fallow	33	117	2	16	168	69.64
Cropland-1	0	0	19	0	19	100
Cropland-2	0	0	0	427	427	100
Total	180	122	21	477		
Producer's accuracy (%)	81.67	95.90	90.48	89.52		
Overall accuracy: 88.75%			**Kappa: 0.811**			
C	Reference data					
Classification	**Urban/orchard**	**Fallow**	**Cropland-1**	**Cropland-2**	**Total**	**User's accuracy (%)**
Urban/orchard	144	29	0	18	191	75.39
Fallow	36	91	2	13	142	64.08
Cropland-1	0	1	19	0	20	95
Cropland-2	0	1	0	446	447	99.78
Total	180	122	21	477		
Producer's accuracy (%)	80	74.59	90.48	93.50		
Overall accuracy: 87.5%			**Kappa: 0.786**			

Integers in tables represent the number of pixels that belongs to a certain classification condition.

did not set any predefined coverage criterion for data selection. Therefore, the actual coverage of fallow, urban, and orchard can overlap (also indicated in Fig. 5), and hence misclassifications are acceptable.

Discussion

Phenological Stages Identification

Remote sensing phenological stages that used to extract land-cover characteristics are often identified by discriminating the time points that separate them. However, it is difficult to choose a mathematically ideal method [11]. Furthermore, as these points are timings that represent dynamic vegetation growth conditions, it is difficult to directly evaluate the results from field phenology observations, because the two kinds of data are not measured at the same spatial scale and often represent different ground phenological events [8]. Therefore, we turned to the sustained vegetation growth trend that phenological stages inherently

exhibited, and time points were thus identified. In this identification process, as the theoretical increase/decrease threshold is not to give a precise quantitative value, the process can be flexible when applied to large scale analysis. Besides, this method can adjust itself according to the maximum and minimum values of the time series data of each pixel. Although time points were only used for later land-cover characteristic identification in this study, they can also be used in land surface phenology research.

Land-cover Characteristics Identification

Parameters were extracted to reflect land-cover characteristics. In this study, the time of highest vegetation biomass (MT) was detected first, and the VI value that represented maximum vegetation coverage (VI_{max}) was identified accordingly. Compared with commonly used methods [10], MT was extracted based on temporal features of time series data rather than by using a single maximum value, this makes it more resistant to variations caused by noise. In ecosystems where growth patterns change as a result of disturbances (either in coverage or phenology caused by events such as insect defoliation, windfall, and wildfire), the time points and therefore MT would change consequently; hence, MT can also be used as an indicator of disturbances. This is especially convenient if an irregular growing season was caused by such a disturbance. When MT is combined with VI_{max}, subtle vegetation damages can be more evident. However, the time points of greenup and dormancy are sensitive to the start of spring and the end of autumn, which make these time points vulnerable to inter-annual meteorological variations. In order to obtain a more stable inter-annual result of MT, adjustments such as use of meteorology data or reference area are recommended.

AIR and VI_{gm} can portray ecosystem traits that represent how ecosystems grow. The trait difference between ecosystems with different vegetation composition is especially evident when coverage differences are minimized (Fig. 6). Though effort has been made to discriminate land-cover types that have different species composition by comparing growing season NDVI [20], this method can further explore the temporal features of ecosystems. Therefore, this method has potential for monitoring land-cover changes caused by species variation (such as species invasion and vegetation succession). Similar parameters extracted from vegetation biomass decrease period can also be used for detecting how vegetation senescence. This kind of information could help us to understand ecosystem changes in more detail, and help us to further explore ecosystem processes and functions, as well as the causes of the ecosystem changes.

Land-cover Classification

As the extracted parameters incorporated both spectral and temporal features, land-cover characteristics can be better explored. Results of this study show that this kind of land-cover classification can achieve relatively satisfying results in practice. Classification schemes that include these parameters will facilitate land-cover mapping in complicated situations, such as in regions where the differences between land-cover types are subtle, or in areas with irregular growing seasons.

The Performance of Noise Reduction Methods

Although an 8-day composition scheme is adopted in MODIS products, the presence of cloud remains a problem in retrieving land-cover characteristics in our study area. Therefore, the performance of noise reduction methods affects the ultimate results. Our results show that the asymmetric Gaussian method and double logistic function performed better than the Savitzky-Golay filter, and that some apparent discrepancies exist in the Savitzky-Golay filter. For example, in Fig. 5C some VI_{max} values in urban are obviously larger than the maximum values of raw data (such as 0.380 of VI_{max} corresponds to 0.201 of maximum raw data). This indicates larger errors in the noise reduced data, and indicates that the Savitzky-Golay filter is less robust in areas where vegetation is sparse and noises are frequent. It further confirms a conclusion obtained by [15] that the asymmetric Gaussian method and double logistic function can maintain the integrity of signals and that the Savitzky-Golay filter is sensitive to noise. Our results also give a direct illustration that the Savitzky-Golay filter is not suitable to deal with noise contaminated data at the seashore.

Conclusion and Outlook

In this study, we tried to identify land-cover characteristics based on the consideration of sustained vegetation growth trends. During this process, an improvement was also made by simulating a method used for determining the hyperspectral red edge position. Our results show that this method can capture ecosystem growth patterns and more detailed ecosystem traits such as species growing strategy and ecosystem growth status. This method has a potential in land-cover dynamic studies related to vegetation coverage and composition changes (such as ecosystem damage evaluation, invasive species monitoring, and vegetation succession validation), and also in land surface phenology monitoring. When combined with auxiliary data, such as soil properties, or carbon fluxes between land surface and atmosphere, improvement in the understanding of human-environment interactions and influence of changes in one ecosystem on another can be conceived.

Author Contributions

Conceived and designed the experiments: MQZ BZ. Performed the experiments: MQZ HQG XX TTZ ZTO. Analyzed the data: MQZ BZ. Contributed reagents/materials/analysis tools: XX. Wrote the paper: MQZ HQG BZ TTZ ZTO.

References

1. Turner BL, Lambin EF, Reenberg A (2007) The emergence of land change science for global environmental change and sustainability. Proc Natl Acad Sci U S A 104: 20666–20671.
2. Rotenberg E, Yakir D (2011) Distinct patterns of changes in surface energy budget associated with forestation in the semiarid region. Glob Change Biol 17: 1536–1548.
3. Luo Y, Weng E (2011) Dynamic disequilibrium of the terrestrial carbon cycle under global change. Trends Ecol Evol 26: 96–104.
4. Lupo F, Linderman M, Vanacker V, Bartholome E, Lambin EF (2007) Categorization of land-cover change processes based on phenological indicators extracted from time series of vegetation index data. Int J Remote Sens 28: 2469–2483.
5. Geerken RA (2009) An algorithm to classify and monitor seasonal variations in vegetation phenologies and their inter-annual change. ISPRS J Photogramm Remote Sens 64: 422–431.
6. Martínez B, Gilabert MA (2009) Vegetation dynamics from NDVI time series analysis using the wavelet transform. Remote Sens Environ 113: 1823–1842.
7. Duchemin B, Goubier J, Courrier G (1999) Monitoring phenological key stages and cycle duration of temperate deciduous forest ecosystems with NOAA/AVHRR data. Remote Sens Environ 67: 68–82.
8. White MA, de Beurs KM, Didan K, Inouye DW, Richardson AD et al. (2009) Intercomparison, interpretation, and assessment of spring phenology in North America estimated from remote sensing for 1982–2006. Glob Change Biol 15: 2335–2359.
9. Zhang XY, Tarpley D, Sullivan JT (2007) Diverse responses of vegetation phenology to a warming climate. Geophys Res Lett 34:
10. Pettorelli N, Vik JO, Mysterud A, Gaillard JM, Tucker CJ et al. (2005) Using the satellite-derived NDVI to assess ecological responses to environmental change. Trends Ecol Evol 20: 503–510.
11. Reed BC, White M, Brown JF (2003) Remote sensing phenology. In:Schwartz MD editor Phenology: An integrative environmental science. Dordrecht: Kluwer Academic Publishers. 365–381.
12. Jiang ZY, Huete AR, Didan K, Miura T (2008) Development of a two-band enhanced vegetation index without a blue band. Remote Sens Environ 112: 3833–3845.
13. Jönsson P, Eklundh L (2002) Seasonality extraction by function fitting to time-series of satellite sensor data. IEEE Trans Geosci Remote Sens 40: 1824–1832.
14. Beck P, Atzberger C, Hogda KA, Johansen B, Skidmore AK (2006) Improved monitoring of vegetation dynamics at very high latitudes: A new method using MODIS NDVI. Remote Sens Environ 100: 321–334.
15. Hird JN, McDermid GJ (2009) Noise reduction of NDVI time series: An empirical comparison of selected techniques. Remote Sens Environ 113: 248–258.
16. Chen J, Jönsson P, Tamura M, Gu ZH, Matsushita B et al. (2004) A simple method for reconstructing a high-quality NDVI time-series data set based on the Savitzky-Golay filter. Remote Sens Environ 91: 332–344.
17. Eklundh L, Jönsson P (2009) Timesat 3.0 software manual.
18. Jönsson P, Eklundh L (2004) TIMESAT - a program for analyzing time-series of satellite sensor data. Comput Geosci 30: 833–845.

19. Cho MA, Skidmore AK (2006) A new technique for extracting the red edge position from hyperspectral data: The linear extrapolation method. Remote Sens Environ 101: 181–193.

20. Senay GB, Elliott RL (2002) Capability of AVHRR data in discriminating rangeland cover mixtures. Int J Remote Sens 23: 299–312.

21. Reed BC, Brown JF, Vanderzee D, Loveland TR, Merchant JW et al. (1994) Measuring phenological variability from satellite imagery. J Veg Sci 5: 703–714.

22. Zhang XY, Friedl MA, Schaaf CB, Strahler AH, Hodges J et al. (2003) Monitoring vegetation phenology using MODIS. Remote Sens Environ 84: 471–475.

6

Desiccation Risk Drives the Spatial Ecology of an Invasive Anuran (*Rhinella marina*) in the Australian Semi-Desert

Reid Tingley*, Richard Shine

School of Biological Sciences A08, University of Sydney, Sydney, New South Wales, Australia

Abstract

Some invasive species flourish in places that impose challenges very different from those faced in their native geographic ranges. Cane toads (*Rhinella marina*) are native to tropical and subtropical habitats of South and Central America, but have colonised extremely arid regions over the course of their Australian invasion. We radio-tracked 44 adult cane toads at a semi-arid invasion front to investigate how this invasive anuran has managed to expand its geographic range into arid areas that lie outside of its native climatic niche. As predicted from their low physiological control over rates of evaporative water loss, toads selected diurnal shelter sites that were consistently cooler and damper (and thus, conferred lower water loss rates) than nearby random sites. Desiccation risk also had a profound influence on rates of daily movement. Under wet conditions, toads that were far from water moved further between shelter sites than did conspecifics that remained close to water, presumably in an attempt to reach permanent water sources. However, this relationship was reversed under dry conditions, such that only toads that were close to permanent water bodies made substantial daily movements. Toads that were far from water bodies also travelled along straighter paths than did conspecifics that generally remained close to water. Thus, behavioural flexibility—in particular, an ability to exploit spatial and temporal heterogeneity in the availability of moist conditions—has allowed this invasive anuran to successfully colonize arid habitats in Australia. This finding illustrates that risk assessment protocols need to recognise that under some circumstances an introduced species may be able to thrive in conditions far removed from any that it experiences in its native range.

Editor: Carlos A. Navas, University of Sao Paulo, Brazil

Funding: RT was funded by a Natural Sciences and Engineering Research Council of Canada Postgraduate Scholarship, an Endeavour International Postgraduate Research Scholarship, and a University of Sydney International Postgraduate Award. Additional funding was provided by Caring for Our Country and the Australian Research Council. The funders had no role in study design, data collection and analysis, decision to publish, or preparation of the manuscript.

Competing Interests: The authors have declared that no competing interests exist.

* E-mail: reid.tingley@gmail.com

Introduction

Human activities have introduced species to areas that lie far outside of their native geographic ranges. Only a small proportion of the global fauna and flora have ever been introduced to a novel environment [1,2], but many of these translocated species flourish and spread widely in their new ranges [1,3–6]. This is surprising, in that we might expect organisms that have evolved under one set of selective forces (reflecting biotic and abiotic challenges confronted in the native range) to be poorly suited to a different environment that poses a novel suite of challenges [7]. Presumably for this reason, invasion success is highest in ecologically generalized organisms that are introduced to places that experience similar climates to their native ranges [6,8–11]. Yet, some invaders flourish in places that impose challenges very different from those faced in their ancestral ranges [12]. For example, spotted knapweed (*Centaurea maculosa*), an invasive neophyte in North America, inhabits drier conditions in its invaded range than in its native European range [13]. Invasive fire ants (*Solenopsis invicta*) also occupy different climatic niches in their native and invaded ranges [14]. Thus, the absence of 'suitable' environmental conditions at a given introduction point may not prevent an invader from thriving at that location once introduced.

One of the most remarkable climatic niche shifts during a biological invasion involves the cane toad, *Rhinella marina*. Cane toads are native to well-watered regions of tropical and subtropical Central and South America. However, this species has not only managed to invade the driest inhabited continent on earth (Australia), but increasingly has penetrated into extremely arid parts of that continent (figure 1), expanding its range to encompass more than 1.2 million km^2 within seven decades of its introduction in 1935 [15]. Nevertheless, cane toad eggs and larvae are obligately aquatic, and the skin of adult toads acts as a free water surface [16]; thus, water availability is likely the primary factor limiting the range expansion of cane toads in arid systems. Even in the Australian wet tropics (where this constraint is less severe), cane toads select diurnal shelter sites that minimize rates of evaporative water loss, and require frequent access to water to rehydrate [17,18]. The toad's dependence on open water also endangers many components of the native arid-zone fauna, in that both toads and native species are concentrated around water bodies for extensive periods of the year in arid systems. This high degree of sympatry may present a major threat for native predators in particular, which have no shared evolutionary history with toads, and thus are extremely vulnerable to toad toxins [19–21].

Here we present the results of a radio-tracking study conducted at the expanding edge of the cane toad's range in semi-arid Queensland. Our aim was to understand how cane toads deal with the harsh hydric conditions posed by the Australian semi-desert. Specifically, we investigated whether desiccation risk influences

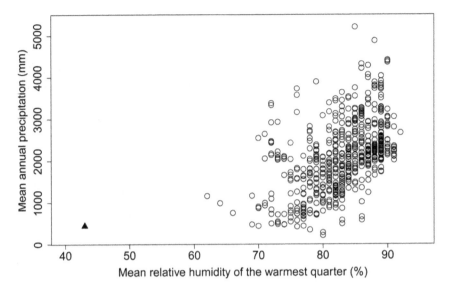

Figure 1. Cane toads in Longreach, QLD (triangle) occupy drier environments than in their native range (circles). Toad occurrence data were taken from museum records (R Tingley & M Kearney, submitted). Climate variables were extracted from the CRU CL 2.0 dataset, which is based on the 1961–1990 normal [39,40].

patterns of toad shelter site selection and daily rates of movement. Identifying the proximate causes of toad distribution and dispersal in arid regions of Australia may clarify the future impacts of toads through space and time, and may also help focus control efforts to eradicate or slow the spread of toads in arid systems.

Methods

Ethics Statement

Permits for this research were provided by the University of Sydney Animal Care and Ethics Committee (permit L04/4-2009/3/4999).

Study Area

Radio-telemetry was conducted at a cattle station (Whitehill) 30 km south-west of Longreach, Queensland. The first cane toad arrived at Whitehill Station in 2007, although successful reproduction was not observed until 2008. Vegetation within the study area predominately consists of paddocks of Mitchell grass (*Astrebla* spp.). Artificial dams constructed for pastoralism provide the only sources of permanent water throughout the year.

The study area lies on the edge of the Australian arid zone (mean annual rainfall = 447.2 mm, mean annual number of days with rain ≥1 mm = 33.3), and experiences high temperatures year-round (mean monthly maximum temperature = 31.4°C; http://www.bom.gov.au). However, weather conditions during our study were unusually humid (rainfall during tracking period = 326 mm *vs.* 159 mm based on long-term normal). Thus, our study was conducted under conditions that maximized toad dispersal opportunities.

Radio-telemetry

Between 01-Jan and 03-Mar-2010, 44 adult cane toads (n = 29 males, mean ± S.E. snout-urostyle length = 111 mm±1.27, mass = 167 g±6.98; n = 15 females, 124 mm±3.05, 289 g±27.1) were fitted with single-stage radio-transmitters (Sirtrack Ltd., Havelock North, NZ). Radio-transmitters were attached to a flexible metal chain around each toad's waist, and did not impede movement or amplexus (transmitter mass ≤3.5% of toad body mass). Toads were released immediately after capture, and located

every 1 to 3 days (median = 1 day) for a total of 2 to16 days (median = 7 days). Toad locations were recorded with a handheld GPS unit (c. 5 m accuracy, Garmin Ltd., Oregon, USA). We calculated daily movement rates by dividing the straight-line distance between successive toad locations by the number of days that had elapsed since last capture. Adult cane toads move about only at night, so our calculations of daily movement rates are based only on diurnal shelter site locations [22].

What Factors Drive Cane Toad Dispersal?

We investigated whether three abiotic variables influenced daily movement rates of radio-tracked toads after accounting for snout-urostlye length (SUL): (i) mean nightly temperature, (ii) number of days since rain, and (iii) distance to the nearest water body. These three variables were selected based on personal observations, and the results of previous studies [23–25]. Mean nightly temperature data (averaged from sunset to sunrise, using data on times from: http://www.usno.navy.mil/USNO/astronomical-applications/data-services/rs-one-year-world) were collected with hygrochon data loggers (Dallas Semiconductor, Texas, USA) placed around the perimeter of each water body. Rainfall data were gathered with a rain gauge located in the middle of the study site.

To determine whether abiotic variables affected (log) movement rates of toads, we used linear mixed effects models and an information-theoretic approach. Akaike's information criterion (AIC) and Bayesian information criterion (BIC) were used to rank the goodness of fit of all possible models containing SUL and the three abiotic variables. These two measures of model parsimony provide upper and lower bounds on model complexity [26]: use of AIC typically results in overly complex models, whereas BIC often selects models that underfit the data (because the penalty for additional parameters is stronger in BIC than in AIC). To facilitate interpretation and reduce the chance of over-fitting models, only two-way interactions between variables were considered. A random effect was also included in all models to account for clustering of movement rates within individual toads; SUL and all abiotic variables were treated as fixed effects. Squared Pearson correlations between observed and predicted movement rates were used to measure the explanatory power of candidate models.

We also investigated whether the straightness of toad movement paths was influenced by the mean distance to the nearest water body. Path straightness was calculated by dividing the net displacement between each toad's initial and final location by the sum of daily movement distances [22]. Perfectly straight paths have a value of 1, whereas paths that start and end at the same location have a value of 0. Only toads that were captured at least four times (n = 33) were included in these analyses. For each toad, we calculated the mean distance to the nearest water body across all locations. We then used a t-test to compare the straightness of movement paths between toads that were close to (<18.6 m), or far from (≥18.6 m) water bodies, based on the median distance to water bodies across all toads.

What Factors Drive Cane Toad Shelter Site Selection?

We used plaster casts of an adult cane toad in the water-conserving posture to quantify desiccation rates in diurnal shelter sites. To mimic the total absorptivity of a cane toad, powder paint was mixed with the plaster before allowing each model to set [27]. A layer of flexible rubber (Performix Plasti Dip, New South Wales, AU) was applied to the undersides of all models to prevent water exchange between models and the substrate. These plaster models lose water at a rate similar to that of live cane toads under field conditions [27]. Models were submerged in water for 24 hours, weighed with a spring scale, and subsequently placed in pairs of shelter sites (used and randomly selected) for *c.* 24 hours before being reweighed. Random locations were selected by spinning the dial of a compass and randomly choosing a distance between 1 and 10 m from each used shelter site. Canopy cover and the maximum height of the closest vegetation were also measured to investigate whether differences in rates of evaporative water loss between sites could be attributed to these characteristics. Canopy cover was measured by placing a plaster model of a toad in each retreat site. Looking from above, we visually estimated the percent of the model that was covered by vegetation.

Differences in rates of evaporative water loss from plaster models placed in used *vs.* random shelter sites were analysed with a paired t-test. We used linear regressions to investigate whether canopy cover and vegetation height influenced rates of evaporative water loss. All statistical analyses were conducted in R© 2.12.0 [28].

Results

Toad movement rate decreased as the number of days since rain increased (i.e., decreased with increasing aridity), but this correlation was modified by an interaction with the distance to the closest water body. Under wet conditions, toads that were far from water moved further than did conspecifics that remained close to water, but this relationship was reversed under dry conditions (figure 2). This interaction between rainfall and distance to water explained 24.1% of the variation in toad movement rate.

We also found support for a quadratic relationship between daily movement rates of toads and mean nightly temperature. Movement rate increased with increasing temperature until *c.* 25°C, but began to slightly decrease above this threshold (figure 3). Inclusion of a quadratic temperature term explained an additional 4.40% of the variance in toad movement rate (pseudo-R^2 = 28.5%). A model containing a quadratic term for temperature and an interaction between the number of days since rain and distance to water also had reasonably high support according to

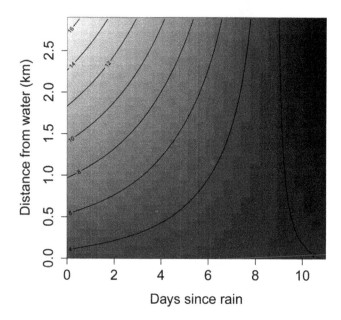

Figure 2. Hydric constraints on cane toad movement rates. Shown is the relationship between log-movement rate, the number of days since rain, and the distance to the closest water body. The interaction between these variables was estimated using a tensor product smooth term within a generalized additive mixed model (estimated degrees of freedom = 3.01). Contours and shading represent different values of the response variable (log-movement rate, m day^{-1}).

AIC and BIC (table 1). These findings were robust to the inclusion of snout-urostyle length, which had a weak (pseudo-R^2 = 28.6%; table 1) negative effect on daily movement rates (i.e., larger toads moved shorter distances).

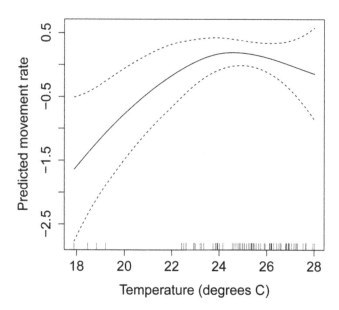

Figure 3. Predictions of cane toad movement rate as a function of mean nightly temperature. This relationship was estimated using a generalized additive mixed model (estimated degrees of freedom = 1.79). Predictions are shown on the scale of the linear predictor, with larger values indicating farther movements.

Table 1. Highest-ranked models of cane toad movement rate.

Rain	Dwater	Rain*Dwater	Temp²	SUL	ΔAIC	wAIC	ΔBIC	wBIC
−0.139	0.00460	−0.000521	3.29–4.74		0	0.666	0	0.525
−0.131	0.00464	−0.000527	3.68–4.64	−0.00828	1.56	0.306	4.77	0.0480
−0.138	0.00483	−0.000546			6.99	0.0200	0.556	0.397

Shown are unstandardized parameter estimates, the difference in Akaike (AIC) and Bayesian information criterion (BIC) between each model and the highest ranked model (ΔAIC and ΔBIC, respectively), and the Akaike (wAIC) and Bayesian weights (wBIC) of each model. Rain = number of days since rain, Dwater = distance to closest water body, Temp² = quadratic effect of mean nightly temperature, SUL = snout-urostyle length.

Desiccation risk also influenced the tortuosity of toad movement paths. Toads that were far from water bodies travelled along straighter paths (mean straightness ± S.E. = 0.641±0.0751) than did conspecifics that generally remained close to water (mean straightness ± S.E. = 0.415±0.0778; $t_{30.9}$ = 2.09, P = 0.0449).

Patterns of toad shelter site selection were driven by desiccation risk as well; plaster models placed in diurnal shelter sites had lower rates of evaporative water loss than did models placed in random locations (mean difference = 38.6 g day^{-1}, $t_{54.0}$ = 13.6, P<0.0001, figure 4A). Toads sheltered in shallow depressions, cracks, or burrows 18.2% of the time, but these subterranean shelters did not provide lower desiccation rates than shelter sites on the surface (evaporative water loss rates in toad shelters above *vs.* below ground: $t_{11.4}$ = −2.02, P = 0.0678). Importantly, differences in rates of evaporative water loss between used and randomly selected shelter sites were not due to the fact that toads sometimes sheltered below ground (mean difference after exclusion of subterranean shelters = 33.6 g day^{-1}, $t_{44.0}$ = 12.8, P<0.0001). Instead, differences in rates of evaporative water loss

between used and random sites were due to variation in canopy cover and vegetation height. Rates of evaporative water loss decreased with increasing canopy cover ($F_{1,\ 108}$ = 130, P<0.0001, R^2-adjusted = 0.542, figure 4B) and vegetation height ($F_{1,\ 108}$ = 32.9, P<0.0001, R^2-adjusted = 0.226, figure 4C).

Discussion

Risk assessment schemes are widely used to predict which alien species should be targeted for quarantine and eradication [29,30]. However, most risk assessments treat alien species as if they are fixed entities, ignoring the fact that microevolution or phenotypic plasticity may allow aliens to invade environments that lie far outside of their ancestral niches [12,31]. In animals, behavioural plasticity has been shown to be particularly important in determining invasion success. For example, introduced vertebrates with larger relative brain sizes (and thus greater behavioural flexibility) have been more successful at establishing viable populations than have smaller-brained species [32–34]. Our results go one step further and demonstrate that

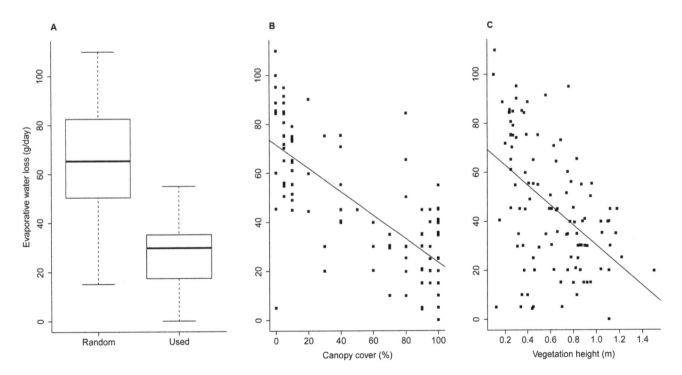

Figure 4. Characteristics of cane toad shelter sites. Plaster models of adult cane toads placed in diurnal shelter sites had lower rates of evaporative water loss than did models placed in random locations (a). These differences between sites were due to (b) variation in canopy cover and (c) vegetation height.

behavioural plasticity can also modify the rate of spread and eventual geographic distribution of an invader. In particular, patterns of cane toad movement and shelter site selection suggest that an ability to exploit spatial and temporal heterogeneity in the availability of surface water has facilitated the range expansion of this anuran invader in the Australian semi-desert.

Unlike some highly specialised arid-zone lineages of anurans, toads (like many forest-origin amphibians) have highly permeable skins, and lack morphological or physiological means to reduce rates of evaporative water loss [19,24]. Our results demonstrate that cane toads in semi-arid systems compensate for this maladaptation by retreating to damp shelter sites by day, and sometimes remaining in these diurnal shelters for days or weeks at a time during prolonged periods of drought. In addition, toads in semi-arid landscapes further minimize desiccation risk by moving long distances between shelter sites only when hydric conditions permit. After it rains, toads which are far from water take advantage of favourable conditions by rapidly moving between shelter sites, presumably in an attempt to reach permanent water sources. However, as the landscape dries out, toads that remain far from water cease moving about, and only toads that are close to permanent water bodies continue to make substantial daily movements. Semi-arid cane toads also reduce the amount of time that they are exposed to desiccating conditions by moving along relatively straight paths when they are far from water bodies. Simulations have shown that such linear dispersal is a highly efficient search strategy for animals attempting to locate suitable habitats in unfamiliar terrain [35]. This suite of water-conserving tactics will be particularly important for toads under conditions of more typical (low) rainfall. In most years, effects of desiccation risk on shelter site selection and movement thus will be stronger than were observed during this study.

Our finding that desiccation risk drives patterns of shelter site selection in cane toads is consistent with previous studies conducted in tropical Queensland [17,18]. Even in these more mesic environments, toads actively selected shelter sites that reduced rates of evaporative water loss. However, our results concerning effects of desiccation risk on daily movement rates differ from those revealed by studies in the Australian tropics. In the wet-dry tropics of the Northern Territory, daily movement rates of toads during the wet season were significantly correlated with abiotic variables [25], but these variables had low explanatory power (R^2 of highest ranked model = 5%). Similarly, in tropical Queensland, a previous radio-tracking study found no significant relationship between abiotic variables and daily movement parameters of cane toads [24]. Thus, desiccation risk appears to have a more marked influence on the movements of cane toads in semi-arid landscapes than in tropical regions. Nonetheless, over 70% of the variation in toad movement rate in our study was unexplained by thermal and hydric constraints. This unexplained variation has several potential sources. For example, radio-telemetry likely underestimated the distances that toads moved between relocations, obscuring relationships between abiotic variables and movement rates. Additionally, expanding the period of data collection to include cooler and drier periods of the year would improve the explanatory power of our models. Future research also could usefully explore the evolutionary underpinnings of the observed movements (e.g., by comparing movement parameters between populations that differ in time since colonization [22]).

Our analyses also revealed that toad movement rates were influenced by mean nightly temperature, although this relationship was much weaker than that between movement rates and the interactive effects of rainfall and distance to water. This finding is consistent with biophysical analyses of the cane toad's fundamental niche, which suggest that activity and movement of toads throughout much of arid Australia are not severely constrained by temperature [36]. In contrast, ambient temperatures determine rates of toad activity and spread in southern areas of the toad's Australian range [31,36].

The results of the current study are of interest not only in understanding how behavioural flexibility allows species to invade novel environments, but also for predicting spatial and temporal variation in the impact of cane toads on native arid-zone fauna. Our finding that desiccation risk constrains the spatial ecology of toads suggests that native species that congregate around water bodies may be especially at risk in arid systems. The likelihood of encounters between toads and native species will be particularly high during droughts, when both toads and native species are forced to remain near water bodies for extensive periods. However, during the wet season, our results illustrate that toads have the ability to spread widely throughout the landscape (>2.8 km from water), and thus encounter (and potentially kill) a wider suite of predators (e.g., goannas, snakes) and prey (e.g., invertebrates) [37]. The effects of cane toads on the native fauna of the arid zone therefore will vary through both space and time, with a high probability of impact around water bodies during periods of drought, and a broader impact across the landscape following rainfall.

Finally, our results also have management implications with respect to the rate of spread and eventual geographic range of cane toads in arid Australia. Toad movement and distribution are largely dictated by the availability of free water, and this dependence will dramatically curtail the toads' ability to invade deeper into the arid zone. However, the provision of abundant artificial water sources for pastoralism will partially overcome this constraint. Decreasing the connectivity of artificial water bodies therefore may offer the most viable management strategy to limit the distribution of toads and slow their rate of spread into arid systems [38].

The cane toad's conquest of semi-arid Australia represents a remarkable example of how invasive species can colonize environments that lie outside of their native climatic niches. Although the degree of climate-match between a species' native and invasive ranges remains a critical predictor of invasion success [6,8–11], this anuran invader is able to flexibly adapt to harsh abiotic conditions by facultatively modifying its behaviour, in order to exploit spatial and temporal heterogeneity in the availability of suitable habitats. Such behavioural plasticity offers a cautionary tale: risk assessment protocols need to recognise that under some circumstances, an introduced species may be able to thrive in conditions far removed from any that it experiences in its native range.

Acknowledgments

First and foremost we would like to thank the Emmott Family for their generous hospitality during the course of fieldwork. CR Tracy, L Schwarzkopf, and D Pike kindly provided assistance with creating plaster models. BL Phillips made helpful comments on an earlier version of this manuscript.

Author Contributions

Conceived and designed the experiments: RT RS. Performed the experiments: RT. Analyzed the data: RT. Contributed reagents/materials/analysis tools: RT. Wrote the paper: RT RS.

References

1. Jeschke JM, Strayer DL (2005) Invasion success of vertebrates in Europe and North America. Proc Natl Acad Sci USA 102: 7198–7202.
2. Tingley R, Romagosa CM, Kraus F, Bickford D, Phillips BL, et al. (2010) The frog filter: amphibian introduction bias driven by taxonomy, body size and biogeography. Global Ecol Biogeogr 19: 496–503.
3. Cassey P (2002) Life history and ecology influences establishment success of introduced land birds. Biol J Linn Soc 76: 465–480.
4. Jeschke JM (2008) Across islands and continents, mammals are more successful invaders than birds. Divers Distrib 14: 913–916.
5. Rodriguez-Cabal MA, Barrios-Garcia MN, Simberloff D (2009) Across islands and continents, mammals are more successful invaders than birds (Reply). Divers Distrib 5: 911–912.
6. Bomford M, Kraus F, Barry SC, Lawrence E (2009) Predicting establishment success for alien reptiles and amphibians: a role for climate matching. Biol Invasions 11: 713–724.
7. Darwin C (1859) The Origin of Species. London: J. Murray.
8. Curnutt JL (2000) Host-area specific climatic-matching: similarity breeds exotics. Biol Conserv 94: 341–351.
9. Blackburn TM, Duncan RP (2001) Determinants of establishment success in introduced birds. Nature 414: 195–197.
10. Hayes KR, Barry SC (2008) Are there any consistent predictors of invasion success? Biol Invasions 10: 483–506.
11. Tingley R, Phillips BL, Shine R (2011) Establishment success of introduced amphibians increases in the presence of congeneric species. Am Nat 177: 382–388.
12. Whitney KD, Gabler CA (2008) Rapid evolution in introduced species, 'invasive traits' and recipient communities: challenges for predicting invasive potential. Divers Distrib 14: 569–580.
13. Broennimann O, Treier UA, Müller-Schärer H, Thuiller W, Peterson AT, et al. (2007) Evidence of climatic niche shift during biological invasion. Ecol Lett 10: 701–709.
14. Fitzpatrick MC, Weltzin JF, Sanders NJ, Dunn RR (2007) The biogeography of prediction error: why does the introduced range of the fire ant over-predict its native range? Global Ecol Biogeogr 16: 24–33.
15. Urban MC, Phillips BL, Skelly DK, Shine R (2007) The cane toad's (*Bufo marinus*) increasing ability to invade Australia is revealed by a dynamically updated range model. Proc R Soc B 274: 1413–1419.
16. Wygoda ML (1984) Low cutaneous evaporative water loss in arboreal frogs. Physiol Zool 57: 329–337.
17. Schwarzkopf L, Alford RA (1996) Desiccation and shelter-site use in a tropical amphibian: comparing toads with physical models. Funct Ecol 10: 193–200.
18. Seebacher F, Alford RA (2002) Shelter microhabitats determine body temperature and dehydration rates of a terrestrial amphibian (*Bufo marinus*). J Herpetol 36: 69–75.
19. Phillips BL, Brown GP, Shine R (2003) Assessing the potential impact of cane toads on Australian snakes. Conserv Biol 17: 1738–1747.
20. Smith JG, Phillips BL (2006) Toxic tucker: the potential impact of cane toads on Australian reptiles. Pac Conserv Biol 12: 40–49.
21. Ujvari B, Madsen T (2009) Increased mortality of naive varanid lizards after the invasion of non-native cane toads (*Bufo marinus*). Herpetol Conserv Biol 4: 248–251.
22. Alford RA, Brown GP, Schwarzkopf L, Phillips BL, Shine R (2009) Comparisons through time and space suggest rapid evolution of dispersal behaviour in an invasive species. Wildlife Res 36: 23–28.
23. Seebacher F, Alford RA (1999) Movement and microhabitat use of a terrestrial amphibian (*Bufo marinus*) on a tropical island: seasonal variation and environmental correlates. J Herpetol 33: 208–214.
24. Schwarzkopf L, Alford RA (2002) Nomadic movement in tropical toads. Oikos 96: 492–506.
25. Phillips BL, Brown GP, Greenlees M, Webb JK, Shine R (2007) Rapid expansion of the cane toad (*Bufo marinus*) invasion front in tropical Australia. Austral Ecol 32: 169–176.
26. Burnham KP, Anderson DR (2004) Multimodel inference: understanding AIC and BIC in model selection. Sociol Methods Res 33: 261–304.
27. Tracy CR, Betts G, Tracy CR, Christian KA (2007) Plaster models to measure operative temperature and evaporative water loss of amphibians. J Herpetol 41: 597–603.
28. R Development Core Team (2010) R: a language and environment for statistical computing. R Foundation for Statistical Computing, Austria.
29. Kolar CS, Lodge DM (2001) Progress in invasion biology: predicting invaders. Trends Ecol Evol 16: 199–204.
30. Kolar CS, Lodge DM (2002) Ecological predictions and risk assessment of alien fishes in North America. Science 298: 1233–1236.
31. Kolbe JJ, Kearney M, Shine R (2010) Modeling the consequences of thermal trait variation for the cane toad invasion of Australia. Ecol Appl 20: 2273–2285.
32. Sol D, Duncan RP, Blackburn TM, Cassey P, Lefebvre L (2005) Big brains, enhanced cognition, and response of birds to novel environments. Proc Natl Acad Sci USA 102: 5460–5465.
33. Sol D, Bacher S, Reader SM, Lefebvre L (2008) Brain size predicts the success of mammal species introduced into novel environments. Am Nat 172: S63–S71.
34. Amiel JJ, Tingley R, Shine R (2011) Smart moves: effects of relative brain size on establishment success of invasive amphibians and reptiles. PLoS ONE 6: e18277.
35. Zollner PA, Lima SL (1999) Search strategies for landscape-level interpatch movements. Ecology 80: 1019–1030.
36. Kearney M, Phillips BL, Tracy CR, Christian K, Betts G, et al. (2008) Modelling species distributions without using species distributions: the cane toad in Australia under current and future climates. Ecography 31: 423–434.
37. Shine R (2010) The ecological impact of invasive cane toads (*Bufo marinus*) in Australia. Q Rev Biol 85: 253–291.
38. Florance D, Webb JK, Dempster T, Worthing A, Kearney MR, et al. (2011) Excluding access to invasion hubs can contain the spread of an invasive vertebrate. Proc R Soc B, In press.
39. New M, Hulme M, Jones PD (1999) Representing twentieth century space-time climate variability. Part 1: development of a 1961–90 mean monthly terrestrial climatology. J Climate 12: 829–856.
40. New M, Lister D, Hulme M, Makin I (2002) A high-resolution data set of surface climate over global land areas. Climate Res 21: 1–25.

7

Macro-Invertebrate Decline in Surface Water Polluted with Imidacloprid: A Rebuttal and Some New Analyses

Martina G. Vijver[1]*, **Paul J. van den Brink**[2,3]

1 Institute of Environmental Sciences (CML), Leiden University, Leiden, The Netherlands, **2** Alterra, Wageningen University and Research centre, Wageningen, The Netherlands, **3** Wageningen University, Wageningen University and Research centre, Wageningen, The Netherlands

Abstract

Imidacloprid, the largest selling insecticide in the world, has received particular attention from scientists, policymakers and industries due to its potential toxicity to bees and aquatic organisms. The decline of aquatic macro-invertebrates due to imidacloprid concentrations in the Dutch surface waters was hypothesised in a recent paper by Van Dijk, Van Staalduinen and Van der Sluijs (PLOS ONE, May 2013). Although we do not disagree with imidacloprid's inherent toxicity to aquatic organisms, we have fundamental concerns regarding the way the data were analysed and interpreted. Here, we demonstrate that the underlying toxicity of imidacloprid in the field situation cannot be understood except in the context of other co-occurring pesticides. Although we agree with Van Dijk and co-workers that effects of imidacloprid can emerge between 13 and 67 ng/L we use a different line of evidence. We present an alternative approach to link imidacloprid concentrations and biological data. We analysed the national set of chemical monitoring data of the year 2009 to estimate the relative contribution of imidacloprid compared to other pesticides in relation to environmental quality target and chronic ecotoxicity threshold exceedances. Moreover, we assessed the relative impact of imidacloprid on the pesticide-induced potential affected fractions of the aquatic communities. We conclude that by choosing to test a starting hypothesis using insufficient data on chemistry and biology that are difficult to link, and by ignoring potential collinear effects of other pesticides present in Dutch surface waters Van Dijk and co-workers do not provide direct evidence that reduced taxon richness and abundance of macroinvertebrates can be attributed to the presence of imidacloprid only. Using a different line of evidence we expect ecological effects of imidacloprid at some of the exposure profiles measured in 2009 in the surface waters of the Netherlands.

Editor: Christopher Joseph Salice, Texas Tech University, United States of America

Funding: These authors have no support or funding to report.

Competing Interests: For transparency reasons, we mentioning the following: PvdB's chair was cofunded between 2008 and 2011 by the following pesticide producers, Bayer, which produces imidacloprid and Syngenta. We feel that this cofunding provides no compete of interest since we don't claim that imidacloprid poses less risks or toxicity than stated in the Van Dijk et al. (2013) as in the current paper we only criticized their methodology. This current work has not been funded. Sponsors thus had no role in study design, data collection and analysis, decision to publish, or preparation of the manuscript.

* E-mail: vijver@cml.leidenuniv.nl

Introduction

The Netherlands is one of the world's foremost agricultural producers, with 2/3 of the total land mass devoted to agriculture or horticulture. Land use is highly intensive in terms of output per hectare or head of livestock [2]. To achieve such high outputs a vast range of agricultural chemicals are used, including fertilizers, veterinary drugs, pesticides and biocides. Different pesticides are used depending on the crop that is grown on the land. There are several routes that pesticides may enter surface waters. Pesticides may be washed into ditches and rivers by rainfall; surface waters can be contaminated by direct overspray or via runoff and leaching from agricultural fields [3]. Emission to surface waters (and thus pesticide residue concentrations) is dictated by many factors such as distance of the crop from the ditch and the mode of application, weather conditions and so on.

Neonicotinoids are the first new class of insecticides to be introduced in the last 50 years. The neonicotinoid imidacloprid is currently one of the most widely used insecticides in the world [4]. Recently, imidacloprid has received much negative attention: The use of certain neonicotoids has been restricted in some countries due to evidence of an unacceptably high risk of toxicity to bees, but this restriction was not in effect in the Netherlands at the time of writing this paper. On April 29, 2013, the European Union passed a two-year ban on the use of three neonicotinoids: European law restricts the use of imidacloprid, clothianidin, and thiamethoxam on flowering plants for two years unless compelling evidence comes out that proves that the use of the chemicals is environmentally safe [5]. This ban is partially, restricted to some applications in specific crops and likely covers 15% of the total use of the three neonicotinoids in the Netherlands [6]. Temporary suspensions had previously been enacted in countries such as France, Germany, Switzerland and Italy. In March 2013, a review of 200 studies on neonicotinoids was published by Mineau and Palmer [7], calling for a ban on neonicotinoid use as seed treatments because of their toxicity to birds, aquatic invertebrates, and other wildlife. The EPA – USA is now re-evaluating the safety of neonicotinoids.

Van Dijk and co-workers [1] aimed to assess the specific relationship between imidacloprid residues in Dutch surface waters, and the abundance of non-target macro-invertebrate taxa.

As also stated by the authors, finding a statistical relationship between those two datasets does not necessarily reflect causality, because there could be other factors (e.g. other pesticide residues, other local habitat factors) which drive observed patterns of abundance. We have some fundamental criticisms on the way the data were analysed and the results were interpreted, and we feel that this can be challenged by existing data. Therefore as a response to the paper of Van Dijk et al [1], and by using additional data, we explore their two key assumptions: 1) residues of pesticides other than imidacloprid, that are collinear with imidacloprid exposure either do not exist or have negligible effects on macroinvertebrate abundance and 2) that imidacloprid concentrations can be extrapolated successfully over 160 days and at a 1 km^2 spatial scale.

Materials and methods

Data collection and treatment

Data on pesticides concentrations in surface water in the Netherlands were obtained from the Dutch Pesticides Atlas. [8]. This is an online tool from which Dutch monitoring data can be collected and processed into a graphic format. Here, data of all pesticide active ingredients and metabolites (n = 634) collected in 2009 were used, since this data set is contiguous with the data used by Van Dijk et al. [1]. Only one year was selected since it can be expected that the correlations between pesticide occurrences will be year-specific, so this correlation should also be assessed for each year specifically. The 2009 dataset covered 302111 individual measurement records of which 19693 measurements exceeded the reporting limit (LOR). The measurements were performed on 4816 samples obtained from 723 different locations. The sample by pesticide dataset is characterised by missing values (90% of entries) and below LOR values (9% of all entries). This is a result of the fact that every water manager has his own suite of pesticides that is sampled, measured and evaluated. The selection of this suite of pesticides is based on the crops and land-use in their region. This selection of pesticides to be monitored improves the efficiency of the monitoring efforts of the individual water managers but yields a data set that has missing values and with many < LOR values when the data of multiple water managers are combined into one. To obtain frequency distributions of the imidacloprid concentrations, data from 2010 and 2011 have also been used.

Environmental quality standards (EQS) of all pesticides were as follows: for imidacloprid the annual average-EQS value (AA-EQS) is 0.067 µg/L (database value set 2-6-2010), and the maximum allowable concentration (MAC-EQS) is 0.2 µg/L (database value set 2-6-2010) as specified by the European Water Framework Directive. In addition, in the Netherlands, the maximum permissible concentration (MPC) of 0.013 µg/L is an important additional criterion (database value set 8-10-2008).

For all samples in which a pesticide could not be detected or quantified, the database substitutes a value of lower than the LOR. The values of reporting limits vary across samples (unique location x time). In our calculations these measurements below LOR are set as zero. We chose to do so, as choosing any other value below LOR would be arbitrary. Moreover, if not taking zero as a value, any other chosen value will result in relatively high toxicity at intensively measured surface waters even if the pesticides are not applied in that area since all measurements results in a lowest value possible of being below the LOR. These types of assumptions are inherent when working with data sets based on monitoring efforts.

Collinearity of imidacloprid concentrations with concentrations of other pesticides

Collinearity refers to a linear relationship between two explanatory variables, meaning that one can be linearly predicted from the others with a non-trivial degree of accuracy. Collinearity was determined on the data set of 2009 measurements restricted to all samples with at least one measurement above the LOR. The reduced data contained measured values for 18% of the samples, of which 8% of the total were measurements above the LOR. In order to assess the correlation between the concentrations of different pesticides we needed a sample by pesticide matrix with as little missing values as possible. From this gappy database, the largest closed data sets were extracted using Principal Component Analysis [9]. For this, measured values in the database were coded as one and missing data by zero. After running the PCA, the species-by-substance matrix was sorted, based on the scores of the substances and samples on the first principal component. Using this approach, it was possible to extract closed data sets by extracting groups of samples with the same score on the first principal component. Four data sets could be extracted that contained more than 100 samples in which the same pesticides were measured. One data set did not include imidacloprid and was not taken into account. The remaining three matrices contained 114, 108 and 191 samples, 27, 51 and 54 pesticides, with 11, 11 and 13% of the measurements above the LOR for data set 1, 2 and 3, respectively. All sampling points of data set 1 were within the provinces of Utrecht and Gelderland while all sampling points of data set 2 and 3 were located in the province of South Holland.

The log((1000 * conc) +1) transformed pesticide concentration values were analysed with Principal Component Analysis (PCA) using the Canoco5 computer programme [10], (see Zafar et al. [11] for the rationale of the transformation]. The pesticide data were centred and standardised for each pesticide. The graphical pictures based on orthogonal coordinate systems describe optimal variance in a dataset. Points that are clustered near each other have a strong correlation. PCA [9] transforms data to a new coordinate system such that the greatest variance by any projection of the data comes to lie on the first coordinate (called the first principal component), the second greatest variance on the second coordinate [12].

Calculating multi substance PAF

The potential affected fraction (PAF) is a common way to express ecotoxicological risks [13]. Following this approach, measured pesticides concentrations were translated into PAF using the species sensitivity distribution (SSD) approach. Toxicity data for each pesticide was obtained from De Zwart [14], and based on acute median effect concentrations (EC50) as derived in the laboratory (database eTox, RIVM as described in [14]). The eTox database consists mainly of data entries from the ECOTOX EPA database. The SSD for imidacloprid is given in Figure 1, and includes 41 different species from 7 different taxonomic groups. Underlying data including references are given in Table S1 of the Supplementary Information. The full database used for the multi substance PAF (msPAF) calculations contained data of 496 different pesticides with 75 different modes of action. To quantify the ecological impacts due to imidacloprid concentrations amongst all other pesticide concentrations as measured in the surface waters, the msPAF was calculated. Firstly, all concentrations of individual pesticides measured over one month per location were aggregated using the maximum measured value. Secondly individual pesticide concentrations were compared to the toxicity data resulting in the PAF. Thirdly, pesticides were grouped based on their mode of action. The PAF's of the pesticides with a similar mode-of-action were added using a concentration addition equation. In this equation, each substance concentration is divided by its effect concentration, ECxa, i.e., the concentration of a that represents a standard effect expressed as EC50 for endpoint x. This gives: Emix (Cmix) = (Ca/ECxa) + (Cb/ECxb) + ... In which Emix(Cmix) is the summed ratio of the mixture components at the exposure concentration of each chemical (Cx). Fourthly, the different pesticides groups with dissimilar mode-of-action were added using a response addition equation. In response addition, the toxicity of the substances in the mixture can be predicted from the product of the fractional effects of the mixture components. This gives Emix (Cmix) = 1 − ((1 − E(Ca)) * (1 − E(Cb)) * ...... In which Emix(Cmix) is the calculated effect of the mixture, Ca the exposure concentration of substance a, and E(ca) the effect of substance a at concentration Ca.

Both models for mixture toxicity are described in Hewlett and Plackett [15]. Chemicals with an unknown mode-of-action were treated according to a unique mode-of-action. As a result an msPAF value per month per monitoring location was derived. In this study we reported the maximum msPAF of the year 2009. The quantification of the relative contribution of imidacloprid on the total chemical pressure as expressed by msPAF was based on acute toxicity data as insufficient chronic toxicity data were available in the literature.

Pairwise combinations of samples taken within 1 km and 160 days

Datasets on imidacloprid concentrations and abundances of macroinvertebrates were linked to each other by Van Dijk and co-workers [1] by using the criteria ≤1 km distance and ≤ 160 days

of time difference. We performed pairwise comparisons of imidacloprid measurements to determine whether imidacloprid concentrations at sites that meet these criteria, matched successfully. Therefore, all imidacloprid measurements were extracted from the 2009 data set. All sampling sites were first ranked on their x coordinate and the difference in distance with the next sample was assessed (using Pythagoras theorem). All site combinations which yielded a difference less than 1 km were extracted. The same procedure was performed using a ranking based on the y-coordinate. The site combinations from both queries were combined. This procedure is not exhaustive since two sites that are not ranked next to each other can also be closer to 1 km from each other, but is likely to find most combinations. The imidacloprid concentrations of all samples taken at the paired sites were compared to each other when the samples were taken within 160 days. The result of the comparison were categorised into: 1) two measurements below the LOR, 2) one measurement below and one above the LOR (0% matching), 3) two measurements above the LOR, of which the number of sample pairs that matched 100% (based on one decimal) was also noted. The analysis resulted in 37 pairs of sites containing a total of 260 observations and 584 concentration measurement pairs being evaluated.

Time series of imidacloprid exposure

For each sampling site it was determined how often imidacloprid samples were analysed. For 34 sampling sites 10 or more samples were analysed, of which imidacloprid was not detected in any of the samples at 14 sites (41%), and in less than half of the samples at 28 sites (82%). The concentration dynamics of the

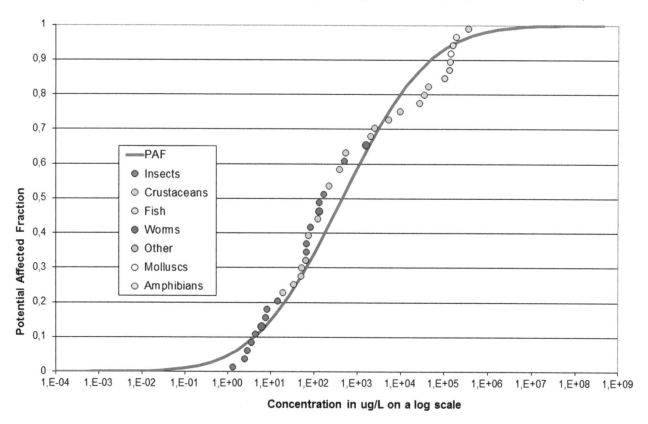

Figure 1. The Species Sensitivity Distribution of imidacloprid based on acute toxicity data. The data consist of 7 different taxomonic groups and 41 species. EPA database downloaded at Oct 23th 2013.

remaining 18% of the sites were plotted to evaluate whether chronic concentrations of imidacloprid may be expected.

Cumulative frequency of maximum imidacloprid concentrations

The measured maximum concentration of each site was compared with threshold concentrations based on the findings of Roessink et al. [16], i.e. the chronic EC10 of the mayfly species *Caenis horaria* and *Cloeon dipterum* (≈ 0.03 µg/L) and the different environmental quality standards. In order to remove the within-site sample dependency, for each sampling site the maximum imidacloprid concentration was extracted. The analysis resulted in 225 negative measurements (below the LOR) and 226 positive measurements (above the LOR).

MPC exceedances of imidacloprid compared to other pesticides

Since only for a restricted number of pesticide AA-EQS and MAC-EQS values have been set in the WFD, we used the (Dutch) MPC standard to compare exceedance frequencies between pesticides. For this comparison, both the magnitude of exceedance as well as the frequency of exceedance was incorporated. Firstly, the exceedance of the MPC of an individual pesticide concentration was derived per measuring location. Secondly, the degree of standard exceedance was weighted according to the following classes: 0 ($\leq$MPC); 1 ($>$ MPC and $\leq$ 2 x MPC), 2 ($>$ 2 x MPC and $\leq$ 5 x MPC) and 5 ($>$ 5 x MPC exceedance). Thirdly, the exceedance classes were summed over all measuring locations per year. Fourthly, pesticides were ranked on the basis of the weighted number of monitoring sites at which the MPC for the compound was exceeded, i.e. corrected for the number of monitoring sites by taking the percentage of sites that show an exceedance of the MPC. Compounds monitored at fewer than ten sites were ignored.

Results and Discussion

For many locations pesticide concentrations have been found to exceed the MPC in 2009 (see Fig. 2). Figure 2 shows that throughout the entire country more than one pesticide exceeds their respective quality standard, so this exceedance is not a common regionally problem. The maximum amount of pesticides exceeding their MPC in one sample is 35. From this it can be concluded that a single pesticide is not likely to drive solely the macro-invertebrate quality, rather all pesticides exceeding the quality standards should be considered.

Collinearity of imidacloprid concentrations with other pesticides

Figure 3A clearly shows that imidacloprid exposure is highly correlated with all chemicals placed on the right, lower side of the diagram, like carbendazim and DEET and to a lesser extend with the large group of chemicals which have a high loading with the horizontal axis, which explains almost double the amount of variance compared to the vertical axis. The results of the second data set (Fig. 3B) show that imidacloprid is placed in the centre of a large group of pesticides placed in the middle of the diagram, since it was measured only in a few samples (7% of the total). The results of the third data set shows a high occurrence of imidacloprid above the LOR (78% of all samples), with concentrations strongly collinear with those that have a high loading on the horizontal axis which explains almost triple the amount of variance of the vertical one (Fig. 3C). The results of the first and third data set show that the contribution of imidacloprid toxicity in surface waters cannot

Figure 2. Number of pesticides exceeding the MPC in 2009. All monitoring locations in the Dutch surface waters with one (yellow); two till five pesticides concentrations (orange); and $>$ five different pesticides (red) exceeding their MPC-values are depicted. Locations were measurements were performed but no exceedances were found are depicted in white.

easily be separated from the toxicity arising from other co-occurring pesticides, or indeed any other co-occurring chemical or physical stressing agent.

The correlations derived from the PCA-plots (Fig. 3) can also be explained from the fact that the active ingredient imidacloprid currently has several authorizations in 38 different products (database ctgb.nl [17], accessed 21-5-2013). The professional use ranges from the use in crops grown in glasshouses such as all different vegetables and in open systems for different bulbs of flowers, potatoes and sugarbeets. Imidacloprid is also registered for use in fruit trees including apple and pear trees. Generally, more than one pesticide is used to protect a specific crop from pest attack. Thus, depending on the land use type, imidacloprid is invariably emitted to surface waters in combination with other pesticides that are authorized to be used on those crops.

Imidacloprid contribution in the msPAF

The potentially affected fraction of the aquatic species by the measured pesticides is higher than 5% in 11 locations (reflecting 1.2 % of all monitoring sites) in the Netherlands in the year 2009. The maximum level that we determined based on the msPAF was 23% in the province of South-Holland. Imidacloprid contributed

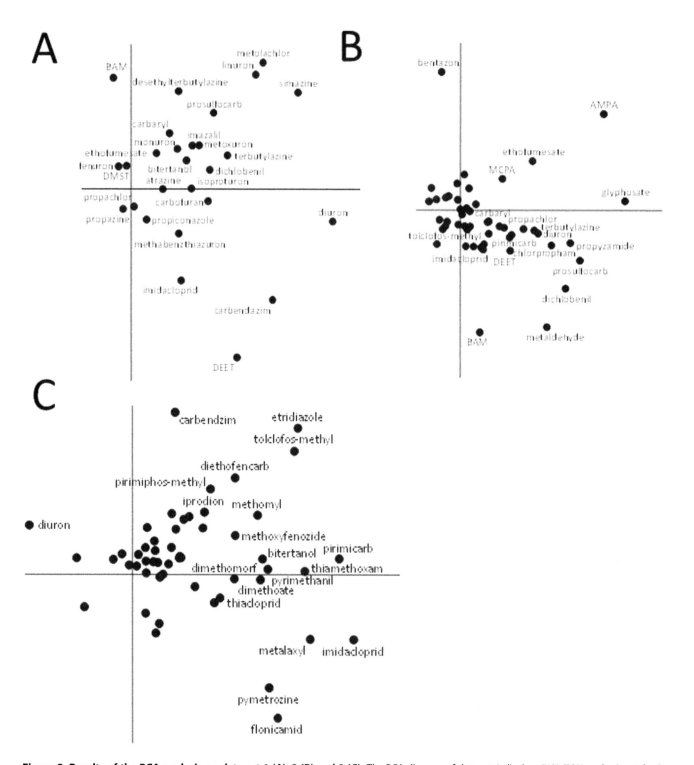

Figure 3. Results of the PCA analysis on data set 1 (A), 2 (B) and 3 (C). The PCA diagram of data set 1 displays 51% (33% on horizontal axis and 18% on vertical one) of the variation in chemical concentrations between the sites while 34% is displayed for data set 2 (21% on horizontal axis and 13% on vertical one) and 38% for data set 3 (28% on horizontal axis and 10% on vertical one).

in 8 out of 11 cases to this potential risk (Table 1). The relative contribution compared to other pesticides as measured at the same location at the same sampling time is rather modest and varied with a maximum of 21% at one location. Note that this calculation was based on acute toxicity data only, so likely is an underestimation of the potential risks that include both acute as chronic effects. From Table 1, it can be deduced that depending on

location, the contribution of specific individual active ingredients differs.

Pairwise combinations of samples

Imidacloprid measurements performed within a time window of 160 days which were taken at sites closer than 1 km from each other were compared. By this pairwise analysis we investigate if

Table 1. Contribution of imidacloprid to the msPAF at locations where msPAF > 5%.

x-coordinate	y-coordinate	Province	Total msPAF of measured pesticides (%)	Relative contribution of imidacloprid to the total msPAF of measured pesticides (%)
N 51 46 39.9	E 4 16 36.7	South Holland	22.53	0
N 52 1 29.6	E 4 30 24.7	South Holland	13.85	7.59
N 51 43 11.8	E 4 16 1.5	Zealand	12.48	0.002
N 51 52 33.5	E 4 10 26.2	South Holland	10.11	0
N 51 46 38.6	E 4 33 19.3	South Holland	9.91	0.009
N 51 45 0.4	E 4 25 46.2	South Holland	9.44	0
N 52 31 7.8	E 4 40 36.5	North Holland	9.25	0.014
N 51 57 10.2	E 4 15 8.8	South Holland	7.09	21.04
N 51 50 20	E 4 35 16.7	South Holland	6.61	0.001
N 51 21 52	E 4 2 10.1	Zealand	6.36	11.49
N 52 41 42.6	E 6 53 54.9	Drenthe	5.64	0.011

selected pairs of imidacloprid concentrations match with each other, and subsequently can be used to accurately link biological effect data and imidacloprid concentrations. Table 2 shows that in 39% of the comparisons there was no match in the presence of imidacloprid above the LOR, while only in 23% of the cases imidacloprid was present above the LOR in both samples. The remaining 38% of comparisons showed two measurements below the LOR. So when imidacloprid is found in at least one of the samples there is a large probability (62%) of not finding imidacloprid in the other site, which hampers the extrapolation of imidacloprid over a time window of 160 day and over a distance of 1 km (Table 2). We, therefore, conclude that the criteria used by Van Dijk et al. [1] to link chemical with biological observations result in a large probability (46%) of linking a site where imidacloprid was detected with a site, where the biological sample was taken, where actually no imidacloprid could be detected. The alternative, i.e. the first measurement being below the LOR and the second one above also has a relatively high probability (34%) (Table 2). Especially in a water-rich country such as the Netherlands, that has more than 350.000 km of ditch systems [18], it should be noted that sampling locations taken within 1 km, not necessarily have a hydrological connection with each other.

Imidacloprid dynamics

The concentration dynamics of imidacloprid (reflecting the concentrations of imidacloprid at the sampling locations with 10 or more samples taken in 2009 and with detection above the LOR in

at least 50% of those samples) are shown in Figure 4. In all but two (Fig. 4B and 4C) of these sampling sites the 28d, EC10 values for *C. horaria* and *C. dipterum* are exceeded for a period longer than 28 days, so at these sites chronic effects of imidacloprid exposure on mayflies can be expected. Also all standards are exceeded for some time in most of the sampling sites, with Fig. 4G showing the largest exeedence for a site near Boskoop in the province of South Holland. It should be noted that these 7 sites only constitute a small percentage (18%) of the total number of sites with 10 or more observations, so likely these exposure patterns represent the worst-cases of the exposure patterns at sites with 10 or more observations. Since we don't know whether there is a bias to measure imidacloprid more intensively at sites where exposure is expected we cannot extrapolate this to the whole population of sites.

Maximum concentrations of imidacloprid

Figure 5 shows the cumulative frequency of the all concentration measurements on the maximum level of imidacloprid for the years 2009, 2010 and 2011. The below LOR measurements are indicated at the 0.001 µg/L level and constituted 50, 53 and 55% of the maximum concentrations in 2009, 2010 and 2011, respectively. The results in Figure 4 show that peak concentrations of imidacloprid in the Dutch surface waters often exceeds the chronic effect concentrations of mayfly as determined in the chronic single species studies by Roessink et al. [16], as well as the three standards. In 2011 the MPC, 28d, EC10, AA-EQS and

Table 2. Result of the comparison of imidacloprid concentrations in samples taken in 2009 at sampling sites closer than 1 km and within 160 days.

Category	# sample pairs	% of total comparisons	% when 1st observation is above LOR	% when 1st observation is below LOR
Two below LOR	217	38		66
One below and above LOR	223	39	46	34
Two above LOR	134	23	54	
100% matching measurements	10	1.7		

LOR = analytical reporting limit.

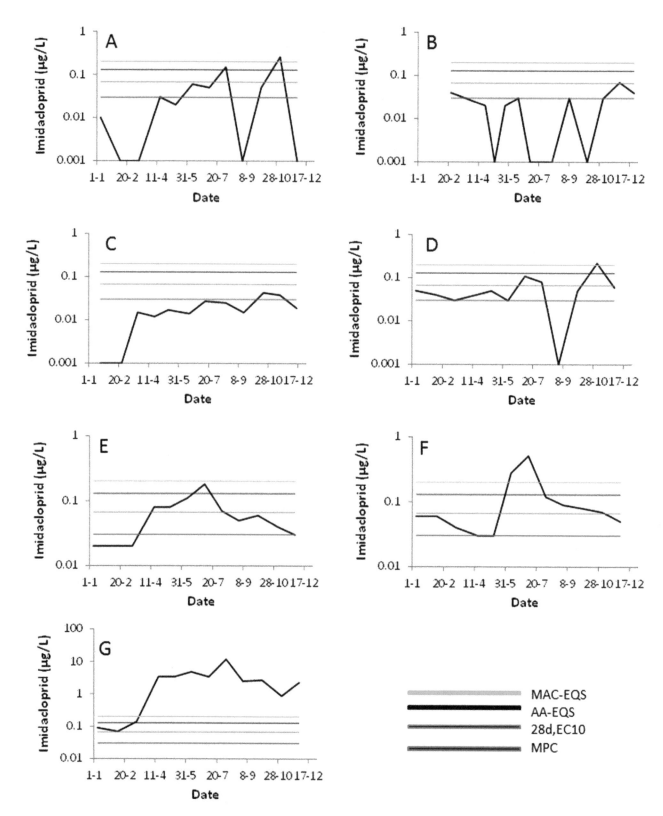

Figure 4. Concentration dynamics at the selected sampling sites (see text for procedure). The sampling sites 4A through 4G have X,Y coordinates of 108313,456412, 105888,455853, 103707,455196, 105927,453177, 170370,518957, 106781,503700 and 105079,453602, respectively. The horizontal lines denotes the MAC-EQS, the AA-EQS, the 28d, EC10 value for the mayflies C. horaria and C. dipterum (Roessink et al., 2013) and the MPC (top to bottom).

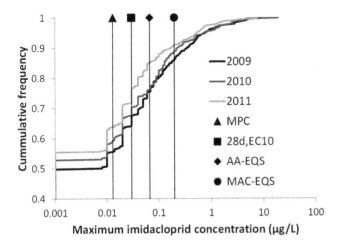

Figure 5. The cumulative frequency of the maximum imidacloprid concentrations of the sampling sites in 2009, 2010 and 2011, together with three standards and the 28d, EC10 of Cloeon dipterum and Caenis horaria.

MAC-EQS threshold values are exceeded by 36, 28, 15 and 9% of the maximum concentrations at the sampling sites, respectively. Since the Hazardous Concentration 5% based on 96h,EC10 values of 0.083 µg/L [16] corresponds more or less with the AA-EQS, acute effects of imidacloprid exposure cannot be excluded at a relatively large proportion of the sites ($\approx$15%). The maximum concentration is of course not a good predictor for the time weighted average concentration of 28d which should ideally be compared with the chronic threshold value of 0.03 µg/L. Still, when combining the results of the time-series (Fig. 3) and the exceedance of this threshold value by the maximum concentrations (Fig. 4) chronic effects of imidacloprid on insects like mayflies may be expected at a vast proportion of sites, with 28% being the most conservative estimate and 5% being the best guess. This 5% is calculated by multiplying the 28% chance of exceeding the threshold value by the maximum concentration and 15% chance of having above LOR measurements at more than 50% of the samples taken at a particular site where imidacloprid is measured at least 10 times. The comparison of the standards with the ecotoxicological threshold value for mayflies also suggests that the

MAC-EQS and AA-EQS are not fully protective for acute and chronic effects on insect taxa, respectively.

Exceedances of environmental quality standards

As stated in the Van Dijk et al [1] paper, in 2009 imidacloprid frequently exceeds quality standards for surface waters: 111 and 62 times for the AA-EQS and the MAC-EQS respectively [8,18]. In addition to the probability of exceeding a standard, also the magnitude of exceedance is important since it is likely that at higher magnitudes the ecological effects are more severe and maybe even last longer. Table 3 shows the compounds that exceeded the MPC most frequently in 2009, ranked according to degree of exceedance.

Imidacloprid was predicted to have a relatively large impact on the ecosystems compared to other pesticides, and gained third place in the Top 10 pesticides violating the environmental quality standards in respect to frequency and magnitude of exceedance. The number of measurements is high, as is also the number of locations from which the samples are taken. This means that monitoring is quite intensive for this compound, and surely covers many different surface waters belonging to different water managers and covering the geographical distribution of the different water types in the Netherlands. Although less intensively measured – a factor 5 to 10 – Table 3 also shows that other pesticides exceed the MPC more often. Thus although imidacloprid poses a significant ecological risk to surface waters in the Netherlands, it is not the only potential cause of degradation in macroinvertebrate abundance, as many other pesticides mentioned in Table 3 also exceed the MPC frequently (and in cases by orders of magnitude) and thus undoubtedly contribute to overall stress regime. It is a common flaw in ecological studies to selectively interpret individual causal agents within stressor regimes as the sole cause of observed phenomena, leading to erroneous conclusions.

Conclusion

Imidacloprid is one of several pesticides that can be detected in surface waters draining agricultural areas at levels frequently exceeding environmental quality standards. Despite this, we show here that key assumptions made by Van Dijk et al. [1] specifically relating to imidacloprid toxicity are not supported by observational data and, therefore, their assessment is unsuitable to determine threshold levels of effects. Specifically, the validity of

Table 3. Top10 pesticides exceeding the MPC in the Netherlands in the year 2009.

Pesticides name	No. of monitoring sites	% Exceedance	No. of measurements	% Exceedance
Captan	38	47	194	13
desethyl-terbuthylazin	63	37	299	10
Imidacloprid	451	44	2133	28
Triflumuron	24	21	142	4
Dicofol	24	17	142	3
Omethoaat	31	16	169	3
Foraat	51	14	313	2
Captafol	15	27	29	14
Fipronil	69	12	230	7
Pyraclostrobin	66	17	341	7

No. = number. The ranking of pesticides is based on frequency and magnitude of exceedances.

two assumptions: 1) that imidacloprid levels are not correlated with toxic levels of other pesticides residues and 2) that chemical exposure data can be extrapolated over a 1 km distance and 160 day time window are here shown to be highly questionable. The ecological status of field sites can be attributed to a complex suite of stressors resulting from a range of anthropogenic practices in the highly managed landscape of the Netherlands, of which pesticides are just one factor, and imidacloprid only one of many pesticides being applied, albeit an important one in terms of ecological risks. We therefore propose that any risk assessment should base the ecological threshold values not solely on field observations but also largely rely on the results of controlled experiments, since these types of experiments allow a full control of separating the imidacloprid stress from other stressors.

Supporting Information

Table S1   Acute toxicity values of imidacloprid (source eTox database, EPA database downloaded Oct 23th

References

1. Van Dijk TC, Van Staalduinen MA, Van der Sluijs JP (2013) Macro-invertebrate decline in surface water polluted with imidacloprid. PLOS ONE 8 (5) e62374.
2. Vijver MG, De Snoo GR (2012) Overview of the state-of-art of Dutch surface waters in the Netherlands considering pesticides. (chapter 9) In: The impact of pesticides, M. Jokanovic (ed.) AcedemyPublish.org, WY, USA. ISBN: 978-0-9835850-9-1.
3. Vijver MG, Van 't Zelfde M, Tamis WLM, Musters CJM, De Snoo GR (2008) Spatial and Temporal Analysis of Pesticides Concentrations in Surface Water: Pesticides Atlas. J Environ Sci Health Part B 43: 665–674.
4. Yamamoto I (1999) "Nicotine to Nicotinoids: 1962 to 1997". In Yamamoto, Izuru; Casida John. Nicotinoid Insecticides and the Nicotinic Acetylcholine Receptor. Tokyo: Springer-Verlag. pp. 3–27 ISBN: 443170213X.
5. McDonald-Gibson C (29 April 2013). *The Independent*. Retrieved 1 May 2013.
6. Van Vliet J, Vlaar LNC, Leendertse PC (2013) Toepassingen, gebruik en verbod van drie neonicotinoïden in de Nederlandse land en tuinbouw. CLM 825- 2013. Available: www.clm.nl. Accessed 2013 May 5.
7. Mineau P, Palmer C (2013) The impact of the nation's most widely used insecticides on birds. Neonicotinoid Insecticides and Birds. American Bird Conservancy. Available: http://www.abcbirds.org/abcprograms/policy/toxins/neonic_final.pdf.
8. Dutch pesticides atlas website. Available: http://www.bestrijdingsmiddelenatlas.nl, version 2.0. Institute of Environmental Sciences (CML) at Leiden University and Waterdienst of the Dutch Ministry of Infrastructure and Environment. Accessed 2013 Oct 23.

2013). Legend: Species selected for the toxicity test were given with their scientific name and with their species group. Toxicity data were given as log10 effect concentrations at which 50% of the organisms showed adverse effects. The scientific papers from which those data are collected are given.

Acknowledgments

The authors thank Donald Baird for his critical comments and language suggestions. We thank Dick de Zwart for providing the eTox database. All pesticides measurements compared to the different EU and MPC quality standards can be found and freely downloaded at www.bestrijdingsmiddelenatlas.nl [8].

Author Contributions

Conceived and designed the experiments: MGV PJB. Performed the experiments: MGV PJB. Analyzed the data: MGV PJB. Contributed reagents/materials/analysis tools: MGV PJB. Wrote the paper: MGV PJB.

9. Jolliffe IT (2002) Principal Component Analysis, Series: Springer Series in Statistics, 2nd ed., Springer, NY. ISBN 978-0-387-95442-4.
10. Ter Braak CJF, Šmilauer P (2012) Canoco reference manual and user's guide: software for ordination, version 5.0. Microcomputer Power, Ithaca, USA, 496 pp.
11. Zafar MI, Belgers JDM, Van Wijngaarden RPA, Matser A, Van den Brink PJ (2012) Ecological impacts of time-variable exposure regimes to the fungicide azoxystrobin on freshwater communities in outdoor microcosms. Ecotoxicol 21:1024–1038.
12. Van Wijngaarden RPA, Van den Brink PJ, Oude Voshaar JH, Leeuwangh P (1995) Ordination techniques for analyzing response of biological communities to toxic stress in experimental ecosystems. Ecotoxicol 4: 61–77.
13. Posthuma L, Suter GW II, Traas TP (eds) (2002) Species Sensitivity Distributions in Ecotoxicology. Lewis Publishers, Boca Raton, FL, USA.
14. De Zwart D (2005) Ecological effects of pesticide use in the Netherlands: Modeled and observed effects in the field ditch. Integrated Environmental Assessment and Management 1:123–134.
15. Hewlett PS, Plackett RL (1959) A unified theory for quantal responses to mixtures of drugs: non-interactive action. Biometrics 15:591–610.
16. Roessink I, Merga LB, Zweers HJ, Van den Brink PJ (2013) The neonicotenoid imidacloprid shows high chronic toxicity to mayfly nymphs. Environ Toxicol Chem 32: 1096 – 1100.
17. Statistics Netherlands. Available: http://www.statline.cbs.nl. Accessed 2013 May 21.
18. De Snoo GR, Vijver MG (eds) (2012) Bestrijdingsmiddelen en waterkwaliteit. Universiteit Leiden, 180 pp., ISBN: 978-90-5191-170-1.

8

When Dread Risks Are More Dreadful than Continuous Risks: Comparing Cumulative Population Losses over Time

Nicolai Bodemer[1,2]*, **Azzurra Ruggeri**[1], **Mirta Galesic**[1,2]

1 Center for Adaptive Behavior and Cognition, Max Planck Institute for Human Development, Berlin, Germany, **2** Harding Center for Risk Literacy, Max Planck Institute for Human Development, Berlin, Germany

Abstract

People show higher sensitivity to dread risks, rare events that kill many people at once, compared with continuous risks, relatively frequent events that kill many people over a longer period of time. The different reaction to dread risks is often considered a bias: If the continuous risk causes the same number of fatalities, it should not be perceived as less dreadful. We test the hypothesis that a dread risk may have a stronger negative impact on the cumulative population size over time in comparison with a continuous risk causing the same number of fatalities. This difference should be particularly strong when the risky event affects children and young adults who would have produced future offspring if they had survived longer. We conducted a series of simulations, with varying assumptions about population size, population growth, age group affected by risky event, and the underlying demographic model. Results show that dread risks affect the population more severely over time than continuous risks that cause the same number of fatalities, suggesting that fearing a dread risk more than a continuous risk is an ecologically rational strategy.

Editor: Enrico Scalas, Università del Piemonte Orientale, Italy

Funding: No current external funding sources for this study.

Competing Interests: The authors have declared that no competing interests exist.

* E-mail: bodemer@mpib-berlin.mpg.de

Introduction

Imagine two different risky events: One threatens to kill 100 people at once; the other threatens to kill 10 people every year over a period of 10 years. The first event represents a *dread risk*, a rare event that kills many people at once, such as a pandemic, an earthquake, or a terrorist attack. The second event represents a *continuous risk*, a relatively frequent event that kills many people over a longer period of time, such as diabetes, air pollution, or car accidents. Which of the two risks is more severe? Both events kill the same number of people and differ only with respect to the time frame. Yet, people react much more strongly to dread risks than to continuous risks, in terms of both perception and avoidance behavior [1–3]. For instance, in reaction to the 9/11 terrorist attacks (a typical dread risk), many Americans avoided air travel and switched to their cars without considering that the risk of dying in a car accident (a continuous risk) is larger than the risk of an airplane terrorist attack, and even of dying in an airplane accident in general [4]. The avoidance of flying and the elevated use of cars increased the number of fatal highway crashes after the 9/11 attacks [1,2,5].

People's higher sensitivity to dread risks compared with continuous risks is often considered a bias: If the continuous risk causes the same number of fatalities, it should not be perceived as less dreadful. In this paper we offer an alternative explanation to the assumption of biased minds and argue that a stronger reaction to dread risks is ecologically rational, because dread risks actually cause a larger cumulative reduction in the population size.

Different hypotheses have been proposed to explain why people fear dread risks more than continuous risks. First, the psychometric paradigm [3] suggests that high lack of control, high catastrophic potential, and severe consequences account for the increased risk perception and anxiety associated with dread risks. Second, people might lack knowledge about the statistical information underlying risks [6], in particular about the large number of fatalities caused by continuous risks. Third, because people estimate the frequency of a risk by recalling instances of its occurrence from their social circle or the media, they may overvalue relatively rare but dramatic risks and undervalue frequent, less dramatic risks [7,8]. This is further supported by findings that people generally overestimate low probabilities and underestimate high probabilities [9–11], although the observed pattern can be partially explained with a regression-to-the-mean effect [8]. Fourth, according to the preparedness hypothesis, people are prone to fear events that have been particularly threatening to survival in human evolutionary history [12]. Given that in most of human evolutionary history people lived in relatively small groups, rarely exceeding 100 people [13,14], a dread risk, which kills many people at once, could potentially wipe out one's whole group. This would be a serious threat to individual fitness, as being in a group reduces predation risk, helps with finding food and hunting, and increases survival chances when injured [15,16]. In line with this hypothesis, Galesic and Garcia-Retamero [17] found that people's fear peaks for risks killing around 100 people and does not increase if larger groups are killed.

A different perspective reveals that dread risks lead to significantly worse short- and medium-term consequences than continuous risks, even if they do not eliminate a whole group. Thus, we focus not only on the overall number of immediate fatalities, as in previous accounts, but also on (a) the population size over time, and (b) the role of the age group that is affected by the risky event. Note that a fatal event strikes twice: it kills a number of people immediately, and it reduces the number of future offspring by reducing the number of their potential parents. A risk that affects children and young adults will have stronger negative effects on future group growth than a risk that affects group members who are past their reproductive period. Dread risks such as pandemics, terrorist attacks, or nuclear accidents are more likely to strike children and young adults compared to many continuous risks such as diabetes, cancer, heart attack, or household accidents, which affect primarily older people [18]. For example, the H1N1 pandemic in 2009 was more likely to infect younger people, whereas older people were relatively immune, probably due to previous exposure to a similar virus strain [19].

We hypothesize that dread risks cause larger cumulative losses on the population level than continuous risks. More specifically, we hypothesize that the number of people-years lost because of a dread risk is larger than the number of people-years lost because of a continuous risk, in particular when the event affects the younger age groups. People-years correspond to the number of people who live 1 year in the population. Hence, by killing a large number of children or young adults at once, dread risks not only deprive the society of their contribution in subsequent years, but they also remove the potential contribution of the offspring the victims could have had if they had survived longer.

To illustrate this hypothesis, consider first a very simplified example. Imagine a population of 40 people, uniformly distributed across four age groups:

Children and adolescents, aged 0–19 years: Pre-fertile generation that may produce offspring in the future.

Young adults, aged 20–39 years: Fertile generation that currently produces offspring.

Older adults, aged 40–59 years: Post-fertile generation.

Elderly adults, aged 60–79 years: Post-fertile generation.

Further assume that the population growth is constant and that every year each young adult produces exactly one offspring. This implies that the number of children at time point i, t_i, corresponds to the number of young adults at time point i-1, t_{i-1}. Moreover, at every t_i a generation shift takes place, so that the number of young adults at t_{i+1} corresponds to the number of children at t_i, and so on for the other groups. Moreover, all elderly adults at t_{i-1} will be dead at t_i. In the absence of any dread risk or continuous risk, the population is constant over time with $N_{total} = 40$ (see Figure 1).

What happens if a dread risk occurs at t_1 that kills 50% of the young adults (i.e., 5 young adults)? At t_1, the total population is reduced to $N_{total} = 35$ ($N_{children} = 10$, $N_{young\ adults} = 5$, $N_{older\ adults} = 10$, $N_{elderly\ adults} = 10$). At t_2 the population is further reduced to $N_{total} = 30$ ($N_{children} = 5$, $N_{young\ adults} = 10$, $N_{older\ adults} = 5$, $N_{elderly\ adults} = 10$), because the number of newborn offspring is smaller due to the fewer young adults. Finally, the population size settles at $N_{total} = 30$, with continuous fluctuation within the respective groups.

What happens if a continuous risk, a disease, occurs at t_1 that kills five young adults over a period of five time steps (one young adult at every t_i, from t_1 to t_5)? Note that the total number of fatalities directly caused by the risk is the same as in the dread risk scenario (i.e., 5). The total population is reduced to $N_{total} = 39$ at t_1

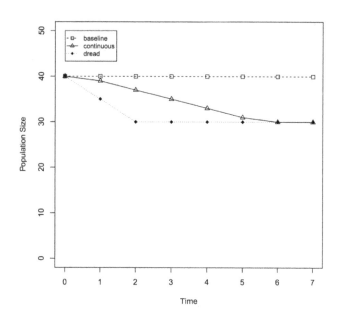

Figure 1. Impact of continuous and dread risk event on cumulative population size. Development of the population size when no risky event is present (baseline), and when a continuous risk (1 individual killed from t_1 to t_5) or a dread risk (5 individuals killed at t_1) event occurs. A dread risk leads to a more immediate impact on cumulative population size that lasts longer compared with the continuous risk.

and continues to decline until t_6, where it finally corresponds to the size of the population hit by the dread risk (see Figure 1).

In sum, the continuous risk takes five more generations to affect the population as severely as the dread risk. The difference in the cumulative losses caused in the population by the dread versus continuous risk, can be calculated by determining the area between the curves representing the difference in the cumulative population sizes of the two conditions (i.e., the difference in people-years over time). In the example in Figure 1, this integral is 20, meaning that the population hit by the dread risk lost 20 people-years more than the population experiencing the continuous risk.

This simple illustration shows that people's tendency to fear dread risks more than continuous risks can be ecologically rational, because dread risks can affect the population more severely in the long run. It is important to note that we do not claim this to be necessarily a universal pattern. We do not exclude there might be situations (i.e., a set of parameters) in which a continuous risk causes stronger cumulative losses than a dread. However, this is the pattern that occurs in all our simulations. In the following, we present results of two sets of more fine-tuned simulations.

Simulation Set 1

In the first set of simulations, we assumed a small population size, similar to groups in which people lived throughout most of evolutionary history [14]. We manipulated whether the population growth rates were constant, increasing, or decreasing, and which age group was exposed to a dread or to a continuous risk.

Methods

We set the total population to 160 people. The individuals were distributed equally across 80 years (i.e., there were 2 individuals for each age at t_0) and across four age groups, as in the illustrative example above. Between conditions, we manipulated (a) whether a dread or a continuous risk occurred, (b) the population growth rate, and (c) which age group was hit by the risk. The risk simulated was either a dread risk that immediately killed 50% of the population of the age group hit, or a continuous risk that killed the same total number of people in the same age group over a period of 10 years. The population growth rate was manipulated by setting the birth rate to either 0.05 (constant population), 0.075 (increasing population), or 0.025 (decreasing population). All individuals would die naturally after their 79th year. The risk hit only children, only young adults, only older adults, or only elderly adults.

In total there were 24 scenarios. Each scenario was simulated 500 times, and we calculated for every time point the average population size within the simulations. We analyzed each scenario by comparing the log difference in cumulative people-years between the dread risk condition and the continuous risk condition after 25, 50, 75 and 100 years.

Results

Figure 2 shows the results for the log difference in cumulative people-years depending on the population growth rate and the hit group after 25, 50, 75, and 100 years. A zero value indicates no difference in cumulative people-years between the dread risk and continuous risk; a negative value indicates a higher loss in cumulative people-years in the dread risk condition, and a positive value a higher loss in the continuous risk condition.

When children and young adults were hit by the risks, the effect was stronger and lasted for the entire 100-year-range simulated. When older and elderly adults were hit, the difference between dread and continuous risks was weaker, decreased over time, and sometimes even became positive.

In sum, the results show that the dread risk affected the cumulative population size more strongly for most scenarios, particularly when it hit children or younger adults. The objective of this first set of simulations was to evaluate the impact of a dread and a continuous risk on small samples that would reflect the sample size of social circles. With a second set of simulations we investigated the effects of such risks on a much larger population of the size of the U.S. population in 2010.

Simulation Set 2

Methods

We set the population size to the actual U.S. population size in 2010 [20] with the respective age distributions and population growth rates. Because the statistics only provided population size for age groups–for instance, 20,201,362 children <5 years old lived in the United States in 2010–we assumed an equal distribution of the children across 0–4 years for reasons of simplicity. As in Simulation Set 1, we manipulated which age group (children, young adults, older adults, elderly adults) was hit by the risk. The risk killed either 20% of the hit group, or the same total number of people over 10 years.

We again ran 500 simulations for each scenario, calculated the averaged population size of the dread risk and continuous risk and plotted the log integrals after 25, 50, 75, and 100 years.

Results

Using real U.S. data, we found support for the findings of the previous simulations. The differences between the cumulative population hit by dread versus continuous risks occurred across all conditions and lasted over, at least, 100 years. Independent of which age group was affected, the dread risk led to a higher loss in

people-years than the continuous risk (Figure 3). Loss was highest when children and young adults were hit by the risk.

Simulation Set 3

The third set of simulations wanted to test whether the results obtained from the first two sets of simulations also hold when an alternative model underlies computation. We used ecology models to define population growth and the impact of a dread and continuous risk. Specifically, we used models described in Lande [21] and Barnthouse [22], who investigated recovery rates of populations after catastrophes. Unlike in simulation sets 1 and 2, here we did not differentiate between the different age groups affected by the dread and continuous risk. Instead, the risky event simply reduced an a priori specified proportion of the population.

Methods

We used a basic growth model of population, given the current population size N_t, the growth rate r and the carrying capacity K (i.e., the carrying capacity represents the maximum population size possible, see [22]):

$$N_{t+1} = N_t + rN_t(\frac{K - N_t}{K}) \qquad (1)$$

We specified that a dread risk kills a particular proportion of the population (see also [21]):

$$D_t = \delta N_t \qquad (2)$$

Immediately after the dread, the remaining population grows as follows:

$$N_{t+1} = (N_t - D_t) + r(N_t - D_t)(\frac{K - (N_t - D_t)}{K}) \qquad (3)$$

In the following time steps, the population again grows according to (1). For the continuous risk, the same total amount of people as in the dread was killed, but over a period of x years:

$$C_t = \frac{D_t}{x} \qquad (4)$$

During these x years, the remaining population grows as follows:

$$N_{t+1} = (N_t - C_t) + r(N_t - C_t)(\frac{K - (N_t - C_t)}{K}) \qquad (5)$$

We use these formulas to simulate changes of population size subject to dread and continuous risks. An analytic solution for time to extinction of a population after random catastrophes is given in Lande [21].

In our simulations, we set the growth rate to either $r = .01$ (increasing population), $r = 0$ (constant population), $r = -.01$ (decreasing population). The initial population size at t_0 was set to 100 people and the carrying capacity to $K = 10,000$. The dread risk eliminated $\delta = .3$ of the population, the continuous risk eliminated the same total number of people over a period of 10 years. We followed the change of population size over a period of

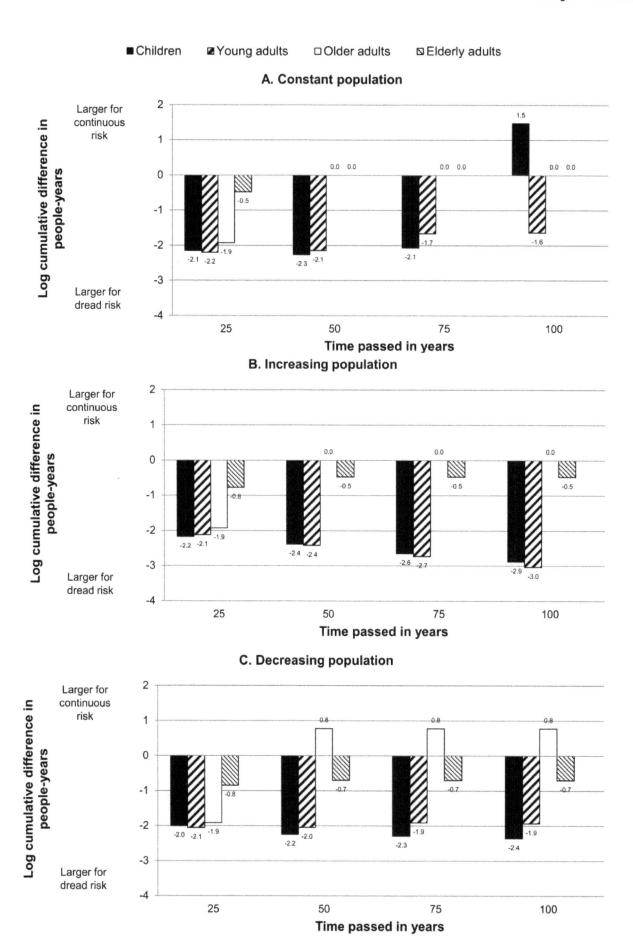

■Children   ▨Young adults   □Older adults   ▨Elderly adults

A. Constant population

B. Increasing population

C. Decreasing population

Figure 2. Log difference in people years lost after continuous and dread risk event (population size: 160). Log difference in people-years lost because of continuous and dread risk, by age group hit by the risk, separately for A. constant, B. increasing and C. decreasing populations. The dread risk killed 50% of a specific age group at once; the continuous risk the same total number of people over a period of ten years. A negative value of the difference indicates that the loss in people-years is larger for the dread risk; a positive value that the loss is larger for the continuous risk. Results show that dread risks lead to larger losses in people-years across time compared with continuous risks, in particular when children and young adults are affected.

100 years. The two risks could occur at any time within such time frame, but always occurred simultaneously, that is, they hit the population at the same *t*. Note that in simulation 1 and simulation 2 the risky events always occurred at the beginning of the time period. We simulated every condition 10,000 times and calculated the average integral, signaling the difference in cumulative people-years after the entire period of 100 years.

Results

In line with the findings of the previous simulations, we found that, over the time period considered, the dread risk resulted in higher losses in cumulative people-years than the continuous risk. The integral for the increasing population was −906, for the constant population −355, and for the decreasing population −137. Hence, even when using another model to test the impact of dread and continuous risks and even when not differentiating between different age groups, the results of the first two sets of simulations still hold.

Discussion

People's stronger reaction to dread risks compared with continuous risks is often perceived as a bias. This result proposes a new perspective against which the current hypotheses account-

ing for people's perception and reaction to dread risks might be reconsidered.

We showed through three different sets of simulations that this is in fact an ecologically rational strategy. The effect of dread risks compared with continuous risks is amplified twice: First by killing more people at a specific point in time, and second by reducing the number of children and young adults who would have potentially produced offspring. Hence, this effect is particularly strong when children and young adults are hit which is often the case for dread risks (e.g., earthquakes, terrorist attacks, pandemics). This result is also in line with findings suggesting that people are more concerned about risks killing younger, and hence more fertile, groups [23]. Moreover, when using a population ecology model and without specifying the age group affected by the respective risky events, the conclusion still holds.

Where does the fear of dread risk come from? Although our study was not designed to provide this answer, we can speculate about some possible answers to this question. According to the preparedness hypothesis mentioned at the beginning, people may be prone to fear risks that threaten their whole group. This trait may be a product of either individual or group selection. Because individual fitness depends and improves with group living [24], in particular in conditions of scarce population density that prevailed throughout much of human evolutionary history [25], an

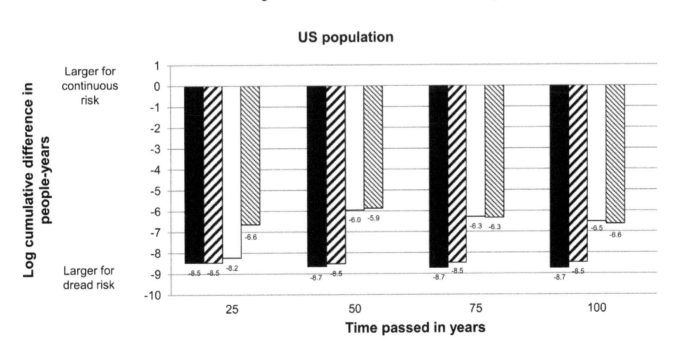

Figure 3. Log difference in people years lost after continuous and dread risk event (US population). Log difference in people-years lost because of continuous and dread risk, based on the US population. The dread risk killed 20% of a specific age group at once; the continuous risk killed the same total number of people over a period of 10 years. Results show that the dread risk leads to a larger loss in people-years over time across all age groups. The loss was largest when children and young adults were affected.

individual might profit from developing alertness to events that threaten to kill her group. An argument could also be made for group selection [26]. Groups that were more alert to dread risks and therefore managed to avoid them, suffered less from dread risks' devastating long-term consequences, which in turn would make them less vulnerable to other groups. Besides the evolutionary arguments, it is also possible that people learn to fear and avoid risks that appear to be particularly dangerous in their current environment. However, at this point we can only speculate about these explanations and further studies could be designed to address them.

There are important practical implications of this finding. For instance, from a public policy perspective, an appropriate reaction to dread risks would be to stimulate increase in birth rates and/or immigration to counterbalance the stronger loss in population size.

In sum, people's fear and stronger risk perception of dread risk, compared to continuous risks, should not be considered an irrational bias, an emotional overreaction to a dramatic event. In fact, people's intuition seems to capture the objective severity of the two different risks.

Acknowledgments

We thank Henrik Olsson for his helpful comments and Anita Todd for editing the paper.

Author Contributions

Conceived and designed the experiments: NB AR. Performed the experiments: NB AR. Analyzed the data: NB AR MG. Contributed reagents/materials/analysis tools: NB AR MG. Wrote the paper: NB AR MG.

References

1. Gigerenzer G (2004) Dread risk, September 11, and fatal traffic accidents. Psychol. Sci. 15: 286–287.
2. Gigerenzer G (2006) Out of the frying pan into the fire: Behavioral reactions to terrorist attacks. Risk Anal 26: 347–351.
3. Slovic P (1987) Perception of risk. Science 236: 280–285.
4. Sivak M, Flannagan MJ (2003) Flying and driving after the September 11 attacks. Am Sci 91: 6–8.
5. Gaissmaier W, Gigerenzer G (2012) 9/11, Act II: A Fine-grained analysis of regional variations in traffic fatalities in the aftermath of the terrorist attacks. Psychol Sci 23: 1449–1454.
6. Gigerenzer G, Mata J, Frank R (2009) Public knowledge of benefits of breast and prostate cancer screening in Europe. J Nat Cancer Inst 101: 1216–1220.
7. Lichtenstein S, Slovic P, Fischhoff B, Layman M, Combs B (1978) Judged frequency of lethal events. J Exp Psych HLM 4: 551–578.
8. Hertwig R, Pachur T, Kurzenhäuser S (2005) Judgments of risk frequencies: Tests of possible cognitive mechanisms. J Exp Psychol LMC 31: 621–642.
9. Preston MG, Baratta P (1948) An experimental study of the auction-value of an uncertain outcome. Am J Psychol 61: 183–193.
10. Tversky A, Kahneman D (1992) Advances in prospect theory: Cumulative representation of uncertainty. J Risk Uncertainty 5: 297–323.
11. Preclec D (1998) The probability weighting function. Econometrica 66: 497–527.
12. Öhman A, Mineka S (2001) Fears, phobias, and preparedness: Toward an evolved module of fear and fear learning. Psychol Rev 108: 483–522.
13. Hill KR, Walker RS, Bozicevic M, Eder J, Headland T et al. (2011) Co-residence patterns in hunter-gatherer societies show unique human social structure. Science 331: 1286–1289.
14. Lee RB, DeVore I (1968) Problems in the study of hunters and gatherers. In: Lee RB, DeVore I, editors. Man the hunter. Chicago IL: Aldine. 3–12.
15. Dunbar RIM, Schultz S (2007) Evolution in the social brain. Science 317: 1344–1347.
16. Krause J, Ruxton GD (2002) Living in groups. Oxford NY: Oxford University Press, 2002.
17. Galesic M, Garcia-Retamero R (2012) The risks we dread: A social circle account. PLoS ONE 7(4): e32837.
18. Statistisches Bundesamt (2011) Gesundheit-Todesursachen in Deutschland. Statistisches Bundesamt Fachserie 12 Reihe 4.
19. European Center for Disease Prevention and Control (2009) 2009 influenza (H1N1) pandemic. ECDC Risk Assessment Version 7.
20. Howeden LM, Meyer JA (2011) Age and sex composition: 2010. Washington, DC: US Census Bureau.
21. Lande R (1993) Risks of population extinction from demographic and environmental stochasticity and random catastrophes. Am Nat 142: 911–927.
22. Barnthouse LW (2004) Quantifying population recovery rates for ecological risk assessment. Environ Toxicol Chem 23: 500–508.
23. Wang XT (1996) Evolutionary hypotheses of risk-sensitive choice: Age differences and perspective change. Ethol Sociobiol 17: 1–15.
24. Kokko H, Johnstone RA, Clutton-Brock TH (2001) The evolution of cooperative breeding through group augmentation. Proc Roy Soc B Bio 268: 187–196.
25. Hassan FA (1981) Demographic archaeology. New York: Academic Press.
26. Nowak MA, Tarnita CE, Wilson EO (2010) The evolution of eusociality. Nature 466: 1057–1062.

9

Human Health Risk Assessment Based on Toxicity Characteristic Leaching Procedure and Simple Bioaccessibility Extraction Test of Toxic Metals in Urban Street Dust of Tianjin, China

Binbin Yu[1], Yu Wang[2], Qixing Zhou[1]*

1 Key Laboratory of Pollution Processes and Environmental Criteria (Ministry of Education), College of Environmental Science and Engineering, Nankai University, Tianjin, China, **2** Department of Agricultural and Biological Engineering, University of Florida, Gainesville, Florida, United States of America

Abstract

The potential ecological and human health risk related with urban street dust from urban areas of Tianjin, China was quantitatively analyzed using the method of toxicity characteristic leaching procedure (TCLP) and simple bioaccessibility extraction test (SBET). In the study, Hakason index, Nemerow index (*P*), the hazard index (*HI*) and the cancer risk index (*RI*) were calculated to assess the potential risk. The sequence of potential ecological risk based on Hakason index was arsenic (As) > cadmium (Cd) > lead (Pb) > copper (Cu) > chromium (Cr), in particular, As and Cd were regarded as high polluted metals. While the results of extraction of TCLP were assessed using *P*, the sequence was As > Pb > Cd > Cr > Cu, which mean that As and Pb should be low polluted, and Cd, Cr and Cu would barely not polluted. For human health, total carcinogenic risk for children and adults was 2.01×10^{-3} and 1.05×10^{-3}, respectively. This could be considered to be intolerable in urban street dust exposure. The sequence in the hazard quotient (*HQ*) of each element was As > Cr > Pb > Cu > Cd. The *HI* value of these toxic metals in urban street dust for children and adults was 5.88×10^{-1} and 2.80×10^{-1}, respectively. According to the characters of chemistry, mobility, and bioavailability of metals in urban street dust, we estimated the hazards on the environment and human health, which will help us to get more reasonable information for risk management of metals in urban environment.

Editor: Qinghua Sun, The Ohio State University, United States of America

Funding: This work was financially supported by the National Natural Science Foundation of China as a key project (grant No. 21037002) and the Ministry of Science and Technology, People's Republic of China as a 863 project (grant No. 2012AA101403-2). The funders had no role in study design , data collection and analysis, decision to publish, or preparation of the manuscript.

Competing Interests: The authors have declared that no competing interests exist.

* E-mail: zhouqx@nankai.edu.cn

Introduction

Along with rapid urbanization, many suburban lands are being converted to residential use, streets, commercial and industrial zones. The high population density leads to an increasing level of urban environmental pollution, industrial discharges, traffic emissions, and waste from municipal activities cause the major anthropogenic troubles [1–4]. Street dust is an important factor of urban pollution, and road activities often cause air pollution and adverse human health effects such as lung cancer, hypertension and cardiovascular diseases [5–9]. Due to the accumulation of metals in street dust with atmospheric deposition by sedimentation interception, human health may be adversely affected by air pollution if metal concentrations reach a level of being considered as toxic pollutants [10].

Many studies reported that the total concentration of heavy metals in street dust was regarded as an indicator of urban air pollution affecting urban environmental quality. More air-environmental scientists described that not only the total concentration of heavy metals in street dust, but also the proportion of their mobile and bioavailable forms were important to environmental and human health risk of metal pollutants [11–

16]. Metals in urban street dust occur in variable forms, such as adsorbed or exchangeable. The forms of toxic metals could affect their mobility and bioavailability including uptake by living organisms, and result in potential risk to the environment and residents. To evaluate long-term impacts of toxic metals on urban environment and risks to residents, it was necessary to investigate chemical forms, mobility, and bioavailability of toxic metals in urban street dust.

Tianjin (39°07′ N, 117°12′ E) is a mega-city in northern China, which is adjacent to Beijing and Hebei Province. It is a municipality along the coast of the Bohai Gulf. There are four distinct seasons, characterized by cold, windy, dry winters affected by the vast Siberian anticyclone, and hot, humid summers due to the East Asian monsoon. The mean temperature is 11.6–13.9°C and the annual average wind speed is 2–5 m s^{-1}. Rapid expansion in Tianjin has made the city become one of the most densely populated regions in China, with the population size of over 1.27 million. Urbanization increases the population density and lots of human activities such as traffic, industry, commerce, petrol combustion, and waste disposal. Of the contaminants in the urban area, toxic metals have caused serious concern of both researchers and governments for their characteristics of accumulation and degradation-resistant. As a crucial component of urban ecosystems, urban dust is subjected to continuous accumulation of contaminants [17–19], especially toxic trace metals [9,20], lots of investigations have been done on metal contamination in urban areas [15,21–25].

In this study, we provided valuable information about toxic metals in urban dust, including chemical characters, mobility, and bioavailability of toxic metals in urban environment. According to the information, we estimated the hazards of toxic metals in street dust, in particular, adverse effects on the environment and human health. The key scientific problems related to street dust pollution were addressed as follows: (1) the spatial distribution of toxic metals in street dust; (2) the mobility and bioavailability of toxic metals; (3) the risk index (RI) and Nemerow index (P) based on chemical behaviors and mobility of toxic metals; (4) the hazards index and total cancer risk based on oral bioavailability of toxic metals; and (5) some environmental management advice to reduce dust pollution.

Materials and Methods

Sampling (No specific permits were required for this study)

Urban street dust samples were collected on pavements next to main roads at periods when no rain had occurred during the previous week, brushes and trays were used for sampling. Along the central line in Tianjin, we collected 87 samples (39°09' N to 39°22' N, 117°14' E to 117°27' E). To uniform samples, the studying area was divided into regular grids of 1×1 km^2 and avoiding serious polluted areas (such as hospitals, gas stations, and bus stations). Each sample was at least 200 g and composed of 5–7 subsamples. Collected samples were put into clean polythene bags and taken to the laboratory as soon as possible. Each urban street dust sample was thoroughly mixed, sieved through a 63 μm nylon sieve, then one part was air-dried, other part was stored at 4°C. The detailed sampling information was depicted in Fig. 1.

Sample analyses

The pH values of urban street dust samples were measured with Milli Q water with a solid-to-solution ratio of 1:2.5 (w/v), organic matter (OM) contents were measured by the wet digestion method, the metals were measured by the microwave assisted acid digestion (US EPA 1996). 0.2 gram of the street dust sample was mixed with 8 mL 65% nitric acid, 5 mL 30% hydrogen peroxide and 3 mL 30% hydrofluoric acid and put into a microwave oven. The digestion procedure was described as follows: stage-1 (10 min to reach 150°C), stage-2 (10 min to reach 180°C) and stage-3 (15 min at 200°C). After cooling below 100°C, sample digestion solutions were evaporated near to dryness, added to 10 mL Milli Q water. Solutions were stored in 10 mL polyethylene vials at 4°C till analysis. Blank and control samples were used to check the accuracy and quality. The contents of arsenic (As), lead (Pb), cadmium (Cd), copper (Cu) and chromium (Cr) were measured by the inductively coupled plasma spectrometry (ICP-MS, DRC-e), the recovery of the metallic elements was in the 85.1–106.8% range.

Toxicity characteristic leaching procedure (TCLP)

Leaching of hazardous metals from street dust samples were examined by means of a TCLP [26]. The pH values of dust samples were beyond 5, the extraction solution (HOAC, pH = 2.88±0.05) was used, the solid - to - liquid ratio was 1:20, the temperature was 23±2°C, and the agitation time was 18 h in a rotary tumbler. After extraction, the leaching solutions were filtered through a glass fiber filter (0.45 μm). The filtrates were analyzed by ICP-MS immediately.

Simple bioaccessibility extraction test (SBET)

Oral bioavailability of metals in street dust samples were measured by SBET [27–29]. Samples were extracted with glycine (0.4 M; pH = 1.5±0.05, pre-adjusted with concentrated hydrochloric acid), the solid-to-liquid ratio was 1:100. Samples were extracted by rotating the samples end-over-end at 37°C for 1 h. The mixture solutions were centrifuged and the supernatant was filtered through 0.45 μm cellulose acetate filter. The pH values of the mixed-solution should be within 0.5 pH units of the starting pH, otherwise the procedure has to be redone. The filtrates were stored in a refrigerator at 4°C until analyzed by ICP-MS in one week.

Pollution risk of total concentrations of urban street dust

Most researchers used potential ecological risk index (RI) as a method to assess the degree of metals pollution in soil. The method was originally introduced by Hakanson [30], according to the toxicity of heavy metals, assessed the response of metals to the environment.

$$C_f{}^i = C^i \div C_n{}^i \tag{1}$$

$$E_i = T_i \times C_f{}^i \tag{2}$$

$$RI = \sum_{i=1}^{m} E_i \tag{3}$$

Where $C_f{}^i$ is the single metal pollution factor, C^i is the concentration of a metal in urban street dust, $C_n{}^i$ is a reference value for a metal, in this study, which is the soil background value in Tianjin [31]. E_i is the monomial potential ecological risk factor, T_i is the response coefficient for the toxicity of the single heavy metal, which for Cd is 30, As is 10, Pb and Cu are 5, Cr is 2 [30], RI is calculated as the sum of all five risk factors for heavy metals in urban street dust. Different categories of metal pollution about E_i and RI were delineated in Table 1.

Assessment of pollution risk using the results of extraction of TCLP

To evaluate the toxicity of the mobile and soluble parts of urban street dust, via a TCLP test, Nemerow index (P) [32] was used to assess the degrees of dust contamination. The pollution index was defined as the ratio of metal concentration to geometric means of background concentration of the corresponding metal:

$$P_i = C_i / S_i \tag{4}$$

and

$$P = \sqrt{\frac{(C_i/S_i)^2{}_{max} + (C_i/S_i)^2{}_{ave}}{2}} \tag{5}$$

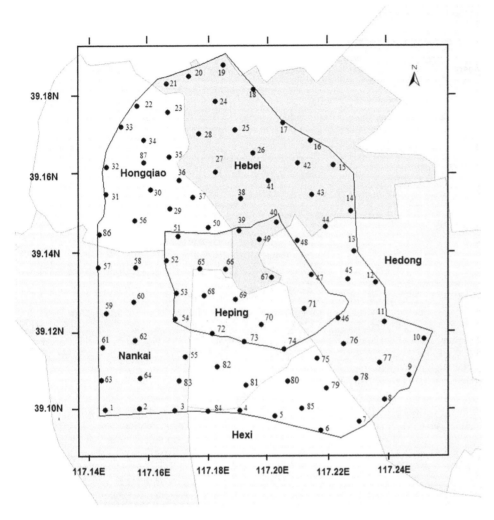

Figure 1. The sampling sites of urban street dust in Tianjin, China.

Where P_i is the evaluation score corresponding to each sample, C_i is the value of extraction of TCLP, S_i is the maximum concentration $(mg \cdot L^{-1})$ of a contaminant for toxicity characteristic, As for 5.0, Pb for 5, Cd for 1.0, Cu for 15, and Cr for 5 [33], P is the metal integrated pollution index, $(C_i/S_i)_{max}$ is the maximum and $(C_i/S_i)_{ave}$ is the average. There are five categories of metal pollution about P were displayed in Table 2.

Potential human health risk of metals in urban street dust

In order to assess both non-carcinogenic and carcinogenic risk for children and adults from ingesting urban street dust, the chronic daily intakes (CDI) of toxic metals and potential risks were used. According to the Exposure Factors Handbook [34], CDI $(mg \cdot kg^{-1} \cdot day^{-1})$ of toxic metals via dust can be calculated using the following equation:

Table 1. Potential ecological risk categories based on E_i and RI values[a].

E_i	Ecological risk categories of single metal	RI value	Ecological risk categories of the environment
$E_i < 40$	Low contamination	$RI < 150$	Low risk
$40 \leq E_i < 80$	Moderate risk	$150 \leq RI < 300$	Moderate risk
$80 \leq E_i < 160$	Considerable risk	$300 \leq RI < 600$	Considerable risk
$160 \leq E_i < 320$	High risk	$RI \geq 600$	Serious risk
$E_i \geq 320$	Serious risk		

[a][30].

Table 2. Potential ecological risk categories based on the P values[a].

P value	Ecological risk category
P≤0.7	Safe
0.7<P≤1.0	Warily level
1.0<P≤2.0	Low contamination
2.0<P≤3.0	Moderate contamination
P>3.0	Serious contamination

[a][31].

$$CDI = C \times CF \times \frac{E_F \times E_D \times Ing_R}{BW \times AT} \qquad (6)$$

Where C is the exposure site concentration $(mg \cdot kg^{-1})$, in this study, C is the mean concentration of SBET; E_F is the exposure frequency of 350 days·year^{-1} [35–36]; E_D is the exposure duration, in this study, 6 years for children and 24 years for adults; Ing_R is the ingestion rate at 200 mg·day^{-1} for children and 100 mg·day^{-1} for adults [34,37]; BW is average body weight, 15 kg for children and 60 kg for adults [35,38], and AT is average time (for non-carcinogens, AT = E_D×365 days; for carcinogens, AT = 70×365 days); CF is conversion factor, $1×10^{-6}$ kg·mg^{-1}.

The potential health risk of each element was calculated with equations (7–8). While Eq.7 is for non-carcinogenic risk, Eq.8 is for carcinogenic risk.

$$HQ = \frac{CDI \times RBA}{RfD_O} \qquad (7)$$

and

$$CR = (CDI \times RBA) \times CSF_O \qquad (8)$$

Where oral reference dose (RfD_o) and cancer slope factor (CSF_o) were obtained from regional screening levels [39], relative bioavailability (RBA) is the ratio of SBET/total contents. RfD_o for Pb has not established in US EPA, in this study, RfD_o for Pb is $3.5×10^{-3}$ mg·kg^{-1}·day^{-1} calculated from the provisional tolerable weekly Pb intake limit (25 µg·kg^{-1}-body weight) recommended by FAO/WHO for adults [24,40]. The toxicity of Cr depends on its valence state (Cr^{6+} and Cr^{3+}), in this study, Cr^{6+} represented total Cr. The RBA, RfD_o and CSF_o values were listed in Table 3.

Though interactions between some metals might result in their synergistic manner, all metal risks were additive. The hazard index (HI) and total risk [41] were estimated with the following equations:

$$HI = \sum_{i=1}^{m} HQ \qquad (9)$$

and

$$TotalRisk = \sum_{i=1}^{m} CancerRisk \qquad (10)$$

If HI exceeds 1.0, there is a chance that non-carcinogenic effects may occur, with a probability which increases with an increase in the value of HI; and then, if HI is less than 1.0, it believed that there is no significant risk of non-carcinogenic effects [42]. Cancer risks (CR) estimates the incremental individual lifetime cancer risk for simultaneous exposure to several carcinogens, the acceptable or tolerable risk for regulatory purposes is in the range of $1×10^{-6}$–$1×10^{-4}$.

Results

Dust properties and metals spatial distribution

The urban street dust samples from Tianjin cover a wide range of pH and OM. The pH values varied from 7.50 to 10.77 with average of 8.28, OM from 0.56% to 4.25% with average of 2.32%. The concentrations of As, Pb, Cd, Cu and Cr in urban street dust varied from 17.18 to 203.78, 20.64 to 155.67, 0.22 to 1.38, 20.65 to 187.78, and 30.85 to 224.60 mg·kg^{-1}, respectively, with the average concentration of 101.41, 60.11, 0.45, 55.47 and 71.85 mg·kg^{-1}, respectively (Table 4). The sequence in the contents of toxic metals in urban street dust was As > Cr > Pb > Cu > Cd. The spatial distributions of metals (As, Pb, Cd, Cu and Cr) in urban areas of Tianjin were depicted in Fig. 2. As was strong contaminated in most sites of the urban areas, which was higher than Class III of National Soil Standards [43]. Pb and Cr were lower than Class II of National Soil Standards, while Cd and Cu were lower than Class II of National Soil Standards in most urban areas and only a few of sites were higher than Class III of National Soil Standards where near the sites were high population density, including stations, commercial zones and construction fields.

Metal mobility and relationship with dust properties

The mobility of toxic metals depends on their chemical forms. In this study, according to the method of TCLP extraction, the range of extraction efficiency of toxic metals varied widely among urban street dust, As is 0.24% to 21.22%, Pb is 0.04% to 26.73%, Cd is 0.24% to 75.35%, Cr is 0.54% to 5.64% and Cu is 0.15% to 7.12%. The sequence of average extraction efficiency of toxic metals is Cd (12.99%) > Pb (4.11%) > Cr (2.70%) > As (2.07%) > Cu (1.40%).

The mobility of toxic metals in urban street dust may be affected by metal-particles and other physiochemical properties of dust, such as pH, and OM. The stepwise multiple regression analysis showed that there were significant correlations among pH, OM, the total concentration and the TCLP-extraction (Cd, Pb and Cu) concentration (Table 5). Whereas for As and Cr in street dust, not obvious relationship was found.

Metal bioaccessibility and relationship with dust properties

Many factors can affect metal bioaccessibility in dust, such as the interactions of metals, properties and constituents of dust. The range of bioaccessibility of metals varied widely among urban street dust, As is 4.94% to 37.60%, Pb is 4.49% to 33.22%, Cd is 8.81% to 43.13%, Cr is 6.70% to 80.70% and Cu is 4.21% to 36.92%. The sequence of average bioaccessibility of metals is Cr (22.63%) > Cd (21.73%) > Cu (19.01%) > Pb (15.04%) > As

Table 3. Health risk from heavy metals in urban street dust (n = 87).

Contaminant	C^a (95% UCL) mg·kg^{-1}	RfD_o^b (mg/kg-day)	CSF_o^b (mg/kg-day)$^{-1}$	Chemical daily intake (mg/kg-day)		Child		Adult	
				Children	Adult	Carcinogenic risk	HQ	Carcinogenic risk	HQ
As noncanc.	101.41	3.00E-04		1.35E-03	6.76E-04		4.56E-01		2.28E-01
As canc.	101.41		1.50E+00	1.11E-04	5.56E-05	1.69E-05		8.44E-06	
Cr noncanc.	71.85	3.00E-03		9.58E-04	4.79E-04		6.60E-02		3.30E-02
Cr canc.	71.85		5.00E-01	7.87E-05	3.94E-05	8.14E-06		4.07E-06	
Cd (diet)	0.45	1.00E-03		6.00E-06	3.00E-06		1.33E-03		6.67E-04
Cu	55.47	4.00E-02		7.40E-04	3.70E-04		3.22E-03		1.61E-03
Pb^c	60.11	3.50E-03		8.01E-04	4.01E-04		3.25E-02		1.63E-02

a 95% upper confidence limit of the mean concentrations.
b [39].
c [40,24].

(14.42%). The bioaccessibility of metals (Pb 59%, and Cu 58%) in Hong Kong urban soils (pH 6.6, SOM 4.4%, clay 7%, and sand 76%) [16] was higher, those (As 27.3%, Pb 71.7%, Cr 5.6%, and Cu 40.4%) almost all higher except Cr in urban roadside soils from Xuzhou, China [44].

Metal bioavailability could be affected by the ingested heavy metal-particles and other physiochemical properties of dust, such as pH, and OM. The stepwise multiple regression analysis showed that there were significant correlation relationships among pH, OM, the total metal concentration and the SBET-extraction (As, Pb, Cd, and Cu) concentration (Table 6). Whereas for Cr in street dust, no obvious relationship was found.

Ecological risk assessment based on *RI* and *P*

To get better image of urban street dust pollution and related risks, the Hakanson's method and the Nemerow's method were applied in this study. The ecological risk assessment results of metals in urban street dust were summarized in Table 7. It was found that the sequence of risk indices (E_i) of single metal was As > Cd > Pb > Cu > Cr. As and Cd are considerable contaminated in urban street dust, the values of E_i were higher than 100, and less than 160. Pb, Cu and Cr were low contaminated, the values of E_i were lower than 40. According to the character of metals, the overall potential ecological risk of the observed metals was quantified. *RI* was calculated as the sum of all the five risk factors. *RI* in urban street dust was 283.72, which mean that the metals in urban street dust were contaminated in the moderate degree ($150 \leq RI < 300$). The sequence of *P* index of each toxic metal was As > Pb > Cd > Cr > Cu (Table 7) in urban street dust. The ecological risk of As and Pb was low, whose value exceeded 1.0, and was less than 2.0. The ecological risk of Cd, Cu and Cr was barely not polluted, whose value was lower than 0.7. From the results, the potential ecological risk of the toxic metals in the extraction of TCLP was lower than that in the total concentrations, and As was the major contaminant in street dust.

Human health risk assessment according to oral bioavailability

Metals are usually non-degradable in the environment, and homeostasis mechanisms are unknown. Thus, biological life would be threaten by any high levels of metals [45]. In an individual lifetime, the incremental risk probability of carcinogens is estimated as a result of exposure to the potential carcinogens. In this study, only As and Cr were assessed through the ingestion exposure modes of urban street dust, for children, As was 1.50×10^{-3}, and Cr was 5.12×10^{-4}, compared to adults, As was 7.49×10^{-4}, and Cr was 2.56×10^{-4}. The total cancer risk for children was 2.01×10^{-3}, and that for adults was 1.05×10^{-3}. The cancer risk was higher than 1×10^{-4}, which was considered to be high potential risk and indicated that the carcinogenic risk of As and Cr in urban street dust exposure cannot be tolerable.

Non carcinogenic risks of a metal in urban street dust is potential, and non-carcinogenic toxicity can occur with time, which is not expressed as the probability of an individual metal with an adverse effect. In this study, the HQ sequence of toxic metals was As > Cr > Pb > Cu > Cd (4.56×10^{-1}, 6.60×10^{-2}, 3.25×10^{-2}, 3.22×10^{-3} and 1.33×10^{-3}, respectively, for children; and 2.28×10^{-1}, 3.30×10^{-2}, 1.63×10^{-2}, 1.61×10^{-3} and 6.67×10^{-4}, respectively, for adults) (Table 4). So the *HQ* values for toxic metals in this study were all lower than 1.0, which means that it was at the safe level. Non-carcinogenic risks were safe for children and adults. The hazards index (*HI*) for toxic metals to residents through the daily ingestion of street dust was 5.88×10^{-1}

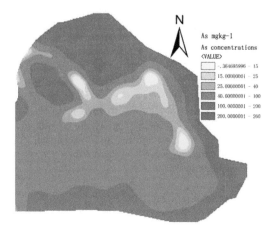

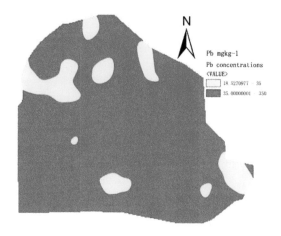

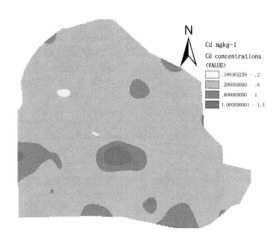

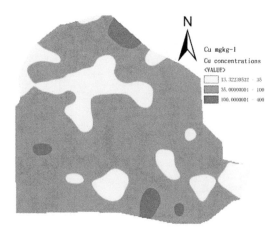

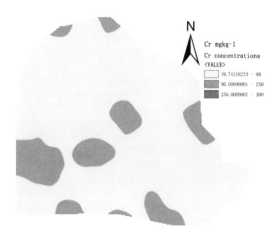

Figure 2. Spatial distribution of metals (As, Pb, Cd, Cu and Cr) in urban street dust of Tianjin, China.

Table 4. A comparison of metal average concentration (mg kg^{-1}) in street dust from some cities.

City (Ref.)	As	Pb	Cd	Cr	Cu
Tianjin	101.41	61.11	0.45	71.85	55.47
Beijing[a]	-	61	1.2	85.6	42
Shanghai[b]	8.01	236.62	0.97	264.32	257.63
Shenyang[c]	-	75.29	0.42	-	51.26
Nanjing[d]	13.4	103	1.10	126	123
Xi'an[e]	10.62	230.52	-	167.28	94.98
Hongkong[f]	66.8	120	-	124	110
National Standard- Class I[g]	15	35	0.20	90	35
National Standard- Class II[h]	25	350	0.60	250	100
National Standard- Class III[i]	40	500	1.0	300	400

[a][22].
[b][21].
[c][44].
[d][24].
[e][50].
[f][51].
[g][43] Environmental quality standard for soils in China (National Environmental Protection Agency of China, 1995), and values in Class I are threshold levels of nationwide natural background.
[h][43] Values in Class II are threshold values established to protect agricultural production and maintain human health.
[i][43] Values in Class III are establbished to maintain normal growth of plants, particularly the trees.

for children, and 2.80×10^{-1} for adults. *HI* was less than 1.0, which means that human exposure to urban street dust was safe.

Discussion

Street dust pollution in urban areas has an important impact on the environment, human health and life quality of residents. Meanwhile, the health of children and adults was affected by inhaling or ingesting street dust with the high metal contamination. It is important to identify the origin and distribution of toxic metals in street dust. The concentration of As in street dust was at the highest level, then followed by Cr, Pb, Cu and Cd. In most urban areas, it contaminated with As. Only the urban areas close to stations, commercial zones and construction fields, it contaminated with Cd and Cu. Compared with the average concentration of metals in street dust from different cities (Table 4), the average concentration of Pb, Cd, Cr and Cu in Tianjin was nearly same as that in Beijing, but lower than that in Shanghai [46], and the average concentration of Pb, Cu and Cd was almost at the same level in Shenyang [47].

Tianjin has a high demand on winter heating, from November to next March. As is known to be serious concentrated in coals, and coal combustion released many metals, such as As, Cr, Pb and Cu. Particularly, vehicle emissions are a major contributor to other

metals in urban street dust, including Pb, Cu and Cd. Vehicle loadings were about 1.76 million in urban areas of Tianjin till 2011. In this study, Cd, Cu and Pb were the main source of auto transport activities, such as vehicle exhaust emissions. Without doubt, there are also a number of other sources of metals in urban street dust, including disintegration of vehicle brakes and tires, atmospheric deposition, road surface wear, municipal solid waste incineration, and residential heating. Besides, many residential buildings are close to streets, which result in the frequent exposure of inhabitants to urban street dust. To reduce the accumulation of toxic metals in street dust, we suggest that it should use clean energy at the heating period and increase green areas in urban areas.

The results of the potential ecological risk for urban street dust using Hakason index and Nemerow index methods were different. The *RI* values of As, Cd, Cu and Cr were higher than *P*, only the risks of Pb were the same. From the results, *RI* could characterize the sensitivity of the toxic metals, and moreover, represent ecological risk resulted from the all contamination in local ecosystems. *RI* estimated the risk in the chemical forms of toxic metals. The categories, concentrations and toxic characteristics of metals could affect the values of E_i and *RI*, and cause the different results. The *P* index was used to estimate the mobility of toxic

Table 5. Regression linear analysis of TCLP-extractable contents, total contents and selected dust properties (pH, and OM).

	Regression equation	R	Sig.
As	C = 3.758−0.296pH+0.143OM−0.003Total	0.304	0.044
Cd	C = −0.059+0.004pH+0.001OM+0.197Total	0.547	0
Cr	C = 9.340−0.069pH+1.402OM+0.040Total	0.366	0.007
Cu	C = 5.310−0.300pH+1.678OM+0.053Total	0.660	0.000
Pb	C = 7.939−0.786pH+2.582OM+0.019Total	0.622	0

Table 6. Regression linear analysis of SBET-extractable contents, total concentrations and selected dust properties (pH, and OM).

	Regression equation	R	Sig.
As	C = −0.338+1.000pH+0.366OM+0.015Total	0.506	0
Cd	C = 0.071−0.009pH+0.020OM+0.131Total	0.692	0
Cr	C = 9.340−0.069pH+1.402OM+0.040Total	0.366	0.007
Cu	C = 5.310−0.300pH+1.678OM+0.053Total	0.660	0.000
Pb	C = 7.939−0.786pH+2.582OM+0.019Total	0.622	0

metals in urban street dust. As and Pb were the major contaminants for the mobility and toxicity, then followed by Cd, Cr and Cu. Despite the concentration of Pb was not higher than Class III of the Chinese National Soil Standards, the P index of Pb was still at the low risk due to the low concentrations of Pb in most urban areas. The results of assessment of chemical forms and mobility of metals were different, because the mobility of metals could be reduced by organic matters in dust samples. In other words, the chemical forms and mobility of metals should be coupled in order to get better assessment of their potential ecological risk. Meanwhile, the properties of urban street dust also affect the mobility and bioaccessiblity of metals in street dust. Toxic metals such as As, Pb, Cd, Cr and Cu could continuously accumulate in urban environment due to their non-biodegradability and the longer residence time. Thus, the local government should pay more attention to the action of reckless and unconscious pollution of the environment, and work out some particular management strategies to achieve better urban environmental quality.

Metals may accumulate in fatty tissues of human bodies, have middle and long-term health risks, and can adversely affect their physiological functions, disrupt the normal functioning of internal organs, or act as cofactors in other diseases [48,49]. Meanwhile, metals could deposit in the circulatory system of people, and each element (As, Cr, Pb, Cu, or Cd) has a distinctive toxicological picture and a particular distribution in human bodies. For instance, Pb accumulates primarily in the liver and in the kidney, while Cd concentrates in the kidney. All these elements originate, following a chronic exposure, complex systemic alterations so that it seems to be quite short-sighted to reduce their toxicological effects to the central nervous system and the cardiovascular system. Metals in urban soils might be transferred to human bodies via ingestion, dermal contact, or breathing, especially to children due to the "hand to mouth" activity during outdoor activities in playgroud and recreational areas [50,51]. The total carcinogens risk was 2.01×10^{-3} for children, and 1.05×10^{-3} for adults. In other words, they were both higher than 1×10^{-4}. With the results, we considered to be high potential risk and indicated that carcinogenic risk of As and Cr in urban street dust

exposure cannot be tolerable. HI for toxic metals were 5.88×10^{-1} for children, and 2.80×10^{-1} for adults, which means that exposure to urban street dust might be safe. Despite HI for children could be safe, low tolerance to toxins and the ingestion of dust through hand-to-mouth pathways, the hazards to children cannot be ignored. The ingestion of dust appears to be the main exposure to street dust, which results in health risk for As, Pb, Cd, Cr and Cu in street dust from anthropogenic sources. But if exposure frequency or ingestion rate is increased, street dust can still result in non-carcinogenic adverse effects on children. Therefore, the potential health risk for children cannot be ignored due to the exposure to the street dust.

Urban street dust receipt large amounts of metals originating from a variety of sources including materials, industrial waste, vehicle emissions, coal burning and other anthropogenic activities. Many metals could be remained in urban street dust for a long time, which may lead to further potential threat to ecosystems and human health. It is well known that chemical speciation of metals should be considered in addition to total concentration when evaluating the potential risk of metals in urban street dust. Therefore, we should consider the mobility and bioavailability of metals in urban street dust in order to offer an insight into potential ecological risk of urban street dust to the environment and human health.

Statement

Our sampling sites belong to local government departments, do not involve the use of private land. The whole sampling was finished in the public land, did not involve the destruction of the vegetation and biological species.

Author Contributions

Conceived and designed the experiments: QZ YW. Performed the experiments: BY. Analyzed the data: BY. Contributed reagents/materials/analysis tools: BY. Wrote the paper: BY. Guided the work as the supervisor and revised the manuscript many times: QZ. Improved the experimental scheme and made an elementary polishing and correction in English: YW.

Table 7. The potential ecological risk indexes.

Contaminant	E_i	Risk degree	RI	P	Risk degree
As	108.81±7.18	Considerable	283.72 Moderate degree	1.40	Low
Pb	15.03±0.67	Low		1.36	Low
Cd	148.54±6.59	Considerable		0.51	Clean
Cu	9.63±0.48	Low		0.16	Clean
Cr	1.71±0.08	Low		0.40	Clean

References

1. Wong CSC, Li XD, Thornton I (2006) Urban environmental geochemistry of trace metals. Environmental Pollution 142(1): 1–16.
2. Liu R, Zhou QX, Zhang LY, Guo H (2007) Toxic effects of wastewater from various phases of monosodium glutamate production on seed germination and root elongation of crops. Frontiers in Environmental Science & Engineering in China 1(1): 114–119.
3. Zhou QX, Wang ME (2010) Adsorption-desorption characteristics and pollution behavior of reactive X-3B red dye in four Chinese typical soils. Journal of Soils and Sediments 10(7): 1324–1334.
4. Zhang L, An J, Zhou QX (2012) Single and joint effects of cadmium and galaxolide on zebrafish (Danio rerio) in feculent water containing bedloads. Frontiers in Environmental Science & Engineering in China 6(3): 360–372.
5. Al-Khashman OA (2004) Heavy metal distribution in dust, street dust and soils from the work place in Karak Industrial Estate, Jordan. Atmospheric Environment 38(39): 6803–6812.
6. Arslan H (2001) Heavy metals in street dust in Bursa, Turkey. Journal of Trace and Microprobe Techniques 19(3): 439–445.
7. Chen B, Shand CA, Beckett R (2001) Determination of total and EDTA extractable metal distributions in the colloidal fraction of contaminated soils using SdFFF-ICP-HRMS. Journal of Environmental Monitoring 3(1): 7–14.
8. Harrison RM, Laxen DPH, Wilson SJ (1981) Chemical associations of lead, cadmium, copper, and zinc in street dusts and roadside soils. Environmental Science & Technology 15(11): 1378–1383.
9. Li XD, Poon CS, Liu PS (2001) Heavy metal contamination of urban soils and street dusts in HongKong. Applied Geochemistry 16(11–12): 1361–1368.
10. Ferreira-Baptista L, Miguel DE (2005) Geochemistry and risk assessment of street dust in Luanda, Angola: a tropical urban environment. Atmospheric Environment 39(25): 4501–4512.
11. Chang EE, Chiang PC, Lu PH, Ko YW (2001) Comparison of metal leachability for various wastes by extraction and leaching methods. Chemosphere 45: 91–99.
12. Ge Y, Murray P, Hendershot WH (2000) Trace metal speciation and bioavailability in urban soils. Environmental Pollution 107(1): 137–144.
13. Kim JY, Kim KW, Ahn JS, Ko I, Lee CH (2005) Investigation and risk assessment modeling of As and other heavy metals contamination around five abandoned metal mines in Korea. Environmental Geochemistry and Health 27(2): 193–203.
14. Poggio L, Vrscaj B, Schulin R, Hepperle E, Marsan FA (2008) Metals pollution and human bioaccessibility of topsoils in Grugliasco (Italy). Environmental Pollution 157(2): 680–689.
15. Lu SG, Bai SQ (2010) Contamination and potential mobility assessment of heavy metals in urban soils of Hangzhou, China: relationship with different land uses. Environmental Earth Sciences 60(7): 1481–1490.
16. Lu Y, Gong ZT, Zhang GL, Burghardt W (2003) Concentration and chemical speciations of Cu, Zn, Pb and Cr of urban soils in Nanjing, China. Geoderma 115(1–2): 101–111.
17. Tang L, Tang XY, Zhu YG, Zheng MH, Miao QL (2005) Contamination of polycyclic aromatic hydrocarbons (PAHs) in urban soils in Beijing China. Environment International 31(6): 822–828.
18. Thornton I, Farago ME, Thums CR, Parrish RR, McGill RAR, et al. (2008) Urban geochemistry: research strategies to assist risk assessment and remediation of brownfield sites in urban areas. Environmental Geochemistry and Health 30(6): 565–576.
19. Zhou QX, Luo Y (2011) Pollution Eco-chemistry (in Chinese). Science Press, Beijing, China 541pp.
20. Luo XS, Yu S, Li XD (2011) Distribution, availability, and sources of trace metals in different particle size fractrions of urban soils in Hong Kong: Implications for assessing the risk to human health. Environmental Pollution 159(5): 1317–1326.
21. Shi G, Chen Z, Bi C, Li Y, Teng J, et al. (2010) Comprehensive assessment of toxic metals in urban and suburban street deposited sediments (SDSs) in the biggest metropolitan area of China. Environmental Pollution 158(3): 694–703.
22. Zheng N, Liu JS, Wang QC, Liang ZZ (2010) Health risk assessment of heavy metal exposure to street dust in the zinc smelting district, Northeast of China. Science of the Total Environment 408(4):7 726–733.
23. Mahanta MJ, Bhattacharyya KG (2011) Total concentrations, fractionation and mobility of heavy metals in soils of urban area of Guwahati, India. Environmental Monitoring and Assessment 173(1–4): 221–240.
24. Hu X, Zhang Y, Luo J, Wang TJ, Lian HZ, et al. (2011) Bioaccessibility and health risk of arsenic mercury and other metals in urban street dusts from a mega-city, Nanjing, China. Environmental Pollution 5(159): 1215–1221.
25. Wang G, Oldfield F, Xia DS, Chen FH, Liu XM, et al. (2012) Magnetic properties and correlation with heavy metals in urban street dust: A case study from the city of Lanzhou, China. Atmospheric Environment 46: 289–298.
26. US EPA (1992) Toxicity characteristic leaching procedure. http://www.epa.gov/wastes/hazard/testmethods/sw846/pdfs/1311.pdf.
27. Oomen AG, Hack A, Minekus M, Zeijdner E, Cornelis C, et al. (2002) Comparison of five in vitro digestion models to study the bioaccessibility of soil contaminants. Environmental Science & Technology 36(15): 3326–3334.
28. US EPA (2007) Guidance for evaluating the oral bioavailability of metals in soils for use in human health risk assessment.
29. US EPA (2008) Standard operating procedure for an in vitro bioaccessibility assay for lead in soil. http://www.epa.gov/superfund/bioavailability/pb_ivba_sop_final.pdf.
30. Hakanson L (1980) An ecological risk index for aquatic pollution control. A sedimentological approach. Water Research 14(8): 975–1001.
31. China National Environmental Monitoring Centre. The element background values of soil in China [M], Beijing: China Environmental Science Press.
32. Nemerow NL, Hisashi S (1970) Benefits of water quality enhancement. Syracuse University, Syracuse, NY, Report No. 16110 DAJ, prepared for the US EPA.
33. Sun YF, Xie ZM, Li J, Xu JM, Chen ZL, et al. (2006) Assessment of toxicity of heavy metal contaminated soils by the toxicity characteristic leaching procedure. Environmental Geochemistry and Health 28(1–2): 73–78.
34. US EPA (1997) Exposure factors handbook. EPA/600/P-95/002F. Washington, DC: Environmental Protection Agency, Office of Research and Development.
35. BMEPRI (2007) Guidance of Site Environmental Assessment. Municipal Environmental Protection Bureau, Beijing.
36. Peng C, Chen WP, Liao XL, Wang ME, Ouyang ZY, et al. (2011) Polycyclic aromatic hydrocarbons in urban soils of Beijing: status, sources, distribution and potential risk. Environmental Pollution 159: 802–808.
37. Calabrese EJ, Kostecki PT, Gilbert CE (1987) How much dirt do children eat? An emerging environmental health question. Comments Toxicol 1: 229–241.
38. Shi GT, Chen ZL, Bi CJ, Wang L, Teng JY, et al. (2011) A comparative study of health risk of potentially toxic metals in urban and suburban road dust in the most populated city of China. Atmospheric Environment 45: 764–771.
39. US EPA (2010) Region 9, Regional Screening Levels Available online at. http://www.epa.gov/region9/superfund/prg/index.html.
40. Ostapczuk P, Valenta P, Rützel H, Nürnberg HW (1987) Application of differential pulse anodic stripping voltammetry to the determination of heavy-metals in environmental samples. Science of the Total Environment 60: 1–16.
41. US EPA (2007) Available: http://www.epa.gov/superfund/bioavailability/bio_guidance.pdf.
42. US EPA (2001) Risk Assessment Guidance for Superfund: Volume III d Part A, Process for Conducting Probabilistic Risk Assessment. US Environmental Protection Agency, Washington, D.C. EPA 540-R-02-002.
43. GB 15618-11995. Available: http://www.nbepb.gov.cn/UploadFiles/lan2/200575165235193.pdf.
44. Wang XS, Qin Y, Chen YK (2007) Leaching Characteristics of Arsenic and Heavy Metals in Urban Roadside Soils Using a Simple Bioavailability Extraction Test. Environmental Monitoring and Assessment 129(1–3): 221–226.
45. Tong STY, Lam KC (2000) Home sweet home? A case study of household dust contamination in Hong Kong. Science of the Total Environment 256: 115–123.
46. Tanner PA, Ma HL, Yu PKN (2008) Fingerprinting metals in urban street dust of Beijing, Shanghai and Hong Kong. Environmental Science & Technology 42(19): 7111–7117.
47. Sun YB, Zhou QX, Xie XK, Liu R (2010) Spatial, sources and risk assessment of heavy metal contamination of urban soils in typical regions of Shenyang, China. Journal of Hazardous Materials 174(1–3): 455–462.
48. Nriagu JQ (1988) A silent epidemic of environmental metal poisoning? Environmental Pollution 50: 139–161.
49. Mehmet Y (2006) Comprehensive comparison of trace metal concentrations in cancerous and non-cancerous human tissues. Current Medicinal Chemistry 13(21): 2513–2525.
50. Han YM, Du PX, Cao JJ, Posmentier ES (2006) Posmentier, Multivariate analysis of heavy metal contamination in urban dusts of Xi'an, Central China. Science of the Total Environment 355: 176–186.
51. Yeung ZLL, Kwok RCW, Yu KN (2003) Determination of multi-element profiles of street dust using energy dispersive X-ray fluorescence (EDXRF). Applied Radiation and Isotopes 58(3): 339–346.

10

Going Beyond the Millennium Ecosystem Assessment: An Index System of Human Dependence on Ecosystem Services

Wu Yang[1]*, **Thomas Dietz**[1,2], **Wei Liu**[1], **Junyan Luo**[1], **Jianguo Liu**[1]

1 Center for Systems Integration and Sustainability, Department of Fisheries and Wildlife, Michigan State University, East Lansing, Michigan, United States of America,
2 Environmental Science and Policy Program, Department of Sociology and Animal Studies Program, Michigan State University, East Lansing, Michigan, United States of America

Abstract

The Millennium Ecosystem Assessment (MA) estimated that two thirds of ecosystem services on the earth have degraded or are in decline due to the unprecedented scale of human activities during recent decades. These changes will have tremendous consequences for human well-being, and offer both risks and opportunities for a wide range of stakeholders. Yet these risks and opportunities have not been well managed due in part to the lack of quantitative understanding of human dependence on ecosystem services. Here, we propose an index of dependence on ecosystem services (IDES) system to quantify human dependence on ecosystem services. We demonstrate the construction of the IDES system using household survey data. We show that the overall index and sub-indices can reflect the general pattern of households' dependences on ecosystem services, and their variations across time, space, and different forms of capital (i.e., natural, human, financial, manufactured, and social capitals). We support the proposition that the poor are more dependent on ecosystem services and further generalize this proposition by arguing that those disadvantaged groups who possess low levels of any form of capital except for natural capital are more dependent on ecosystem services than those with greater control of capital. The higher value of the overall IDES or sub-index represents the higher dependence on the corresponding ecosystem services, and thus the higher vulnerability to the degradation or decline of corresponding ecosystem services. The IDES system improves our understanding of human dependence on ecosystem services. It also provides insights into strategies for alleviating poverty, for targeting priority groups of conservation programs, and for managing risks and opportunities due to changes of ecosystem services at multiple scales.

Editor: Just Cebrian, MESC; University of South Alabama, United States of America

Funding: We gratefully acknowledge financial support from the National Science Foundation (NSF Award Numbers 0709717 and OISE-0729709), the National Aeronautics and Space Administration (NASA Award Number NNX08AL04G), Michigan State University AgBioResearch (http://agbioresearch.msu.edu/), and fellowships from Michigan State University's Environmental Science and Policy Program (http://www.espp.msu.edu/) and Graduate Office (http://grad.msu.edu/). The funders had no role in study design, data collection and analysis, decision to publish, or preparation of the manuscript.

Competing Interests: The authors have declared that no competing interests exist.

* E-mail: yangwu1201@gmail.com

Introduction

The Millennium Ecosystem Assessment (MA) was designed to assess the consequences of ecosystem change and provide scientific information that could aid in sustainably managing ecosystems for human well-being [1]. Although the intended audience was decision-makers, MA also provided a conceptual framework for studying interactions among four key components (i.e., indirect drivers, direct drivers, ecosystem services, and human well-being) of coupled human and natural systems (CHANS) [2], and identified future research needs [3].

Of the interactions between the four components in the MA framework, the linkage between ecosystem services and human well-being is perhaps least understood. The relationship between human well-being and the social factors that influence it has been extensively studied [4,5,6,7,8]. Through efforts like land change science and structural human ecology, the relationships between changes of ecosystems services and factors that influence them have also begun to be documented from local to regional and

global scale [9,10]. It has been recognized that humans substantially depend on ecosystem services, which range from basic provisions of food, fresh water and fuel, through regulation of water and air quality, to cultural services like ecotourism [1,11]. The MA also established that during recent decades, two thirds of these services have degraded or are in decline due to the unprecedented scale of human activities [1]. But what are the consequences of such dramatic degradation to short-term and long-term human well-being? The risks of ecosystem degradation and their consequences for human well-being, including nonlinear or abrupt changes, are poorly quantified. On the one hand, there is a lack of a robust theory that links ecological diversity to ecosystem dynamics and ecosystem services [3]. On the other hand, the scientific community lacks understanding of how and to what extent humans depend on ecosystem services. For instance, it has been widely recognized that the poor are most dependent on ecosystem services and most vulnerable to the degradation of ecosystem services [1]; however, it is generally not known how such dependence differs across time, space and various population groups (e.g., across income levels). To better understand, monitor and manage such dependences, a quantitative approach is urgently needed.

From our perspective, there are at least four reasons to quantify human dependence on ecosystem services. First, the relationship between ecosystem services and poverty seems obvious but the dependence of the poor on ecosystem services is rarely quantified, which leads to a pervasive tendency to overlook it in statistics, poverty assessments and natural resource management decisions [12]. A quantitative measurement of such dependence and its integration into decision making could reverse inappropriate strategies that could otherwise lead to further marginalization of the poor and increased pressure on ecosystem services.

Second, benefits provided by ecosystem services are often unequally distributed across different population groups and there may be trade-offs among groups. With better understanding of the distribution of benefits from ecosystem services (e.g., fuelwood, clean water, non-timber forest products, and tourism) across different population groups, conservation and development programs may be better designed to guide the flow of benefits from ecosystem services to target priority population groups. The Wolong Nature Reserve of China provides a compelling example. Most benefits obtained from ecotourism flow to the outside tourism development companies rather than local households [13,14], a common phenomenon in many other areas [15]. Government policies should encourage household relocation closer to tourism facilities and provide more support to local households (e.g., provide training to improve human capital and offer favorable loan opportunities) to enhance their capacity for participating in tourism businesses [13,14]. In doing so, more benefits from tourism would flow to local households and substantially reduce their pressure on provisioning services by which local ecosystems provide fuelwood, bamboo shoots, and traditional Chinese medicine.

Third, a better understanding of such dependence would draw attention to currently unmanaged risks and unrealized opportunities that come with ecosystem change. For example, agricultural supply chains can be tightly dependent on ecosystem services and thus are vulnerable to dramatic ecosystem degradation. Unprecedented human activities would likely lead to more frequent extreme climate events and natural disasters (e.g., storms, floods, droughts, and landslides), cause tremendous destruction to ecosystems and their services, and threaten the livelihoods of those people who are highly dependent on corresponding ecosystem services [1,10]. Yet few managers or policy analysts understand this dependence and related unintended consequences, and even fewer manage the potential risks and opportunities [16,17].

Fourthly, a quantitative measurement of such dependence would improve the understanding of human-nature interactions. One of the major advances and challenges of the CHANS approach for studying human-nature interactions is to construct coupled models by integrating sub-models of both human and natural subsystems [18]. The key of such integration requires good understanding of the interactions between human and natural subsystems. Currently, there are few coupled models integrating drivers, ecosystem services, and human well-being to systematically understand human-nature interactions [3,19,20]. The quantification of human dependence on ecosystem services could potentially serve as a proxy to facilitate such integration and understanding.

The objectives of this study were to (1) propose the conceptual basis of an index of dependence on ecosystem services (IDES) system to measure the degree of human dependence on ecosystem

services; (2) demonstrate the construction of the IDES system with empirical data; and (3) illustrate advantages and applications of the proposed IDES system. Specifically, we first provided the conceptual basis of an IDES system, including an overall index and sub-indices for different categories of ecosystem services based on the widely accepted MA framework. We then delineated the process of estimating the indices at Wolong Nature Reserve. We examined temporal changes of the overall IDES and shifts in of the structure of the IDES system (i.e., changes of sub-indices). We compared the overall index with an alternative indicator (i.e., the commonly used agricultural income share) to illustrate the advantages of our proposed index system. Moreover, we assessed the dependence of the poor on ecosystem services. In particular, we analyzed how households' dependences on ecosystem services differ across different degrees of access to capitals (i.e., natural, human, financial, manufactured, and social capitals). We also evaluated the spatial heterogeneity of the overall IDES.

Methods for Developing an Index System of Human Dependence on Ecosystem Services

2.1 Conceptualization of the index system

The term ecosystem services is defined and used in a variety of ways [1,11,21,22,23,24,25]. Here we aligned with the definitions of the MA as the benefits that people obtain directly or indirectly from ecosystems, including both natural systems or highly managed systems [1]. In particular, we included agricultural products as part of ecosystem services. We acknowledge that some literatures might exclude products from highly managed systems (e.g., agro-ecosystems and constructed wetlands) and restrict ecosystem services to goods and services provided by natural systems only. But since the logic of our analysis is driven by the MA, we felt it appropriate to adhere to the definition by the MA. Our proposed index system of dependence on ecosystem services includes an overall index and three sub-indices. The overall index of human dependence on ecosystem services is defined as the ratio of net benefits obtained from ecosystems to the absolute value of total net benefits that derived from ecosystems and other socioeconomic activities (e.g., migrant work, and small business unrelated to ecosystem services, see Table S1). In addition to the overall index, a sub-index can be calculated for each category of ecosystem services under the MA framework (i.e., provisioning, regulating services, and cultural services) [1]. Because supporting services are the bases for other three types of services, following the common practice in ecosystem service assessment, they are not included in IDES to avoid double accounting. As shown by the definition, the higher value of the overall index or sub-index represents the higher dependence on the corresponding ecosystem services, and thus the higher vulnerability to the degradation or decline of corresponding ecosystem services. The equations for calculating three sub-indices and the overall IDES are given as below.

$$IDES_i = ENB_i / |\sum\nolimits_{i=1}^{3} ENB_i + SNB| \qquad (1)$$

$$IDES = \sum\nolimits_{i=1}^{3} IDES_i \qquad (2)$$

where i is the category of ecosystem services (i.e., provisioning, regulating, and cultural services); $IDES_i$ is the sub-index for category i; ENB_i is the total net benefit obtained from category i ecosystem services; SNB is the total net benefit obtained from socioeconomic activities; IDES is the overall index.

There are four reasons for using net benefits instead of gross benefits. First, ecosystems generate both services and dis-services to humans. Dis-services may include pests and diseases causing reduction in agricultural production and other unintended negative health consequences for organisms including humans [26]. Second, the generation and delivery of ecosystem services may entail costs (e.g., costs of seeds, fertilizers, and pesticides for agricultural products). Using the gross benefits could potentially mislead decision making [27]. One might opt for a program that has the largest increase in gross benefits when another program has a larger yield of net benefits, thereby choosing an inefficient program. Third, using net benefits allows the inclusion of trade-offs between different ecosystem services [28]. Such trade-offs would not be correctly represented if gross benefits are used without considering the costs of delivering those services. Finally, using net benefits facilitates cross-context comparisons. Few previous ecosystem service assessments have evaluated net benefits [29,30,31]. Many previous studies have evaluated only the gross benefits so results from different studies are not comparable because ecosystem dis-services and costs of generating ecosystem services can be substantial and vary considerably across contexts.

Both the sub-indices and overall index can be negative. This is because net benefits are not necessarily positive. Total net benefits from each category and all categories of ecosystem services summed can be negative. The ecological and economic meaning of an index with negative value is that the gross benefit obtained from ecosystem services is lower than the sum of costs for generating the corresponding ecosystem services and costs of ecosystem dis-services. For example, the gross benefits of producing agricultural products may be lower than the total costs of seeds, fertilizers, and pesticides.

2.2 Methods for constructing the index system

The index system is constructed to assess net benefits of a unit of analysis (e.g., household). The procedures for this approach are in some ways similar to that of many Cost-Benefit Analyses (CBA) [32,33] and Ecosystem Service Assessments (ESA) [1,28,30,31] where data from a variety of sources are aggregated into an integrated assessment and where the unit of analysis for which the calculation is done must be specified. For CBA and ESA this is often a region or nation, while here we will work at the household level.

Where markets for the gross benefits and costs exist, assessments are relatively straightforward and simple. It is easy to apply market-based valuation methods such as the market price method, the appraisal method, and the avoided cost method [31,34,35,36]. Otherwise, when market data are not available, nonmarket valuation methods such as the contingent valuation method, the travel cost method, the stated preference method, and the hedonic price method can be used [31,34,36,37].There are also cross-cutting methods, such as the benefit transfer method and unit-day value method, which combine both market-based and nonmarket methods [38,39,40]. Recently, integrated approaches such as the Integrated Valuation for Ecosystem Services and Tradeoffs (InVEST) have focused on assessing ecological production and then applying economic valuation methods [28,41]. A variety of reviews and guidelines have discussed these economic valuation methods in detail (e.g., [32,33,34,42]). A summary and critique of the use of these methods was presented by Bateman [37] and thus we do not discuss the use of these economic valuation methods in detail here. We provided an example of how different types of data could be collected through various economic valuation methods to assess the net benefits for constructing the IDES system. The

following empirical study will demonstrate the integration of different data sources and valuation methods in detail.

Consider a rural household living in a forest area as an example. Costs and benefits from agricultural products and other socioeconomic activities are parts of the household's income and expenditures and could be captured in a survey with relative ease, using best practices for economic surveys. But when benefits or avoided costs that do not involve market transaction (e.g., non-timber forest products such as fruits, herbal medicine, and fuelwood), they are not shown in the household's income and expenditures as conventionally defined and thus are not captured by conventional economic survey methods. If there are established payments for ecosystem services (PES) programs, then the obtained benefits (e.g., payments) and associated costs (e.g., labor costs for monitoring forests) have market values. If such PES programs are not in place, an ESA can be conducted by adding corresponding survey questions, for example by using contingent valuation method (see case studies in [31,43]). An ESA can also be conducted using integrated tools such as InVEST for the entire study area (see case studies in [28,41]) and then disaggregating to the household level (e.g., divided by total number of households in the entire study area or calculated by defining a buffer zone of accessibility to certain ecosystem services based on each house-hold's location, see an example of fuelwood collection in [44]).

Empirical Demonstration of Constructing the Index System

3.1 Description of the demonstration area

Here we provide an example to demonstrate the index system at Wolong Nature Reserve (N 30°45′–31°25′, E 102°52′–103°24′, Fig. 1) in China. We choose Wolong Nature Reserve as our study area for three reasons. First, situated in the transition from Sichuan Basin to the Qinghai-Tibet Plateau, it is within one of the top 25 global biodiversity hotspots endowed with enormous ecosystem services [45,46]. Second, it is one of the earliest nature reserves established in China [47]. Like many other protected areas, there are human residents living inside who depend on many types of ecosystem services. Third, our research team has been conducting studies in this area over the past 18 years and has accumulated extensive datasets and local knowledge that give us a well-grounded basis for testing the IDES concept, methods and applications.

The primary purpose of Wolong Nature Reserve is to protect giant pandas (*Ailuropoda melanoleuca*) as well as regional forest ecosystems and rare plant and animal species [48]. When it was established in 1963, its initial size was ~20,000 ha but was expanded to its current size of ~200,000 ha in 1975 [48]. It is home to ~10% of the total wild giant panda population [48]. Currently, there are ~4900 local human residents, distributed in ~1200 households in two townships (i.e., Wolong and Gengda Townships) within the Reserve (Fig. 1). The majority of local residents are farmers involved in subsistence activities such as cultivating maize and vegetables, raising livestock (e.g., pigs, cattle, yaks, and horses), collecting traditional Chinese medicine, keeping bees, and collecting fuelwood for cooking and heating (Table S1).

In response to the massive droughts in 1997 and floods in 1998, the Chinese government started to implement a series of ecosystem service policies [49,50], including two of the world's largest payments for ecosystem services (PES) programs: the Natural Forest Conservation Program (NFCP) and the Grain-to-Green Program (GTGP) [49]. These PES programs aim mainly to improve regulating services such as soil erosion control, water conservation, carbon sequestration, and air purification. From

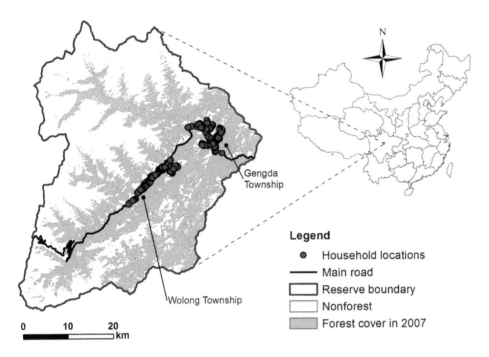

Figure 1. Wolong Nature Reserve in Sichuan Province, southwestern China.

2000, Wolong Nature Reserve started to implement GTGP, NFCP, as well as a local PES program called the Grain-to-Bamboo Program (GTBP) [48,51]. NFCP aims to conserve and restore natural forests through logging bans, afforestation, and monitoring, using PES approach to motivate conservation behavior [49,51]. GTGP and GTBP aim to convert cropland on steep slopes to forest/grassland, and bamboo forest by providing farmers with subsidies, respectively [48,49,51].

3.2 Data and methods

Ethics Statement. The household survey data used here were collected with the permission from the Wolong Administration Bureau of Wolong Nature Reserve. A verbal consent process from interviewees was used due to the low level of education of our interviewees. The verbal consent script was read to the selected interviewees before conducting the survey. Only when they agreed to participate in our survey, we then continued to ask questions in the designed survey instruments. Or else, we did not collect any information but switched to the next selected interviewee. The Institutional Review Board of Michigan State University (http://www.humanresearch.msu.edu/) approved the verbal consent process, verbal consent script, and survey instruments.

Data collection and analyses. For this study we used household survey data to estimate the obtained net benefits from ecosystem services (or equivalently gross benefits and costs) for households. This allows us to construct the IDES estimates for households. Our surveys were conducted in the summer of 1999 and the end of 2007 to obtain data covering activities in 1998 and 2007. We tracked the same randomly sampled 180 households so the data constitutes a panel. Usually the household heads or their spouses were chosen as interviewees because our past experience indicates that they are the decision makers and are most familiar with household affairs [52]. To facilitate cross-context comparisons, we used the categories for household income and expenditure data that are consistent with those of the National Bureau of Statistics of China [53] and thus with standard economic survey

methods. We used the MA classification for ecosystem services to generate sub-indices. It is important to note that it is impractical, if not impossible, to assess all the ecosystem services in a study area. This analysis only attempts to include as many major ecosystem services as possible using the best available data in our study area.

As the term implies direct ecosystem benefits are those that are used directly in generating human well-being. For example, agricultural products are provisioning services that provide direct benefits from agricultural ecosystems. Other services contribute indirectly to human well-being. Sometimes indirect benefits are only one step removed from direct benefits (i.e., first-order indirect benefits) and sometimes they are more distantly linked (i.e., secondary or more distant indirect benefits). For example, local households do not directly partake in the ecotourism activities but they one-step indirectly benefit from the cultural services of ecotourism through providing transportation, food and accommodation services to eco-tourists. But ecotourism may also enhance the development of infrastructure (e.g., road construction), create more job opportunities, and thus provide indirect benefits several steps removed from the cultural services. Generally, the challenge in identifying benefits for CBA is to separate the genuine indirect effects from those that are double accounting [54]. Usually, if there is not a strong rationale, only direct benefits and costs are included to avoid double accounting [33]. However, in our study area, first-order indirect benefits capture an important part of benefits from ecosystem services and the inclusion of them do not cause double-accounting (Table S1). As a first approximation, here we included direct benefits and first-order indirect benefits in our calculations because these captured the majority benefits in our study area (Table S1). We adapted the MA classification for types of related ecosystem services (Table S1) to make it appropriate for our study area. For some specific services, the classification may differ from what would be appropriate in other areas. But this does not affect the comparisons using the overall index and sub-indices of IDES, which are based on the generalizable MA framework.

Some households obtained negative net benefits in agricultural operating income when the total gross agricultural income was lower than total agricultural expenditure, due to pests, diseases, natural disasters (e.g., storms and landslides), and/or low prices of agricultural products. Most income from ecosystem services comes from provisioning services, but some households also have income from ecotourism, which we categorized as benefits related to cultural services, and income from PES programs, which we categorized as benefits related to regulating services (Table S1).

The benefits that households obtained from ecosystems include not only the benefits reflected in their income, but also the avoided costs not reflected in their expenditures. Two major items of avoided costs were assessed here. One is the reduced electricity fees through a subsidized electricity price. Because the conservation of forests also dramatically conserves watersheds in our study area, local households were given a reduction of electricity price of 0.07 yuan per kilowatt-hour in both 1998 and 2007 (yuan: Chinese Currency, 1 USD = 7.52 yuan as of 2007). Thus, the avoided electricity fees could be calculated by multiplying their consumed electricity amount and the reduced price. Another item of avoided cost is from fuelwood collection for energy use. Households would need to pay for alternative energy sources (e.g., electricity, coal) if they do not collect fuelwood. Because households need to spend labor collecting fuelwood, in the past when one household did not have enough laborers in the fuelwood collection season, one might exchange laborers or hire laborers from other households. Thus, the monetary value of collected fuelwood can be estimated as the market value of the labor spent on collecting it. In our household survey, we measured the collected amount of fuelwood and total labor spent in collecting it. We then calculated the shadow price of fuelwood (approximately 0.10 yuan per kilogram in 1998 and 0.20 yuan per kilogram in 2007). Data for each household on each of these sources of net income and avoided costs were then used to construct the index system. The dataset used for this study is provided in the Supporting Information (Data file S1).

3.3 Results of the index system

Table 1 showed the results of net benefits from different sources and the overall IDES and corresponding sub-indices in both 1998 and 2007. Our results showed a dramatic increase of net benefits from all categories of ecosystem services and socioeconomic activities. From 1998 to 2007, the total net benefit from ecosystem services has increased from an average of approximately 1,723 yuan to 12,972 yuan (both values were in present values for 1998). Meanwhile, from 1998 to 2007, the total net benefit from socioeconomic activities also has dramatically increased from an average of approximately 2,456 yuan to 12,350 yuan.

Table 1 also showed that the overall index of households' dependences on ecosystem services has increased from approximately 0.42 in 1998 to 0.61 in 2007. The average overall IDESs were 0.45 in 1998 and 0.61 in 2007, indicating that approximately 45% and 61% of total net benefits to households came from ecosystem services in 1998 and 2007, respectively. Approximately 54% and 63% households had an overall IDES larger than 0.50, and 9% and 16% households had an overall IDES of 1.00 in 1998 and 2007, respectively. Overall these results suggested that most households in our study area were highly dependent on ecosystem services and some were essentially completely dependent on them.

The percent of households obtained positive net benefits from provisioning services were 89% and 85% in 1998 and 2007, respectively. Almost all households benefited from regulating services in both 1998 and 2007. Perhaps most interesting, almost all households in 2007 acquired positive net benefits from regulating services through the PES programs (i.e., NFCP, GTGP, and GTBP). These programs were the major reason for the dramatic increase of net benefits from regulating services. However, almost no household in 1998 and only 11% households obtained positive net benefits from cultural services such as ecotourism.

3.4 Advantages and applications of the index system

Our IDES is better than the agricultural income share in reflecting households' dependences on ecosystem services. Agricultural income share, or the ratio of agricultural income to total income, is a commonly used indicator that can approximately reflect a rural household's dependence on ecosystem services. Although the proposition that the poor are more dependent on ecosystem services is rarely examined quantitatively, it is a widely accepted notion [3,12]. Here, we compared the overall IDES with agricultural income share by examining their relation to overall household income. Our results suggested that the overall IDES were negatively associated with household income in both 1998 and 2007 (Table 2).That is, higher income households make less use of ecosystem services. These results confirmed the common view that low incomes households are more dependent on ecosystem services. However, the association of household income with agricultural income share was significant only in 1998 but not in 2007 (Table 2). These results indicated that our overall IDES was better than the agricultural income share as a measure of rural households' dependences on ecosystem services. In our study area, income had become decoupled from income share from agriculture but not from use of ecosystem services by our second survey.

Comparing the results of the index system in 1998 and 2007 (Table 1), the reasons that IDES is a better measure than agricultural income share are easy to see. In 1998, most of the household income was from agriculture, which was classified as benefits related to provisioning services (Table S1). The overall IDES was almost equivalent to the sub-index for provisioning services (Table 1) and thus was similar to agricultural income share in reflecting a household's dependence on ecosystem services in 1998. In 2007, household income sources became more diverse and included many non-agricultural items (Table S1). Therefore, unlike the overall IDES, agricultural income share no longer accurately reflected households' dependences on ecosystem services.

These results demonstrate some of the uses of the index system. We further illustrated its applications by examining how the benefits of ecosystem services were unequally available to households. Because ecosystem services flow from natural capital, households who possess substantial access to natural capital should obtain more benefits from ecosystems and thus would be more dependent on ecosystem services than those with less access to natural capital. The positive associations between indices of dependence on ecosystem services and the area of cropland supported this argument (Table 3). Although poverty is often defined in terms of low access to financial capital, households in financial poverty often have limited access to human, manufactured, and social capital. Thus, the proposition that the poor depend more on ecosystem services may be generalized from poverty in financial capital to poverty in human, manufactured, and social capital. Table 3 supported such negative associations between indices of high dependence on ecosystem services and a number of measures of low access to different forms of capital.

It should be noted that the relationship between dependence on ecosystem services and lack of access to forms of capital can change rather rapidly. In 1998, the local economy at Wolong Nature Reserve was relatively closed and mainly relied on

Table 1. Net benefits, overall IDES and sub-indices in 1998 and 2007.

Net benefits/Indices	1998		2007	
	Mean (S.D.)	Range (Minimum: Maximum)*	Mean (S.D.)	Range (Minimum : Maximum)
Net socioeconomic benefit (yuan)	2456.38 (3315.50)	(0:16,600)	12,350.10 (21,027.75)	(0:186,046)
Net benefit from provisioning services (yuan)	2308.77 (2506.89)	(−2671:13,676)	8544.97 (14,063.43)	(−2620:107,003)
Net benefit from regulating services (yuan)	77.60 (92.88)	(0:544)	2900.64 (2003.59)	(0:10,448)
Net benefit from cultural services (yuan)	3.33 (44.72)	(0:600)	1526.10 (13,476.34)	(0:177,626)
Total net benefit from ecosystem services (yuan)	2389.71 (2527.22)	(−2620:14,182)	12971.70 (19,043.19)	(−27:181,801)
Sub-index for provisioning services	0.4131 (0.8627)	(−7.0351:0.9973)	0.3754 (0.3131)	(−0.1750:1)
Sub-index for regulating services	0.0340 (0.0714)	(0:0.7518)	0.2112 (0.2026)	(0:1)
Sub-index for cultural services	0.0003 (0.0038)	(0:0.0513)	0.0257 (0.1143)	(0:0.8568)
Overall IDES	0.4473 (0.8237)	(−6.9019:1)	0.6123 (0.3055)	(−0.0015:1)

Notes:
Monetary values for net benefits in 2007 were discounted into present values of 1998 for comparison.
*Negative value of an index means that the gross benefit from ecosystem services is lower than the sum of costs from ecosystem dis-services and costs of generating the corresponding ecosystem services.

agriculture. Since 2000, the NFCP, GTGP, GTBP, and tourism development have led to more non-farm income. The local economy became more open. For example, thousands of tourists visit Wolong Nature Reserve every year to view giant pandas. As a result, local farmers began to grow vegetables selling to outside markets, and some young farmers enrolled their cropland in GTGP and GTBP and migrated to urban areas for work. These were the reasons why some indicators of capital such as the distance to the main road, area of cropland, social ties were not as important in 1998 as they were in 2007 (Table 3). Furthermore, most households in 1998 lived in low quality wooden or stone sheds, while in 2007, with more money available, some of them lived in high quality concrete houses. Households actively strategize how to substitute one form of capital for another in order to achieve access to needed resources and enhance well-being [55].

Discussion and Conclusions

This proposed index system is a step forward in quantifying human dependence on ecosystem services. As mentioned above, it is impractical, if not impossible, to capture all ecosystem services in any study. However, we have implemented the index system with existing household survey data and were able to capture major ecosystem services in our initial estimates. Our results suggest that such an approximation is a viable approach that reveals useful information on human dependence on ecosystem services. The

overall index and its sub-indices can reveal the general pattern of households' dependences on ecosystem services, and the variations across time, space, and different levels of access to multiple forms of capital.

Given the fact that different individuals, households and communities rely on different ecosystem services, in practice it is likely that the measurement of benefits, costs and IDES will depend on relatively detailed data that is collected with an understanding of the local context. To facilitate comparisons across time, space and different institutional levels, it is necessary that different studies use a common platform such as the generally accepted MA framework for classification. While it will usually not be possible to capture all benefits and costs associated with ecosystems services, care should be taken to accurately estimate the most important benefits and costs in the local context in order to construct overall IDES estimates. To avoid misinterpretation in some circumstances (e.g., the effect of Integrated Conservation and Development Projects on changes in households' dependence on ecosystem services), such comparisons should not only focus on the overall IDES index but also consider the sub-indices as well as the structure of dependence on ecosystem services (i.e., the distribution of three sub-indices).

It should be noted that IDES measures the relative importance of ecosystem services, with comparison to other socioeconomic activities, in providing benefits directly and indirectly to humans. If one wants to compare the absolute values of benefits from ecosystem services across different areas, one should use the net

Table 2. Comparison of overall IDES and agricultural income share for their associations with gross household income.

	Household income in 1998	Household income in 2007
Agricultural income share	−0.355***	−0.012
Overall IDES	−0.194**	−0.405***

Notes:
Numbers are Spearman's rhos. Total samples are the same 180 randomly sampled households across years.
*p<0.05;
**p<0.01;
***p<0.001.

Table 3. Regression of sources of variation on overall IDES.

	Variable	IDES 1998	IDES 2007
Natural capital	Area of cropland (Mu, 1 Mu = 1/15 ha)	0.020 (0.014)	0.042*** (0.007)
Human capital	Household size	−0.077 (0.049)	−0.037* (0.016)
	Number of laborers	−0.070 (0.056)	−0.080*** (0.018)
	Average education of adults (year)	−0.032† (0.017)	−0.032*** (0.009)
	Average age of adults (year)	0.011* (0.005)	0.008*** (0.002)
Financial capital	Household income (yuan, log)	0.071 (0.075)	−0.152*** (0.025)
	Per capital income (yuan, log)	0.143 (0.095)	−0.126*** (0.028)
Manufactured capital	Type of house (0 for low quality non-concrete sheds and 1 for high quality concrete house)	−0.022 (0.139)	−0.189*** (0.048)
	Distance to the main road (meter, log)	−0.029 (0.027)	0.042*** (0.010)
Social capital	Social ties to local township and reserve level officials (0: low; 1: high).	0.065 (0.129)	−0.188** (0.065)
Spatial heterogeneity	Township (0: Gengda; 1: Wolong)	0.098 (0.133)	−0.101* (0.047)

Numbers outside and inside parentheses are coefficients and robust standard errors of bivariate regressions, respectively. Dependent variables are overall IDES in 1998 and 2007, respectively.
Notes:
Total samples are the same 180 randomly sampled households across years.
$^{†}p<0.01$;
$^{*}p<0.05$;
$^{**}p<0.01$;
$^{***}p<0.001$.

benefits obtained from ecosystem services, which are also provided through construction of the IDES system.

It should also be noted that there is a substantial difference between being dependent on ecosystem services and being dependent on PES. The PES program compensates some of the forgone benefits that local households enjoyed before the implementation of conservation policies. But the PES program does not necessarily compensate all the forgone benefits. For example, in our study area, the main purpose of payments from NFCP is to protect natural forests. As a result payments from NFCP mostly compensate the forgone provisioning (e.g., timber harvest) services whereby regulating services (e.g., soil erosion control, carbon sequestration, and water conservation) are increased. Cultural services (e.g., recreation and ecotourism) are not included. In our study, we therefore included benefits from ecotourism which are not captured in the PES from the NFCP.

Using the overall IDES, we confirmed the proposition that the poor are more dependent on ecosystem services, and thus are more vulnerable to degradation or decline of the corresponding ecosystem services. More importantly, we generalized this proposition to those disadvantaged groups who possess less access to multiple forms of capital (i.e., human, financial, manufactured, and social capital) and found they too were more dependent on ecosystem services than the affluent.

Although we demonstrated the construction of the IDES system and its applications based only on data from the Wolong Nature Reserve, the conceptual basis of IDES and methodology we used were designed to be generalizable. While we examined households, the unit of analysis could range from individuals to communities, regions, and nations. Our analysis here is a proof of the concept of IDES. Further elaborations are warranted and could potentially improve the estimates of human dependence on ecosystem services.

We believe the IDES index system presented here has some major advantages to advance the understandings of linkages between ecosystem services and human well-being and support decision-making. First, this paper empirically demonstrated how the index system could better reflect human dependence on ecosystem services than the other commonly used indicator (i.e., agricultural income share) at the household level. Second, the index system provides both a composite index and sub-indices. This allows the quantitative analysis of the structure of human dependence on ecosystem services and the quantitative examination of the interwoven linkages between different types of ecosystem services and different components of human well-being. Future studies could combine IDES with indicators of indirect drivers, direct drivers, and human well-being to construct integrated models based on the MA framework to better understand complex interactions among human and natural components (e.g., to assess how human dependence on ecosystem services may affect human well-being). The improved understanding may help to develop theories on the complexity of CHANS and inform decision making in a rapidly changing global environment. Third, the improved understanding of linkages between ecosystem services and human well-being, if integrated into decision-making, may avoid some inappropriate strategies that aggravate the marginalization of disadvantaged groups. This in turn could reduce the pressure these groups place on ecosystems to obtain services critical to them. However, a distinction should be made: dependence on ecosystem services is not equivalent to pressure on ecosystems because there are often sustainable ways to extract ecosystem services. Acquisitions of regulating and cultural services (e.g., air purification and ecotourism) are often non-consumptive, while many uses of provisioning services (e.g., timber, fuelwood) are often consumptive and may or may not be sustainable. The reduction of pressure or impacts on ecosystem services can be realized through reduction of overall dependence on ecosystem services or through a shift of the structure of dependence to different types of ecosystem services such as a shift from high dependence on provisioning services to high dependence on regulating and cultural services. It can also be achieved by extracting provisioning services in ways that do not harm the ecosystem. Fourthly, improved understanding may also enhance the effectiveness and long-term viability of conservation and development programs by targeting priority population groups such as those with limited access to capital and high dependence on provisioning services. Finally, such understanding could draw stakeholders' attention to the unmanaged risks and unrealized opportunities associated with ecosystem service changes. Climate change and other global changes are causing rapid shifts in ecosystem structure and function and may threaten continued flow of services to those most dependent upon them. Taking our study area as an example, conservation and development efforts such as NFCP, GTGP, GTBP, and tourism development have already reduced many households' dependences on provisioning services; however, the very uneven distribution of benefits from ecosystem services may create potential risks and impede the future success of such policies.

References

1. Millennium Ecosystem Assessment (2005) Ecosystems & Human Well-being: Synthesis. Washington, DC.: Island Press.
2. Liu J, Dietz T, Carpenter SR, Alberti M, Folke C, et al. (2007) Complexity of coupled human and natural systems. Science 317: 1513–1516.
3. Carpenter SR, DeFries R, Dietz T, Mooney HA, Polasky S, et al. (2006) Millennium Ecosystem Assessment: Research Needs. Science 314: 257–258.
4. Campbell A (1976) Subjective Measures of Well-Being. American Psychologist 31: 117–124.
5. Diener E, Suh EM, Lucas RE, Smith HL (1999) Subjective well-being: Three decades of progress. Psychological Bulletin 125: 276–302.
6. Diener E (2000) Subjective well-being — The science of happiness and a proposal for a national index. American Psychologist 55: 34–43.
7. Grant S, Langan-Fox J, Anglim J (2009) The Big Five Traits as Predictors of Subjective and Psychological Well-Being. Psychological Reports 105: 205–231.
8. Abdallah S, Thompson S, Marks N (2008) Estimating worldwide life satisfaction. Ecological Economics 65: 35–47.
9. Turner BL, Lambin EF, Reenberg A (2008) Land Change Science Special Feature: The emergence of land change science for global environmental change and sustainability. Proceedings of the National Academy of Sciences, USA 105: 2751–2751.
10. Rosa EA, Dietz T, Diekmann A, Jaeger CC, editors(2010) Human Footprints on the Global Environment: Threats to Sustainability. Cambridge, Massachusetts: MIT Press.
11. Daily GC (1997) Nature's Services: Societal Dependence on natural Ecosystems. Washington, D.C.: Island Press.
12. Shackleton C, Shackleton S, Gambiza J, Nel E, Rowntree K, et al. (2008) Links between Ecosystem Services and Poverty Alleviation: Situation analysis for arid and semi-arid lands in southern Africa. Ecosystem Services and Poverty Reduction Research Programme: DFID, NERC, ESRC.
13. He G, Chen X, Liu W, Bearer S, Zhou S, et al. (2008) Distribution of Economic Benefits from Ecotourism: A Case Study of Wolong Nature Reserve for Giant Pandas in China. Environmental Management 42: 1017–1025.
14. Liu W, Vogt CA, Luo J, He G, Frank KA, et al. (2012) Drivers and Socioeconomic Impacts of Tourism Participation in Protected Areas. PLoS ONE 7: e35420.
15. Kiss A (2004) Is community-based ecotourism a good use of biodiversity conservation funds? Trends in Ecology & Evolution 19: 232–237.
16. Grigg A (2008) Dependency and impact on ecosystem services – unmanaged risk, unrealised opportunity: A briefing document for the food, beverage and tobacco sectors. Fauna & Flora International.
17. Liu J, Yang W (2012) Water Sustainability for China and Beyond. Science 337: 649–650.
18. McConnell WJ, Millington JDA, Reo NJ, Alberti M, Asbjornsen H, et al. (2011) Research on Coupled Human and Natural Systems (CHANS): Approach, Challenges, and Strategies. Bulletin of the Ecological Society of America 92: 218–228.
19. Carpenter SR, Mooney HA, Agard J, Capistrano D, DeFries RS, et al. (2009) Science for managing ecosystem services: Beyond the Millennium Ecosystem Assessment. Proceedings of the National Academy of Sciences of the United States of America 106: 1305–1312.
20. Yang W, Dietz T, Kramer DB, Chen X, Liu J (2013) Going beyond the Millennium Ecosystem Assessment: an index system of human well-being. PLoS One: In press. doi: 10.1371/journal.pone.0064582.
21. Costanza R, d'Arge R, de Groot R, Farber S, Grasso M, et al. (1997) The value of the world's ecosystem services and natural capital. Nature 387: 253–260.
22. Farber SC, Costanza R, Wilson MA (2002) Economic and ecological concepts for valuing ecosystem services. Ecological Economics 41: 375–392.
23. Wallace KJ (2007) Classification of ecosystem services: Problems and solutions. Biological Conservation 139: 235–246.
24. Wallace K (2008) Ecosystem services: Multiple classifications or confusion? Biological Conservation 141: 353–354.
25. Boyd J, Banzhaf S (2007) What are ecosystem services? The need for standardized environmental accounting units. Ecological Economics 63: 616–626.
26. Zhang W, Ricketts TH, Kremen C, Carney K, Swinton SM (2007) Ecosystem services and dis-services to agriculture. Ecological Economics 64: 253–260.
27. Naidoo R, Ricketts TH (2006) Mapping the economic costs and benefits of conservation. PLoS Biology 4: 2153–2164.
28. Nelson E, Mendoza G, Regetz J, Polasky S, Tallis H, et al. (2009) Modeling multiple ecosystem services, biodiversity conservation, commodity production, and tradeoffs at landscape scales. Frontiers in Ecology and the Environment 7: 4–11.
29. Birch JC, Newton AC, Aquino CA, Cantarello E, Echeverria C, et al. (2010) Cost-effectiveness of dryland forest restoration evaluated by spatial analysis of ecosystem services. Proceedings of the National Academy of Sciences, USA 107: 21925–21930.
30. Chang J, Wu X, Liu AQ, Wang Y, Xu B, et al. (2011) Assessment of net ecosystem services of plastic greenhouse vegetable cultivation in China. Ecological Economics 70: 740–748.
31. Yang W, Chang J, Xu B, Peng C, Ge Y (2008) Ecosystem service value assessment for constructed wetlands: A case study in Hangzhou, China. Ecological Economics 68: 116–125.
32. Hanley N, Shogren JF, White B (2001) Introduction to Environmental Economics. New York: Oxford University Press.
33. Boardman AE, Greenberg DH, Vining AR, Weimer DL (2006) Cost-Benefit Analysis: Concepts and Practice. Upper Saddle River, New Jersey: Prentice Hall.
34. Barbier EB (2011) Pricing Nature. Annual Review of Resource Economics 3: 337–353.
35. Chee YE (2004) An ecological perspective on the valuation of ecosystem services. Biological Conservation 120: 549–565.
36. Scott MJ, Bilyard GR, Link SO, Ulibarri CA, Westerdahl HE, et al. (1998) Valuation of Ecological Resources and Functions. Environmental Management 22: 49–68.
37. Bateman IJ, Mace GM, Fezzi C, Atkinson G, Turner K (2011) Economic Analysis for Ecosystem Service Assessments. Environmental & Resource Economics 48: 177–218.
38. Wilson MA, Hoehn JP (2006) Valuing environmental goods and services using benefit transfer: The state-of-the art and science. Ecological Economics 60: 335–342.
39. Ready R, Navrud S (2006) International benefit transfer: Methods and validity tests. Ecological Economics 60: 429–434.
40. Shrestha R, Rosenberger R, Loomis J (2007) Benefit Transfer Using Meta-Analysis in Recreation Economic Valuation. Environmental Value Transfer: Issues and Methods 9: 161–177.
41. Kareiva P, Tallis H, Ricketts TH, Daily GC, Polasky S (2011) Natural Capital: Theory and Practice of Mapping Ecosystem Services: Oxford University Press.

42. Richard TC, Nicholas EF, Norman FM (2001) Contingent Valuation: Controversies and Evidence. Environmental and Resource Economics 19: 173–210.

43. Hanley N, MacMillan D, Wright RE, Bullock C, Simpson I, et al. (1998) Contingent valuation versus choice experiments: Estimating the benefits of environmentally sensitive areas in Scotland. Journal of Agricultural Economics 49: 1–15.

44. He G, Chen X, Bearer S, Colunga M, Mertig A, et al. (2009) Spatial and temporal patterns of fuel collection in Wolong Nature Reserve: Implications for panda conservation. Landscape and Urban Planning 92: 1–9.

45. Myers N, Mittermeier RA, Mittermeier CG, da Fonseca GAB, Kent J (2000) Biodiversity hotspots for conservation priorities. Nature 403: 853–858.

46. Liu J, Daily GC, Ehrlich PR, Luck GW (2003) Effects of household dynamics on resource consumption and biodiversity. Nature 421: 530–533.

47. Liu J, Linderman M, Ouyang Z, An L, Yang J, et al. (2001) Ecological degradation in protected areas: The case of Wolong Nature Reserve for giant pandas. Science 292: 98–101.

48. Wolong Nature Reserve (2005) Development History of Wolong Nature Reserve [in Chinese]; Lai B, editor. Chengdu: Sichuan Science and Technology Press.

49. Liu J, Li S, Ouyang Z, Tam C, Chen X (2008) Ecological and socioeconomic effects of China's policies for ecosystem services. Proceedings of the National Academy of Sciences, USA 105: 9477–9482.

50. Liu J, Ouyang Z, Yang W, Xu W, Li S (2013) Evaluation of Ecosystem Service Policies from Biophysical and Social Perspectives: The Case of China. In: Levin SA, editor. Encyclopedia of Biodiversity (second edition). ed. Waltham, MA: Academic Press. pp. 372–384.

51. Yang W, Liu W, Viña A, Luo J, He G, et al. (2013) Performance and prospects on payments for ecosystem services programs: evidence from China. Journal of Environmental Management: In press. doi: 10.1016/j.jenvman.2013.1004.1019.

52. An L, Liu J, Ouyang Z, Linderman M, Zhou S, et al. (2001) Simulating demographic and socioeconomic processes on household level and implications for giant panda habitats. Ecological Modelling 140: 31–49.

53. National Bureau of Statistics of China (2011) Rural Household Survey Instrument [in Chinese]. Beijing: National Bureau of Statistics of China.

54. De Rus G (2010) Introduction to Cost-benefit Analysis: Looking for Reasonable Shortcuts: Edward Elgar Pub.

55. Chen X, Frank KA, Dietz T, Liu J (2012) Weak ties, labor Migration, and environmental impacts: Toward a sociology of sustainability. Organization and Environment 25: 3–24.

11

Accounting for Ecosystem Alteration Doubles Estimates of Conservation Risk in the Conterminous United States

Randy Swaty[1]*, Kori Blankenship[2], Sarah Hagen[3], Joseph Fargione[3], Jim Smith[4], Jeannie Patton[5]

1 The Nature Conservancy, Marquette, Michigan, United States of America, 2 The Nature Conservancy, Bend, Oregon, United States of America, 3 The Nature Conservancy, Minneapolis, Minnesota, United States of America, 4 The Nature Conservancy, Jacksonville, Florida, United States of America, 5 The Nature Conservancy, Boulder, Colorado, United States of America

Abstract

Previous national and global conservation assessments have relied on habitat conversion data to quantify conservation risk. However, in addition to habitat conversion to crop production or urban uses, ecosystem alteration (e.g., from logging, conversion to plantations, biological invasion, or fire suppression) is a large source of conservation risk. We add data quantifying ecosystem alteration on unconverted lands to arrive at a more accurate depiction of conservation risk for the conterminous United States. We quantify ecosystem alteration using a recent national assessment based on remote sensing of current vegetation compared with modeled reference natural vegetation conditions. Highly altered (but not converted) ecosystems comprise 23% of the conterminous United States, such that the number of critically endangered ecoregions in the United States is 156% higher than when calculated using habitat conversion data alone. Increased attention to natural resource management will be essential to address widespread ecosystem alteration and reduce conservation risk.

Editor: Tamara Natasha Romanuk, Dalhousie University, Canada

Funding: This work was funded by the TNC-LANDFIRE Cooperative Agreement 10-CA-11132543-054 with the U.S. Forest Service and Dept. of the Interior. The funders had no role in study design, data collection and analysis, decision to publish, or preparation of the manuscript.

Competing Interests: The authors have declared that no competing interests exist.

* E-mail: rswaty@tnc.org

Introduction

Conservation assessments at regional, national, and global levels have commonly relied upon data on the magnitude and rate of habitat conversion to crop production or urban uses as an evaluation of conservation risk [1,2,3,4,5,6]. While this approach provides useful information, it neglects the fact that much habitat — while not converted outright— could be highly degraded due to logging, fire suppression, biological invasions, grazing, and other land management practices.

Data to assess the extent of ecosystem alteration have previously not been available at broad scales. Recently, however, a national land-cover assessment of ecosystem alteration based on remote sensing and departure from reference natural vegetation conditions has been conducted for the United States (www.landfire.gov) [7,8]. These data capture human alteration of ecosystem structure and composition through disturbances such as fire suppression, conversion to plantations, logging, and biological invasions from introduced plant species. In many cases, this altered vegetation has reduced habitat value for species of conservation concern [9,10,11]. For example, vegetation structure and composition affect habitat use by grassland birds [12], forest mammal diversity [13,14], grassland arthropod diversity [15,16], and ecosystem services [17,18]. Therefore, conservation risk assessments must consider ecosystem alteration in addition to habitat conversion in order to fully capture impacts to biodiversity and ecosystem services.

We used LANDFIRE's national map of ecosystem alteration to calculate a conservation risk index for ecoregions in the conterminous United States, expanding a previous assessment

based on habitat conversion [19]. We selected ecoregions as the scale of analysis because these geographic units share similar species, ecological dynamics, and environmental conditions and are widely used for conservation planning [20,21].

This analysis provides, for the first time, a comprehensive picture of ecosystem alteration in the United States. Large-scale conservation planning has focused on protecting land from conversion in part because it is relatively easy to map protected and converted areas. Although management practices and associated ecosystem alteration on unconverted lands is arguably of equal or greater importance for conservation, data availability has, until now, limited consideration of ecosystem alteration in large-scale conservation planning.

Materials and Methods

Ecosystem alteration and land conversion were assessed for the conterminous United States using LANDFIRE National Project spatial data (www.landfire.gov). LANDFIRE's measure of ecosystem alteration assesses the difference between estimated reference conditions (historic vegetation structure and composition) and current vegetation [7,8]. Lands classified as urban, agricultural, or barren (Fig. 1A) were excluded in the LANDFIRE analysis. To generate reference conditions that incorporated natural disturbance regimes (e.g. fire, insects, and storms), LANDFIRE used the Vegetation Dynamics Development Tool (VDDT, www.essa.com) and the LANDSUM model [22,23,24] to estimate reference conditions within each of 1,667 Biophysical Settings (BpS; represents dominant vegetation prior to Euro-American settlement based on edaphic and disturbance factors [8]). These models,

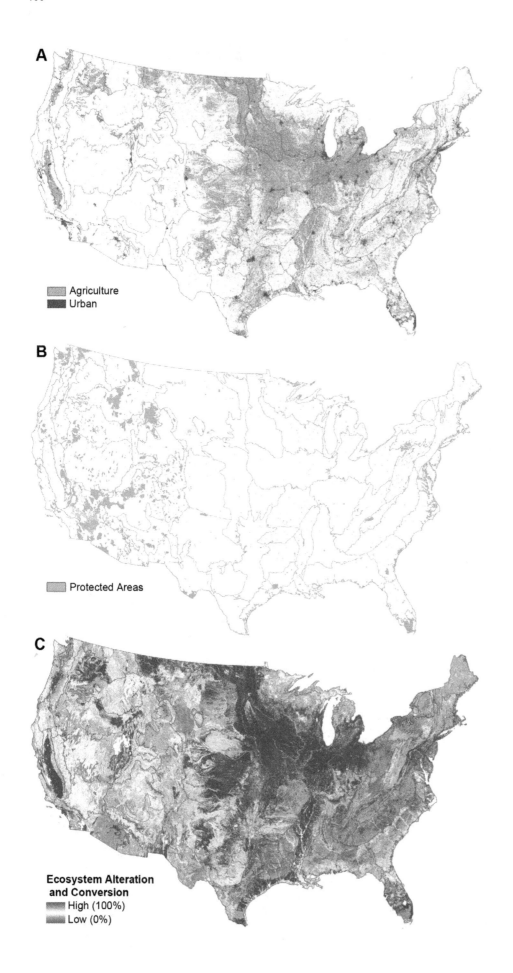

Agriculture
Urban

Protected Areas

**Ecosystem Alteration
and Conversion**
High (100%)
Low (0%)

Figure 1. Mapping components of the ecological conservation risk index. (A) Areas converted to agricultural and urban land use, and (B) protected areas, and (C) ecosystem alteration and conversion (converted lands are considered to be 100% altered). High alteration indicates a substantial shift in vegetation structure and/or composition from reference conditions. Grey lines indicate ecoregional boundaries.

which were tailored for each BpS, predict the average proportion of an ecosystem in each of several (up to five) successional states defined by cover, height, and dominant vegetation. For example, in the Western Cascades Western Hemlock Forest, five successional states were defined as shown in Table 1. All reference vegetation for each BpS was assumed to fall into one of the defined successional states.

Any particular location is expected to transition through successional states over time. Given this dynamic nature of vegetation, it is not possible to assign any particular location to a single reference successional state. Therefore, reference vegetation models were designed to predict the proportion of land cover in different successional classes for the entire extent of a BpS rather than to make fine-grained predictions about land cover.

To map current vegetation type, cover, and height, each 30-meter pixel in the United States was categorized based on remotely sensed data trained using 331,900 ground-truth vegetation plots [8]. The classification system recognized 398 existing vegetation types, 27 cover classes, and 12 height classes. Current land cover was categorized based on the same successional states defined in the reference vegetation analysis, using the three current vegetation data layers (i.e. vegetation type, cover, and height). Current vegetation that did not fall into one of the successional state categories was assigned to one of two alternative "uncharacteristic" states: uncharacteristic native or uncharacteristic exotic. Although one component of the land cover data underwent validation analysis (existing vegetation type [25,26]), the "current vegetation successional state" data layer that we used in our analysis did not.

The degree of ecosystem alteration in each ecosystem was quantified using a similarity index based on the proportion of land cover in different successional states in reference versus current conditions [8]. An alteration metric was computed for each BpS in each Ecological Subsection (hereafter "ecosystems") [27]. This ecosystem alteration index ranges from 0 to 100 (Fig. 1C), with scores of 67 and higher considered to indicate highly altered ecosystems [28]. To assess the sensitivity of our results to this threshold, we also calculated our results using a threshold of 57 and 77. The use of this threshold acts to exclude areas that are not highly altered from subsequent analyses. We note that lands that are not "highly altered" may still be moderately altered and that this alteration may still have detrimental effects on habitat values, wildlife, and ecosystem services. If so, our assessment of conservation risk is conservative. To assess conservation risk at the scale of ecoregions (each of which contains numerous ecosystems), we tabulated the percentage of land covered by ecosystems found to be highly altered within each ecoregion.

As an index of the relative conservation risk at the ecoregional scale, we developed the Ecological Conservation Risk Index (ECRI). The ECRI is an extension of the Conservation Risk Index (CRI), which is calculated as the ratio of percent area converted to percent area protected (Fig. 1B) for a given biome or ecoregion [19]. Although other approaches are available for determining conservation risk for ecosystems [1,4,6], CRI is unique in that it was developed to be applied to ecoregions and the data requirements for its calculation are available at national scales. Because ecosystem alteration may also erode habitat value, we add the percent area highly altered to the percent area converted to calculate ECRI, given by the formula:

$$ECRI = (\% \text{ Converted} + \% \text{ Highly Altered})/\% \text{ Protected}$$

As a comparison, we applied both the CRI and the ECRI to Bailey's ecoregions [29,30] in the conterminous United States. For CRI (or ECRI), ecoregions in which habitat conversion (and high alteration) >20% and CRI (or ECRI) >2 were classified as Vulnerable; those in which conversion (and high alteration) >40% and CRI (or ECRI) >10 were classified as Endangered; and those with conversion (and high alteration) >50% and CRI (or ECRI) >25 were classified as Critically Endangered [19].

Protected Areas were based on the 2009 World Database on Protected Areas (Fig. 1B; [31]). We included both areas designated for biodiversity protection (IUCN categories I–IV, including U.S. National Parks and Wilderness areas) and those designated for multiple management objectives (IUCN categories V–VI) in our analysis. Proposed areas, areas mapped with a point location (whose area could not be calculated), and areas or portions of areas in water were excluded from the analysis.

Results

Approximately 29% of the land area of the conterminous United States has been converted to human use, with roughly 24% (182 million hectares) converted to agriculture and 5% (37 million hectares) converted to urban land use [7]. However, these numbers do not include the widespread occurrence of ecosystem alteration. Our analysis shows an additional 23% of non-converted lands in the conterminous United States have high levels of ecosystem alteration, indicating a significant shift in vegetation structure and composition relative to reference conditions. In total, more than half (52%) of the United States has been highly altered or converted.

Table 1. Definitions for the succession classes in the Western Cascades Western Hemlock Forest.

Succession Class	Vegetation Height	Vegetation Cover	Dominant Vegetation
A	<5 meters	0–60%	fireweed (*Epilobium angustifolium*) and red alder (*Alnus rubra*) with tree seedlings
B	5–50 meters	61–100%	Douglas-fir (*Pseudotsuga* menziesii) and western hemlock (*Tsuga heterophylla*)
C	5–50 meters	20–60%	Douglas-fir (*P. menziesii*) with some shrubs such as salal (*Gaultheria shallon*)
D	>50 meters	20–60%	Douglas-fir (*P. menziesii*)
E	>50 meters	61–100%	Douglas-fir (*P. menziesii*) and Western hemlock (*T. heterophylla*)

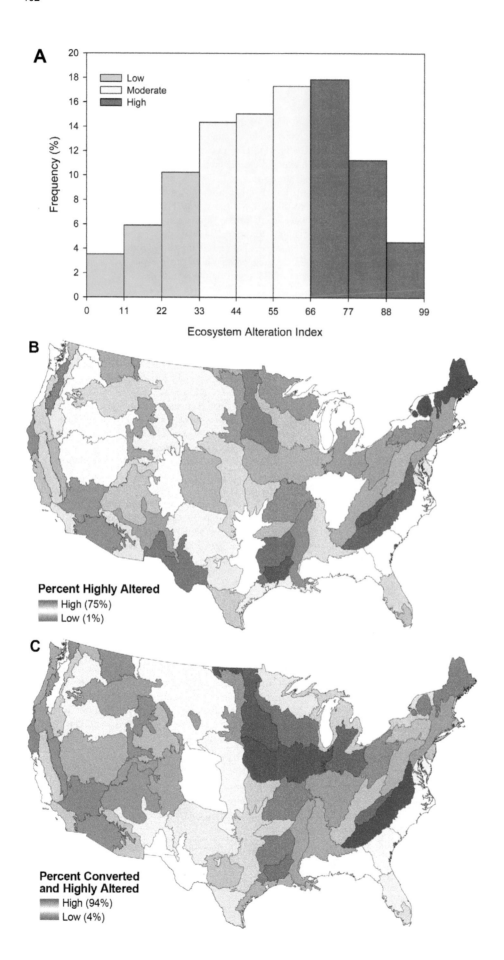

Percent Highly Altered
High (75%)
Low (1%)

**Percent Converted
and Highly Altered**
High (94%)
Low (4%)

Figure 2. Percent highly altered and converted by ecoregion. (A) The frequency distribution of the ecosystem alteration index in the conterminous United States (excluding converted and barren lands). (B) Percent highly altered by ecoregion. (C) Percent highly altered or converted by ecoregion. Grey lines indicate ecoregional boundaries.

In addition to these highly altered lands, many lands are moderately altered. Figure 2 shows the frequency distribution for our ecosystem alteration index on unconverted lands, showing the percent of lands that have low, moderate, and high ecosystem alteration. The average ecosystem alteration index value was 54% (Fig. 2A). However, this alteration was not distributed evenly across the United States, with some areas having a higher percentage of highly altered areas (Fig. 2B). The percent of an ecoregion that was highly altered ranged from a low of 1% in the Northern Tallgrass Prairie to a high of 75% in the Northern Appalachians. When considering both ecosystem conversion and alteration, the percent of an ecoregion that was impacted ranged from 3% in the California North Coast to 94% in the Piedmont (Fig. 2C).

Based on the relationship between the amount of ecosystem conversion and the amount of land protection, the original Conservation Risk Index [19] identified 20 Vulnerable, 9 Endangered, and 9 Critically Endangered ecoregions across the United States (Fig. 3A). When we add in the new ecological alteration data, we find a dramatic increase in critically endangered ecoregions (from 9 to 23, with a range from 17 to 29 critically endangered ecosystems in our sensitivity analysis; Fig. 3B, C, and Table S1). Critically endangered areas included large areas of deciduous forest (from New England to Appalachia) and grasslands (in the central United States) with high levels of ecosystem alteration that went undetected using previous habitat conversion assessments. Overall, the inclusion of ecosystem alteration increased the conservation risk index across the United States such that 35 of the 69 ecoregions increased by one or two risk categories (Fig. 3B, C). The number of ecoregions with increased risk ranged from 22 to 44 in our sensitivity analysis.

Discussion

Our ecological conservation risk assessment (ECRI) reveals ecoregions to be at greater risk than was apparent based on land conversion alone. Over half of the conterminous United States is either converted or highly altered. However, these impacts are not evenly distributed, with some ecoregions receiving a dispropor-tionate share of ecosystem alteration and conversion. Notably, the three ecoregions with the highest percent of land that was highly altered were the Northern Appalachians, West Gulf Coastal Plain, and Southern Blue Ridge. While the vegetation in the Northern Appalachian and Southern Blue Ridge Mountains is only 4–11% converted to row crop or urban uses, current vegetation lacks the tall closed-canopy characteristics of the old growth forests that historically dominated these areas. In the West Gulf Coastal Plain, vegetation has shifted from Wet Longleaf Pine Savanna and Flatwoods (33% of the ecoregion historically) and Upland Longleaf Pine Forest and Woodland (22% of the ecoregion historically) vegetation to 23% uncharacteristic vegetation cover, primarily Loblolly pine (*Pinus taeda*) plantations. Taking this ecosystem alteration into account increased the assessed conser-vation risk to these ecoregions, elevating them to Vulnerable or Endangered status. In total, consideration of ecosystem alteration caused 35 ecoregions to increase one or two risk levels. This highlights the need for significant conservation efforts focused on sustainable vegetation management and landscape-scale vegeta-tion restoration to reduce conservation risk.

Ecosystem alteration can be addressed with improved land management, using management actions that are targeted to the causes of ecosystem alteration. The proximate causes of alteration are characterized by the LANDFIRE ecosystem alteration dataset, which indentifies areas that have altered canopy cover, canopy height, or species composition. Loss of old growth, such as via logging, can be detected by reductions in canopy height and cover and shifts in species composition. Increases in canopy cover and shifts in composition can indicate fire suppression. And increases in "exotic uncharacteristic vegetation" explicitly identify areas that have been invaded by exotic plants (Fig. 4). These signatures of logging, fire suppression, and invasive species provide a national overview of the need for forest protection and improved forestry techniques to restore old growth forest characteristics, prescribed fire to restore natural fire regimes, and regionally specific approaches, such as appropriate grazing practices, to fight invasive species. We illustrate this with three examples: 1) Great Basin Desert Scrub [32], 2) Ozarks Oak Woodland [33], and 3) Western Cascades Western Hemlock Forest [34].

In the Great Basin, invasive species are a leading cause of ecological alteration (Fig. 5A; Fig. 4). Currently, over 25% of the Great Basin Desert Scrub ecosystem is mapped as "Uncharacteristic Exotic" in LANDFIRE (Fig. 4B), presumably due to the invasion of cheatgrass (*Bromus tectorum*), estimated to cover 20,000 km^2 [35]. In the Ozarks Oak Woodlands, fire suppression is a leading cause of ecosystem alteration (Fig. 5B). The Ozarks Oak Woodland ecosystem currently exhibits mostly closed canopy conditions (~80% of land cover) that were less common under reference conditions (~20% of land cover) due to relatively frequent low intensity surface fires across the ecosystem prior to significant European settlement [36,37]. In Western Hemlock Forests of the Western Cascades, logging is a leading cause of ecological alteration (Fig. 5C). Under reference conditions, these Western Hemlock Forests were dominated by tall (>50 m), closed canopy, old growth Douglas fir (*Pseudotsuga menziesii*) and western hemlock (*Tsuga heterophylla*) stands (~70% of land cover). Currently, however, the landscape is dominated by closed canopy young forest stands 5–50 m tall (~82% of land cover), a result of decades of logging [38].

In all three cases, biodiversity conservation is threatened by ecosystem alteration. Cheatgrass invasion of Desert Scrub threatens species including sage grouse (*Centrocercus urophasianus*) and desert tortoise (*Gopherus agassizii*) [39,40]. Fire suppression in the Ozarks threatens savanna-dependent species such as the eastern collared lizard (*Crotaphytus collaris collaris*) [41]. Loss of Western Hemlock old growth forest threatens bird species such as the Marbled Murrelet (*Brachyramphus marmoratus*) [42], mammals such as northern flying squirrels (*Glaucomys sabrinus*) [43], and ectomycorrhizal fungi unique to forests with old-growth characteristics [44].

Although protected areas generally provide abatement from some threats to biodiversity such as development and forest clearing, we found that even within protected areas, 21% of non-converted lands have high levels of ecosystem alteration. This finding suggests that increased attention to management or restoration of vegetation conditions on our public lands is warranted. For example, to address widespread fire suppression in fire-dependant forests, some level of fire regime restoration and fuels treatment will be needed for restoration of both biodiversity and ecosystem services such as carbon storage [45,46,47]. Fire suppression can lead to increased risk of costly catastrophic fires in

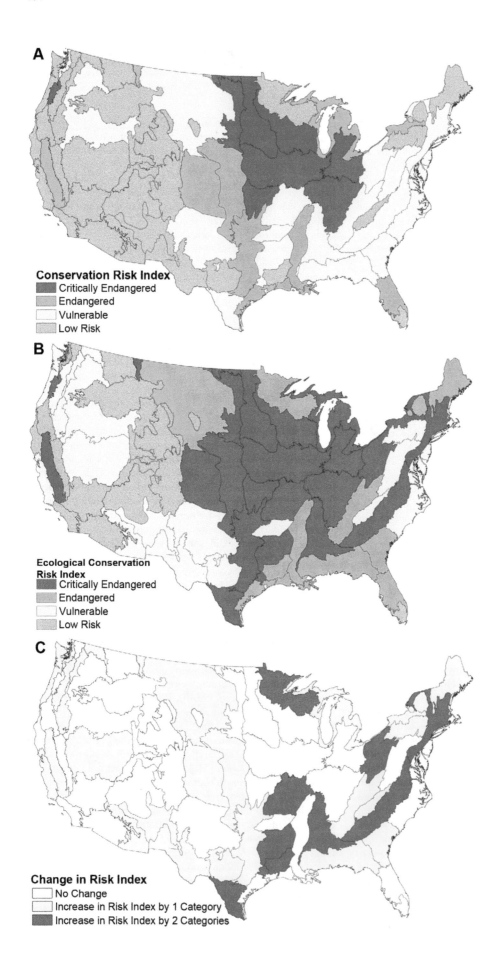

Figure 3. Ecological Conservation Risk Index shows increased risk for ecoregions compared to a Conservation Risk Index that does not include ecosystem alteration. (A) Conservation Risk Index, calculated following [19]. (B) Ecological Conservation Risk Index, which includes ecosystem alteration. (C) Increased risk measured by the Ecological Conservation Risk Index, quantified as the number of risk categories by which each ecoregion increased. Grey lines indicate ecoregional boundaries.

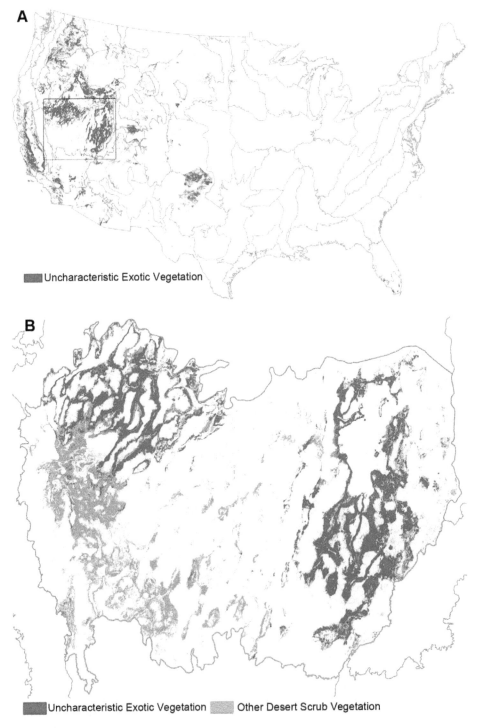

Figure 4. Uncharacteristic exotic vegetation in (A) the United States and (B) the Great Basin ecoregion. The area bordered by a dotted line in panel (A) is magnified in panel (B). Vegetation that is unique when compared to pre-settlement reference conditions is considered uncharacteristic. Uncharacteristic vegetation can be generated by either native or exotic vegetation; here we show the areas dominated by exotic vegetation. Grey lines indicate ecoregional boundaries.

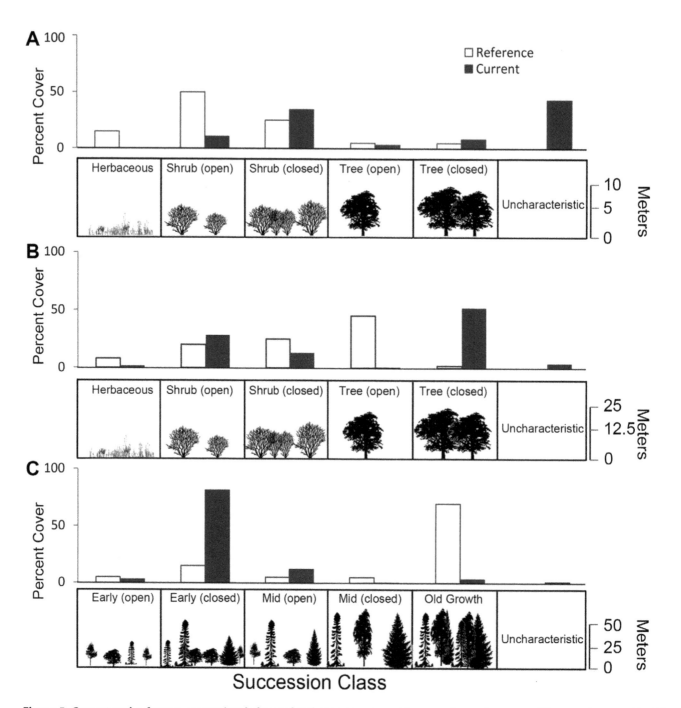

Figure 5. Current and reference successional classes for three ecosystems. Departure from reference conditions can be caused by (A) increases in uncharacteristic vegetation, as in Great Basin Salt Desert Scrub, (B) increases in closed canopy successional classes, as in Ozark Oak Woodland, or (C) increases in early successional classes, as in Cascades Western Hemlock Forest. These vegetation changes are the expected outcomes of biological invasion, fire suppression, and logging, respectively.

many ecosystems [48,49] often due to an unnatural buildup of fuels [50,51]. Ongoing large-scale federal efforts such as Landscape Conservation Cooperatives and US Forest Service forest plan revisions could benefit from the ecosystem alteration information presented here to both assess the need for restoration and to help target management activities. Restoration and management of vegetation within public protected areas may be more feasible than on private lands, which commonly have smaller parcel sizes and typically lack mechanisms for coordinating management across parcels.

With the addition of an ecological alteration dataset to the original Conservation Risk Index based only on land conversion, our analysis provides a more complete picture of the conservation status of ecoregions and can help identify not only areas in need of greater protection, but also areas in need of improved land management. While important, land protection strategies alone will be insufficient to meet conservation risks that we have identified. Successful conservation strategies will also require broader application of ecologically based vegetation management such as: 1) restoration of fire regimes and/or increased use of fire surrogates, 2) forestry techniques that accelerate development of appropriate vegetation structure and composition, 3) invasive species control, and 4) improved grazing practices. Greater resources should be directed to ecosystem management, particularly within the ecoregions at greatest conservation risk as a result of ecosystem alteration.

Acknowledgments

We thank P. Kareiva, J. Hoekstra, P. Doran, T. Hall, and R. Lalasz for constructive comments on the manuscript.

Author Contributions

Conceived and designed the experiments: RS KB JS JF. Performed the experiments: RS KB SH. Analyzed the data: RS KB SH JF. Wrote the paper: RS KB JF JS JP.

References

1. NatureServe (2009) NatureServe conservation status assessments: Methodology for assigning ranks. Arlington, VA: NatureServe.
2. Noss RF, LaRoe ET, Scott JM (1995) Endangered ecosystems of the United States: a preliminary assessment of loss and degradation National Biological Service. pp 1–56.
3. Rodriguez JP, Balch JK, Rodriguez-Clark KM (2007) Assessing extinction risk in the absence of species-level data: quantitative criteria for terrestrial ecosystems. Biodiversity and Conservation 16: 183–209.
4. Nicholson E, Keith DA, Wilcove DS (2009) Assessing the Threat Status of Ecological Communities. Conservation Biology 23: 259–274.
5. Hoekstra JM, Boucher TM, Ricketts TH, Roberts C (2004) Confronting a biome crisis: global disparities of habitat loss and protection. Ecology Letters 8: 23–29.
6. Rodriguez JP, Rodriguez-Clark KM, Baillie JEM, Ash N, Benson J, et al. (2011) Establishing IUCN Red List Criteria for Threatened Ecosystems. Conservation Biology 25: 21–29.
7. LANDFIRE (2007) LANDFIRE 1.1.0 Fire Regime Condition Class (FRCC) layer. Available: http://landfire.cr.usgs.gov/viewer. Accessed 2011 Jul 18.
8. Rollins MG (2009) LANDFIRE: A nationally consistent vegetation, wildland fire, and fuel assessment. International Journal of Wildland Fire 18: 235–249.
9. Noss RF (1990) Indicators for monitoring biodiversity: a hierarchical approach. Conservation Biology 4: 355–364.
10. Tews J, Brose U, Grimm V, Tielborger K, Wichmann MC, et al. (2004) Animal species diversity driven by habitat heterogeneity/diversity: the importance of keystone structures. Journal of Biogeography 31: 79–92.
11. Lindenmayer D, Hobbs RJ, Montague-Drake R, Alexandra J, Bennett A, et al. (2008) A checklist for ecological management of landscapes for conservation. Ecology Letters 11: 78–91.
12. Herkert JR (1994) The effects of habitat fragmentation on midwestern grassland bird communities. Ecological Applications 4: 461–471.
13. Fox BJ, Fox MD (2000) Factors determining mammal species richness on habitat islands and isolates: habitat diversity, disturbance, species interactions and guild assembly rules. Global Ecology and Biogeography 9: 19–37.
14. Sullivan TP, Sullivan DS (2001) influence of variable retention harvests on forest ecosystems II. Diversity and population dynamics of small mammals. Journal of Applied Ecology 38: 1234–1252.
15. Joern A (2005) Disturbance by fire frequency and bison grazing modulate grasshopper assemblages in tallgrass prairie. Ecology 4: 861–873.
16. Schaffers AP, Raemakers IP, Sýkora KV, ter Braak JF (2008) Arthropod assemblages are best predicted by plant species composition. Ecology 89: 782–794.
17. Zavaleta ES (2000) Valuing ecosystem services lost to Tamarix invasion in the United States. In: Mooney HA, Hobbs RJ, eds. Invasive species in a changing world. Washington DC: Island Press. pp 261–300.
18. Chapin FS, Zavaleta ES, Eviner VT, Naylor RL, Vitousek PM, et al. (2000) Consequences of changing biodiversity. Nature 405: 234–242.
19. Hoekstra JM, Boucher TM, Ricketts TH, Roberts C (2005) Confronting a biome crisis: global disparities of habitat loss and protection. Ecology Letters 8: 23–29.
20. Dinerstein E, Olson DM, Graham DJ, Webster AL, Primm SA, et al. (1995) A conservation asessment of the terrestrial ecoregions of Latin America and the Carribean. Washington, D.C.: The World Bank.
21. Groves CL, Valutis L, Vosick D, Neely B, Wheaton K, et al. (2000) Designing a Geography of Hope: A Practitioner's Handbook for Ecoregional Conservation Planning. Special ed. Arlington, Virginia: The Nature Conservancy.
22. Keane RE, Parsons R, Hessberg P (2002) Estimating historical range and variation of landscape patch dynamics: limitations of the simulation approach. Ecological Modelling 151: 29–49.
23. Keane RE, Holsinger M, Pratt SD (2006) Simulating historical landscape dynamics using the landscape fire succession model LANDSUM version 4.0 Gen. Tech. Rep. RMRS-GTR-171CD. In: US Forest Service RMRS, editor. Fort Collins, CO. 73 p.
24. Pratt SD, Holsinger L, Keane RE (2006) Using simulation modeling to assess historical reference conditions for vegetation and fire regimes for the LAND-FIRE prototype project. In: Rollins MG, Frame CK, eds. The LANDFIRE prototype project: nationally consistent and locally relevant geospatial data for wildland fire management Gen Tech Rep RMRS-GTR-175. Fort Collins, CO: U.S. Forest Service, Rocky Mountain Research Station. pp 277–314.

25. LANDFIRE (2010) LANDFIRE national eastern milestone overall quality assessment report. Available: http://www.landfire.gov/dp_quality_assessment.php. Accessed 2011 Jul 18.
26. LANDFIRE (2009) LANDFIRE national western milestone overall quality assessment report. Available: http://www.landfire.gov/dp_quality_assessment.php. Accessed 2011 Jul 18.
27. Cleland DT, Freeouf JA, Keys JEj, Nowacki GJ, Carpenter CA, et al. (2007) Ecological subregions: sections and subsections for the conterminous United States [1:3,500,000]. General Technical Report WO-76. Washington, DC: USDA Forest Service.
28. Hann WJ, Bunnell DL (2001) Fire and land management planning and implementation across multiple scales. International Journal of Wildland Fire 10: 389–403.
29. Bailey RG (1983) Delineation of ecosystem regions. Environmental Management 7: 365–373.
30. Bailey RG (1989) Explanatory supplement to ecoregions map of the continents. Environmental Conservation 16: 307–309.
31. WDPA (2009) World Database on Protected Areas Annual Release 2009. protectedareas@unep-wcmc.org. Accessed 2011 Jul 18.
32. LANDFIRE (2009) LANDFIRE 1.0.0 Vegetation Dynamics Models: Inter-Mountain Basins Mixed Salt Desert Scrub. Available: http://www.landfire.gov/index.php. Accessed 2011 Jul 18.
33. LANDFIRE (2009) LANDFIRE 1.0.0 Vegetation Dynamics Models: Ozark-Ouachita Dry Oak Woodland. Available: http://www.landfire.gov/index.php. Accessed 2011 Jul 18.
34. LANDFIRE (2009) LANDFIRE 1.0.0 Vegetation Dynamics Models: North Pacific Maritime Mesic-Wet Douglas-fir-Western Hemlock Forest. http://www.landfire.gov/index.php.
35. Bradley BA, Mustard JF (2005) Identifying land cover variability distinct from land cover change: Cheatgrass in the Great Basin. Remote Sensing of Environment 94: 204–213.
36. Guyette RP, Muzika RM, Dey DC (2002) Dynamics of an anthropogenic fire regime. Ecosystems 5: 472–486.
37. Yang J, He HS, Shifley SR (2008) Spatial controls of occurrence and spread of wildfires in the Missouri Ozark Highlands. Ecological Applications 18.
38. Nonaka E, Spies TA (2005) Historical range of variability in landscape structure: A simulation study in Oregon, USA. Ecological Applications 15: 1727–1746.
39. Connelly JW, Knick ST, Braun CE, Baker WL, Beever EA, et al. (2011) Greater sage-grouse: ecology and conservation of a landscape species and its habitat. Studies in Avian Biology: http://sagemap.wr.usgs.gov/monograph.aspx.
40. Brooks ML, Berry KH (2006) Dominance and environmental correlates of alien annual plants in the Mojave Desert, USA. Journal of Arid Environments 67: 100–124.
41. Templeton AR, Robertson RJ, Brisson J, Strasburg J (2001) Spatial controls of occurrence and spread of wildfire in the Missouri Ozark Highlands. Ecological Applications 18: 1212–1225.
42. Spies TA, McComb BC, Kennedy RSH, McGrath MT, Olsen K, et al. (2007) Potential effects of forest policies on terrestrial biodiversity in a multi-ownership province. Ecological Applications 17: 48–65.
43. Carey AB (2000) Effects of new forest management strategies on squirrel populations. Ecological Applications 10: 248–257.
44. Smith JE, Molina R, Huso MMP, Luoma DL, McKay D, et al. (2002) Species richness, abundance, and composition of hypogeous and epigeous ectomycorrhizal fungal sporocarps in young, rotation-age, and old-growth stands of Douglas-fir (Pseudotsuga menziesii) in the Cascade Range of Oregon, U.S.A. Canadian Journal of Botany 80: 186–204.
45. Noss RF, Beier P, Covington WW, et al. (2006) Recommendations for integrating restoration ecology and conservation biology in ponderosa pine forests of the southwestern United States. Restoration Ecology 14: 4–10.
46. Hurteau M, North M (2009) Fuel treatment effects on tree-based carbon storage and emissions under modeled wildfire scenarios. Frontiers in Ecology and the Environment 7: 409–414.
47. Wiedinmyer C, Hurteau MD (2010) Prescribed fire as a means for reducing forest carbon emissions in the Western U.S. Environmental Science and Technology 44: 1962–1932.
48. Donovan GH, Brown TC (2007) Be careful what you wish for: the legacy of Smokey Bear. Frontiers in Ecology and the Environment 5: 73–79.

49. Stephens SS, Ruth LW (2005) Federal forest-fire policy in the United States. Ecological Applications 15: 532–542.

50. Agee JK, Skinner CN (2005) Basic principles of forest fuel reduction treatments. Forest Ecology and Management 211: 83–96.

51. Shang BZ, He HS, Crow TR, Shifley SR (2004) Fuel load reductions and fire risk in central hardwood forests of the United States: a spatial simulation study. Ecological Modelling 180: 89–102.

12

Integrated Assessment of Heavy Metal Pollution in the Surface Sediments of the Laizhou Bay and the Coastal Waters of the Zhangzi Island, China: Comparison among Typical Marine Sediment Quality Indices

Wen Zhuang[1,2,3]**, Xuelu Gao**[1]*

1 Key Laboratory of Coastal Environmental Processes and Ecological Remediation, Yantai Institute of Coastal Zone Research, Chinese Academy of Sciences, Yantai, Shandong, China, **2** College of City and Architecture Engineering, Zaozhuang University, Zaozhuang, Shandong, China, **3** University of Chinese Academy of Sciences, Beijing, China

Abstract

The total concentrations and chemical forms of heavy metals (Cd, Cr, Cu, Ni, Pb and Zn) in the surface sediments of the Laizhou Bay and the surrounding marine area of the Zhangzi Island (hereafter referred to as Zhangzi Island for short) were obtained and multiple indices and guidelines were applied to assess their contamination and ecological risks. The sedimentary conditions were fine in both of the two studied areas according to the marine sediment quality of China. Whereas the probable effects level guideline suggested that Ni might cause adverse biological effects to occur frequently in some sites. All indices used suggested that Cd posed the highest environmental risk in both the Laizhou Bay and the Zhangzi Island, though Cd may unlikely be harmful to human and ecological health due to the very low total concentrations. The enrichment factor (EF) showed that a substantial portion of Cr was delivered from anthropogenic sources, whereas the risk assessment code (RAC) indicated that most Cr was in an inactive state that it may not have any adverse effect either. Moreover, the results of EF and geoaccumulation index were consistent with the trend of the total metal concentrations except for Cd, while the results of RAC and potential ecological risk factor did not follow the same trend of their corresponding total metal concentrations. We also evaluated the effects of using different indices to assess the environmental impact of these heavy metals.

Editor: Wei-Chun Chin, University of California, Merced, United States of America

Funding: This study was co-supported by the National Natural Science Foundation of China (41376083), the Department of Science and Technology of Shandong Province (2012GHY11535) and the CAS/SAFEA International Partnership Program for Creative Research Teams (Representative Environmental Processes and Resources Effects in Coastal Zones). The funders had no role in study design, data collection and analysis, decision to publish, or preparation of the manuscript.

Competing Interests: The authors have declared that no competing interests exist.

* E-mail: xlgao@yic.ac.cn

Introduction

Heavy metals in ecosystems have received extensive attention because they are toxic, non-biodegradable in the environment and easy to accumulate and magnify in organisms. Concentrations of heavy metals in aquatic ecosystems have increased considerably due to the inputs of industrial waste, sewage runoff and agriculture discharges [1,2]. In other words, heavy metal pollution may likely go with the rapid economic development [3,4]. The measurements of heavy metals only in the water and in the suspended material are not conclusive due to water discharge fluctuations and low resident time [5]. With a combined action of adsorption, hydrolysis and co-precipitation, only a small part of free metal ions stay dissolved in water, and a large quantity of them get deposited in the sediments [6]. However, when environmental conditions change, sediments may transform from the main sink of heavy metals to sources of them for the overlying waters [1,7]. Therefore the contents of heavy metals in sediments are often monitored to provide basic information for environmental risk assessment [8,9].

In recent decades, various risk assessment indices have been applied to evaluate the environmental risks of metals in marine sediments. Caeiro et al. [10] classified them into three types: contamination indices, background enrichment indices and ecological risk indices. In the present study contamination indices and background enrichment indices were collectively called contamination indices. To assess the metal contamination, the geoaccumulation index (I_{geo}) [5,11] and the enrichment factor (EF) [3,12] are often used. Meanwhile, the risk assessment code (RAC) [13] and the potential ecological risk index (ER) [14] are very popular indices in evaluating the ecological risk posed by heavy metals in sediments. Therefore, these four indices were employed to assess the contamination and ecological risks of the six selected metals (i.e., Cd, Cr, Cu, Ni, Pb and Zn) in the surface sediments of the Laizhou Bay and the coastal waters of the Zhangzi Island (hereafter also referred to as Zhangzi Island for short), China. Numerous sediment quality guidelines (SQGs) also have been developed to deal with environmental concerns. Marine Sediment Quality of China (GB 18668-2002) [9] is one of the SQGs usually used as a general measure of marine sediment contamination in

China, so this guideline was chosen in this study. TEL (threshold effects level) and PEL (probable effects level), which were proved to be effective sediment quality guidelines [15,16], were also used in this study. Many studies have shown that different conclusions or even contradictory conclusions may be drawn by using different risk assessment methods for the same sample or for different elements within the same sample [10,13,17]. Therefore, the relationships among the four index methods and the total concentrations of metals were explored to find out their differences in the environmental risk assessment of heavy metals. Since all the four indices are very popular in evaluating the environmental risks posed by heavy metals in sediments all over the world, we hope to provide useful information about these indices for other researchers to refer to when they carry out similar studies.

To sum up, the purposes of this study are i) to quantify and explain the spatial distribution and fractionations of six heavy metals (Cd, Cr, Cu, Ni, Pb and Zn) in the surface sediments of the Laizhou Bay and the Zhangzi Island; ii) to explore the degree of contamination and the potential ecological risks of these heavy metals to the environment; and iii) to investigate the differences among the used risk assessment indices and the SQGs in the environmental risk assessment of heavy metals.

Materials and Methods

Ethics statement

This study did not involve endangered or protected species and no specific permissions were required for these locations/activities in this study. The specific locations of the present study were shown in Fig. 1.

Study area

The Laizhou Bay (area–7000 km^2, coastline length–320 km, mean depth <10 m, max. depth–18 m) lies in the southern part of the Bohai Sea, accounting for up to 10% of the total area (Fig. 1). It is a semi-closed shallow area with relatively flat seafloor which is formed by the accumulation of riverine suspended matters. There are more than a dozen of rivers running into the Laizhou Bay, among which the Yellow River and the Xiaoqinghe River influence the Laizhou Bay most. Riverine sediment load of the Yellow River and the Xiaoqinghe River began to decrease from the second half of the 20th century, whereas the amount of the contaminants brought by them increased with years [18,19,20]. From the western coast to the eastern coast of the Laizhou Bay, there in turn have Dongying Port, Yangjiaogou Port, Weifang Port and Longkou Port which are important ports in Shandong Province. Due to the abundant seawater resources and underground brine resources in the coastal Laizhou Bay, one of the biggest chemical industrial bases in the world called Weifang Binhai Economic Development Zone is located along its southwestern coast. More than 400 chemical enterprises are located nearby and large amounts of non-purified or insufficiently purified wastewaters are discharged into the Laizhou Bay. The Laizhou Bay can be characterized as a region surrounded by areas with high population growth and rapid economic development in China. The Laizhou Bay used to be one of the most important spawning and breeding grounds for many marine organisms in China. The rapid economic development has brought serious ecological damages to the Laizhou Bay, and fishery resources in the Laizhou Bay are gradually disappearing. The overall sharp decrease of fishery resources and unpredictable nature of the sediment highlight the necessity of the environmental risk assessments of pollutants, especially the heavy metals possessing high affinities for sedimentary materials.

The Zhangzi Island (coastline length–60 km) is located in a national first-class clean sea area in the northern North Yellow Sea which is ~100 km away from Dalian City, Liaoning Province (Fig. 1). The offshore area of the Zhangzi Island is the largest aquaculture base for choice rare seafood in China, which produces conchs, sea cucumbers, scallops, abalones, sea urchins and so on. The only national original seed field for *Patinopecten yessoensis* is located in this area. Fodder-feeding has been banned from seafood farming for many years in this area, and the cultivation of crops on the island is also prohibited. It is thought that natural environment in the Zhangzi Island is generally better than other typical Chinese coastal seas such as the Laizhou Bay. However, sewage (residual feeds, excrement and suspended particles, etc.) discharged during the process of raising seedlings and the frequent operation of ships may result in the deterioration of the local waters and sediment qualities.

Sampling

In this study, a total of 18 surface sediment samples were collected in the Laizhou Bay in October 2011 and 7 surface sediment samples were collected in the Zhangzi Island in November 2011 (Fig. 1). In the Laizhou Bay, sampling sites L1 and L6 were near to the new and old mouths of the Yellow River, respectively; site L18 was near to the estuary of the Xiaoqinghe River. In the Zhangzi Island, the sampling sites were chosen stochastically. Three sites were located in the intertidal zone (Z1–Z3); three were in the coastal waters (Z4–Z6), among which Z4 and Z5 were located in the mariculture areas where sea cucumbers and scallops were farmed, respectively. Site Z7 was in the intertidal zone of an outer island called the Dalian Island. Surface sediment samples (~0–5 cm) were collected by a stainless steel grab sampler and/or a plastic spatula, and were placed in acid-rinsed polyethylene bags. They were transported to the laboratory in a cooler box with ice packs and stored at 4 °C until further treatment.

Analytical methods

The information about the fractionations of metals in the surface sediments was obtained by a sequential extraction procedure reported by Rauret et al. [21]. The four operationally defined geochemical fractions which were separated under this scheme are acid soluble, reducible, oxidizable and residual. The detailed sequential extraction protocol used in this study has been described elsewhere [22].

Previous experiments have shown that sample drying could alter the solid phase distribution of trace elements [23,24]. Furthermore, the elemental concentrations in sediments are highly dependent on the grain size [25,26]. So a drying and grinding treatment could potentially alter the extractability of elements [27]. For the reasons above, wet and unground sediments were used for the sequential extraction procedure in this study to reduce errors.

The mixture of concentrated HF, HNO$_3$ and HClO$_4$ (5:2:1) [22] was used to digest five randomly selected residues instead of the so-called pseudototal digestion with aqua regia used by Rauret et al. [21]. The total metal concentrations in all samples were obtained by the same method used to get the metal concentrations in the residual fraction. The concentrations of metals in residual fractions were estimated by subtracting the metal concentrations obtained in the first three steps of sequential extraction from the total metal concentrations. The sum of the measured values of the four geochemical fractions accounted for 85–110% of the values from the total digestion experiment.

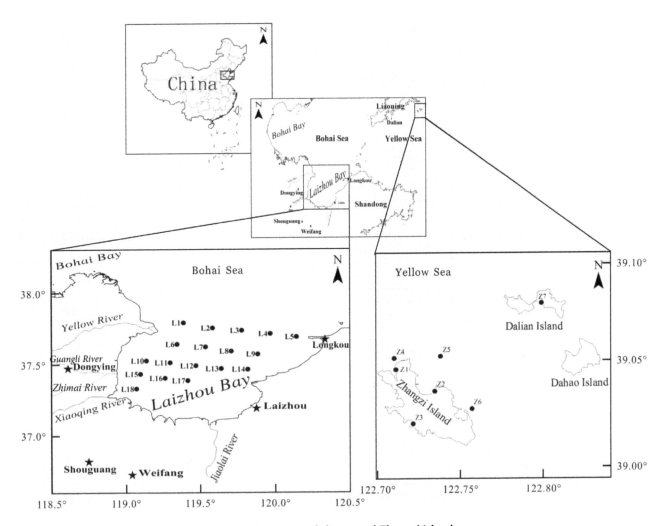

Figure 1. Location of sampling sites in the Laizhou Bay and the coastal Zhangzi Island.

Inductively coupled plasma mass spectrometry (PerkinElmer Elan DRC II) was applied in this work for the determination of Cd, Cr, Cu, Pb, Ni and Zn. In addition, the concentration of Al was analyzed by inductively coupled plasma optical emission spectrometer (PerkinElmer Optima 7000 DV) and the enrichment factor for each element was calculated.

The total organic carbon (TOC) in sediments was obtained by determining the total carbon using an Elementar vario MACRO cube CHNS analyzer after removing the inorganic carbon with 1 M HCl. The substance concentrations of sediments were expressed on the dry weight basis based on the results of moisture contents, which were determined gravimetrically by comparing the weight differences before and after heating an aliquot of sediment at 105 °C until constant weight was obtained. The grain size of samples was analyzed by a Malvern Mastersizer 2000 laser diffractometer capable of analyzing particle sizes between 0.02 and 2000 μm. The percentages of the following three groups of grain sizes were determined: <4 μm (clay), 4–63 μm (silt), and >63 μm (sand) [3,21].

Quality control

The analytical data quality was guaranteed through the implementation of laboratory quality assurance and quality control methods, including the use of standard operating procedures, calibration with standards, analysis of reagent blanks, and analysis of replicates. The precision of the analytical procedures was tested by recovery measurements on the Chinese national geostandard samples (GBW-07333 and GBW-07314). The results were consistent with the reference values, and the differences were all within 10%. The precision of the analytical procedures, expressed as the relative standard deviation (RSD), ranged from 5% to 10%. The precision of the analysis of standard solution was better than 5%. All analyses were carried out in duplicate, and the results were expressed as the mean.

Assessment of sediment contamination and ecological risks

In this study, four different indices were used to assess the degree of heavy metal contamination and ecological risks in the surface sediments of the Laizhou Bay and the Zhangzi Island. For the comparison purpose, the average upper continental crust (UCC) values [29] were chosen as the reference background values in all of the following related indices (Table 1).

1. Enrichment factor (EF). EF is a useful contamination index in determining the degree of anthropogenic heavy metal pollution. The EF for each element was calculated to evaluate anthropogenic influences on heavy metals in sediments using the following formula [3]:

Table 1. The metal guideline values of two different criteria used to distinguish marine sediment quality and the average upper continental crust (UCC) values.

Sediment quality guidelines	Cd	Cr	Cu	Ni	Pb	Zn	Reference
Class I upper limit	0.5	80	35		60	150	[9]
Class II upper limit	1.5	150	100		130	350	[9]
Class III upper limit	5	270	200		250	600	[9]
TEL guideline	0.68	52.3	18.7	15.9	30.2	124	[28]
PEL guideline	4.2	160	108	42.8	112	271	[28]
UCC	0.098	35	25	20	20	71	[29]

Content unit is $\mu g\ g^{-1}$ dry weight for all elements.

$$EF = \frac{(C_x/C_{Al})_{Sample}}{(C_x/C_{Al})_{Background}}$$

Where C_x and C_{Al} denote the concentrations of element x and Al in the samples and in UCC, respectively. In this study, Al was used as the reference element for geochemical normalization, because it represents the quantity of aluminosilicates which is generally the predominant carrier phase for metals in coastal sediments and its natural concentration tends to be uniform [30]. According to EF values, each sample falls into one of the seven tiers: i) $EF<1$ indicates no enrichment; ii) $1<EF<3$ is minor enrichment; iii) $3<EF<5$ is moderate enrichment; iv) $5<EF<10$ is moderately severe enrichment; v) $10<EF<25$ is severe enrichment; vi) $25<EF<50$ is very severe enrichment; and vii) $EF>50$ is extremely severe enrichment [12].

2. Geoaccumulation index (I_{geo}). I_{geo} is also a contamination index which is defined by the following equation:

$$I_{geo} = \log_2(C_n/1.5B_n)$$

Where C_n is the measured concentration of metal n; B_n is the geochemical background concentration of metal n. Correction index 1.5 is usually used to characterize the sedimentary and geological characteristics of rocks and other effects [5]. The geoaccumulation index consists of seven classes: $I_{geo}\leq0$ (Class 0, practically uncontaminated); $0<I_{geo}\leq1$ (Class 1, uncontaminated to moderately contaminated); $1<I_{geo}\leq2$ (Class 2, moderately contaminated); $2<I_{geo}\leq3$ (Class 3, moderately to heavily contaminated); $3<I_{geo}\leq4$ (Class 4, heavily contaminated); $4<I_{geo}\leq5$ (Class 5, heavily to extremely contaminated); $5<I_{geo}$ (Class 6, extremely contaminated) [31].

3. Risk assessment code (RAC). RAC which was originally developed by Perin et al [32] is widely used in ecological risk assessments of heavy metals in sediments. RAC is defined as:

$$RAC = Exc\% + Carb\%$$

Exc% and Carb% are percentages of metals in exchangeable and carbonate fractions (i.e., acid soluble fractions in the present study). According to RAC values, each sample falls into one of the five tiers: i) $RAC\leq1\%$ (no risk); ii) $1\%<RAC\leq10\%$ (low risk); iii)

$10\%<RAC\leq30\%$ (medium risk); iv) $30\%<RAC\leq50\%$ (high risk); v) $50\%<RAC$ (very high risk).

4. Potential ecological risk factor (ER). ER was originally developed by Hakanson [33] and is also an index widely used in ecological risk assessments of heavy metals in sediments. According to this methodology, the potential ecological risk index is defined as:

$$ER^i = Tr^i \cdot C_f^i$$

$$C_f^i = C_o^i/C_n^i$$

ER^i is the potential ecological risk factor for a given element i; Tr^i is the toxic-response factor for element i (e.g., Cd = 30, Cu = Pb = Ni = 5, Cr = 2, Zn = 1); C_f^i, C_o^i and C_n^i are the contamination factor, the concentration in the sediment and the background reference level for element i, respectively. According to Hakanson [33] the following tiers are used for the ER^i value: i) $ER\leq40$ (low risk); ii) $40<ER\leq80$ (moderate risk); iii) $80<ER\leq160$ (considerable risk); iv) $160<ER\leq320$ (high risk); v) $320<ER$ (very high risk).

Sediment quality guidelines

Numerous sediment quality guidelines (SQGs) have been developed to deal with environmental concerns, and two of them were chosen to assess the contamination extent of individual metals in the surface sediments of the Laizhou Bay and the Zhangzi Island (Table 1).

The marine sediment quality of China (GB18668-2002) [9] has defined three grades of marine sediments, in which the contents of five metals (i.e., Cd, Cr, Cu, Pb and Zn) are regarded as parameters used to classify marine sediment quality. According to this criterion, three classes are identified: i) mariculture, nature reserve, endangered species reserve, and leisure activities are suitable; ii) industry and tourism site can be established; iii) only used for harbor.

Threshold effects level (TEL) and probable effects level (PEL) are also sediment quality guidelines which are widely used [28]. TEL is the concentration below which adverse biological effects rarely occur; PEL is the concentration above which adverse biological effects frequently occur.

Based on the fact that heavy metals occur in sediments as complex mixtures, the mean PEL quotient method has been applied to determine the possible biological effect of combined toxicant groups by calculating the mean quotients for a large range of contaminants using the following formula [34]:

$$meanPELquotient = \sum (C_x/PEL_x)/n$$

Where C_x is the sediment concentration of component x, PEL_x is the PEL for compound x and n is the sum of components. Based on the analyses of matching chemical and toxicity data from over 1000 sediment samples from the USA estuaries, the mean PEL quotients of <0.1 have an 8% probability of being toxic, the mean PEL quotients of 0.11–1.5 have a 21% probability of being toxic, the mean PEL quotients of 1.51–2.3 have a 49% probability of being toxic, and the mean PEL quotients of >2.3 have a 73% probability of being toxic [15].

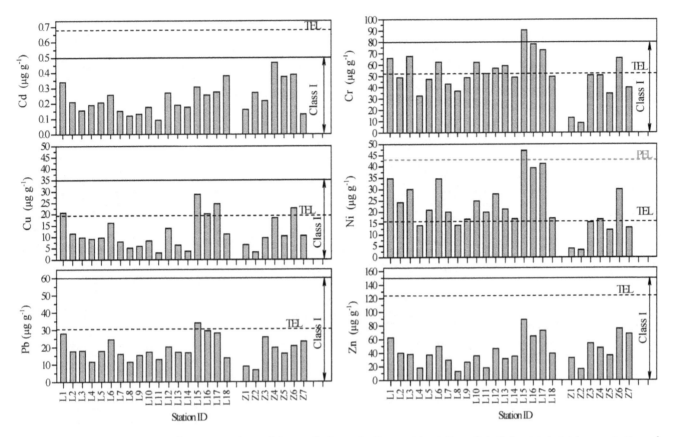

Figure 2. Information of the total concentrations of the studied metals. The spatial variations of studied metals in total concentrations of the surface sediments from the Laizhou Bay and the coastal Zhangzi Island. The horizontal dash lines represent their corresponding TEL or PEL concentrations; the horizontal solid lines represent their corresponding higher boundary values of Class I sediment category of China.

Statistical Analysis

Statistical methods were applied to process the analytical data in terms of the distribution and correlation among the studied parameters. Pearson's correlation coefficient analysis was performed to identify the relationship among heavy metals in sediments and their possible sources. The principal component analysis (PCA) of the normalized variables (Z-scores) was performed to extract significant principal components (PCs) and further reduce the contribution of variables with minor significance. After that factor analysis (FA) was conducted. These PCs were then subjected to varimax rotation to generate varifactors (VFs). The commercial statistics software package SPSS (version 19.0) for Windows was used for statistical analyses mentioned above in the present study.

The agglomerative hierarchical clustering (AHC) analysis was conducted on the normalized data set using Ward's method with Euclidean distances as a measure of similarity to assess the interrelationships among the sampling sites. The XLSTAT software (version 2013) was used in the AHC analysis.

Results and Discussion

Metals in total concentrations

The spatial distribution of heavy metals is shown in Fig. 2 and the related information is summarized in Table 2. Based on the mean concentrations, the target elements in the surface sediments of the Laizhou Bay exhibited the following descending order: Cr $(56.7 \ \mu g \ g^{-1})$ >Zn $(41.5 \ \mu g \ g^{-1})$ > Ni $(25.9 \ \mu g \ g^{-1})$ > Pb $(19.4 \ \mu g \ g^{-1})$ > Cu $(12.0 \ \mu g \ g^{-1})$ > Cd $(0.22 \ \mu g \ g^{-1})$; in the

Zhangzi Island the corresponding result was Zn $(47.1 \ \mu g \ g^{-1})$ > Cr $(37.4 \ \mu g \ g^{-1})$ > Pb $(17.3 \ \mu g \ g^{-1})$ > Ni $(13.5 \ \mu g \ g^{-1})$ > Cu $(11.5 \ \mu g \ g^{-1})$ > Cd $(0.29 \ \mu g \ g^{-1})$.

In the Laizhou Bay, the highest concentrations of Cr $(90.4 \ \mu g \ g^{-1})$, Cu $(28.7 \ \mu g \ g^{-1})$, Ni $(47.1 \ \mu g \ g^{-1})$, Pb $(30.4 \ \mu g \ g^{-1})$ and Zn $(88.6 \ \mu g \ g^{-1})$ were all found in the surface sediments of site L15, which was about 10 km from the estuary of the Guanglihe River, Dongying City; the highest concentration of Cd $(0.38 \ \mu g \ g^{-1})$ was found at site L18 which was about 8 km from the estuary of the Xiaoqinghe River, Weifang City. Relatively higher concentrations of all the six metals studied were also found at sites L1, L6, L16 and L17. L1 and L6 were about 10 km from the new and old mouths of the Yellow River, respectively, indicating the contribution to heavy metal content of terrigenous input. The samples from L17 and L16 had the first and the second highest percentages of fine fractions (clay and silt) which demonstrated that the deposition of fine grained materials physically controls the abundance and distribution of metals in sediments [16]. The information about the grain size and TOC in the sediments of this study has been described in detail in Gao et al. [35].

In the Zhangzi Island, the highest concentrations of Cr $(62.2 \ \mu g \ g^{-1})$, Cu $(22.5 \ \mu g \ g^{-1})$, Ni $(30.0 \ \mu g \ g^{-1})$, Pb $(25.8 \ \mu g \ g^{-1})$ and Zn $(75.4 \ \mu g \ g^{-1})$ were all found in the surface sediments of site Z6, which was about 0.25 km from the coast and had the highest percentage of fine fractions and TOC [35]. The highest concentration of Cd $(0.47 \ \mu g \ g^{-1})$ was found in the surface sediments of site Z4 located in the mariculture area about 0.25 km from the coast where sea cucumbers were farmed. The concentrations of the rest of the metals studied at site Z4 were

Table 2. Heavy metal concentrations in the surface sediments of the Laizhou Bay and the coastal Zhangzi Island; and the related values reported for the surface sediments of other marine areas of China are shown for comparison.

Location	Sampling date		Cd	Cr	Cu	Ni	Pb	Zn	References
Laizhou Bay, China	Oct., 2011	Range	0.09–0.38	32.4–90.4	2.9–28.7	14.1–47.1	11.4–34.0	12.8–88.6	Present study
		Mean	0.22	56.7	12.0	25.9	19.4	41.5	
Coastal Zhangzi Island, China	Nov., 2011	Range	0.13–0.47	8.4–65.6	3.3–22.5	3.2–30.0	6.7–25.8	16.2–75.4	Present study
		Mean	0.29	37.4	11.5	13.5	17.3	47.1	
Coastal Shandong Peninsula (Yellow Sea), China	2007		na[a]	57.8	20.0	31.2	28.4	74.7	[36]
Liaodong Bay, China	2009		na	46.4	19.4	22.5	31.8	71.7	[37]
Coastal East China Sea, China	May, 2009		0.30	84.2	33.1	36.1	28.0	102.4	[38]
Coastal Bohai Bay, China	May, 2008		0.22	101.4	38.5	40.7	34.7	131.1	[3]
Intertidal Bohai Bay, China	May, 2008		0.12	68.6	24.0	28.0	25.6	73.0	[39]
Jinzhou Bay, China	Oct., 2009		26.8	na	74.1	43.5	124.0	689.4	[14]
Laizhou Bay, China	May, 2007		0.081	57.1	13.3	19.4	20.2	59.4	[40]
Laizhou Bay, China	May, 2008		0.11	na	15.0	na	11.7	50.8	[41]
Daya Bay, China	Jan., 2006		0.052	na	20.8	31.2	45.7	113	[22]
North Bohai and Yellow Sea, China	Oct., 2008		0.15	47	13	na	25	60	[42]
Changjiang Estuary, China	Apr. and Aug., 2009		0.26	78.9	30.7	31.8	27.3	94.3	[43]
The Atlantic and Cantabric coasts, Spain	2001–2007		na	na	115.0	na	91.0	230.0	[44]
İzmit Bay, Turkey	Apr. 2002		5.10	75.0	66.4	41.2	104.7	961	[45]
Masan Bay, Korea	2004–2005		1.24	67.1	43.4	28.8	44.0	206.3	[46]
Gironde Estuary, France			0.48	78.4	24.5	31.7	46.8	168.0	[47]
San Francisco Bay, USA	Mar. 2000– Mar. 2001		0.14	19	33	33	19	60	[48]
Jade Bay, Germany	2009–2010		0.25	49	7	10	16	43	[49]

[a]na: not available.
Content unit is $\mu g\ g^{-1}$ dry weight for all elements.

also relatively higher than the other sites in the Zhangzi Island except Z6, which reflected the anthropogenic influence (e.g. fishing operations) on heavy metals.

For the comparison purpose, the average UCC values (Table 1) and related values reported about the surface sediments of some of the marine areas in China and other countries were also shown (Table 2). In the surface sediments from the Laizhou Bay, the mean total contents of Cd, Cr and Ni were clearly higher with respect to their corresponding average values in the UCC; the mean total content of Cu was close to its corresponding average value in the UCC. In the Zhangzi Island, only the values of Cd and Cr were higher than their corresponding average values in the UCC. The average concentrations of Cd, Cr, Ni, Pb and Zn in the Laizhou Bay, Cd, Cu, Ni, Pb and Zn in the Zhangzi Island were within the range identified in the other marine areas listed in Table 2. The average concentrations of Cu in the Laizhou Bay and Cr in the Zhangzi Island were lower than all the values in the other studies listed. All the average concentrations of the studied

metals in the two areas were far below the values of the Jinzhou Bay in China, the Atlantic and Cantabric coasts in Spain, the İzmit Bay in Turkey and the Masan Bay in Korea which were much heavily polluted coastal zones in the world [14,44,45,46]. All the average concentrations of the studied metals in the Zhangzi Island were close to the values of the Jade Bay in Germany where sediment quality was in good condition [49].

Correlation analyses have been widely used in environmental studies. They provide an effective way of revealing the relationships between multiple variables and parameters by which the factors as well as sources of chemical components could be better understood [5,13,38,50]. The correlation matrix for the parameters studied was shown in Table 3. All the metals were significantly correlated with each other in the surface sediments of the Laizhou Bay, suggesting a major common origin in sediments in this area. The wastewater discharged from industrial sources into the surrounding rivers which runs into the Laizhou Bay could be responsible for this [18,19,20]. It has been reported

Table 3. Pearson correlation matrix for the sediment components.

		Cd	Cr	Cu	Ni	Pb	Zn	%Clay	%Silt	%Sand	%TOC
Laizhou Bay	Cd	1	0.472[c]	0.736[a]	0.539[c]	0.602[b]	0.722[a]	0.458	0.216	−0.331	0.381
	p		0.048	0.000	0.021	0.008	0.001	0.056	0.390	0.180	0.118
	Cr		1	0.786[a]	0.932[a]	0.894[a]	0.883[a]	0.535[c]	0.394	−0.473[c]	0.252
	p			0.000	0.000	0.000	0.000	0.022	0.105	0.047	0.312
	Cu			1	0.917[a]	0.924[a]	0.949[a]	0.759[a]	0.608[b]	−0.700[a]	0.530[c]
	p				0.000	0.000	0.000	0.000	0.007	0.001	0.024
	Ni				1	0.962[a]	0.938[a]	0.746[a]	0.593[b]	−0.685[b]	0.488[c]
	p					0.000	0.000	0.000	0.010	0.002	0.040
	Pb					1	0.965[a]	0.735[a]	0.572[c]	−0.668[b]	0.489[c]
	p						0.000	0.001	0.013	0.002	0.040
	Zn						1	0.708[a]	0.526[c]	−0.628[b]	0.484[c]
	p							0.001	0.025	0.005	0.042
Coastal Zhangzi Island	Cd	1	0.456	0.653	0.501	0.072	0.036	0.716	0.783[c]	−0.774[c]	0.525
	p		0.303	0.112	0.252	0.878	0.939	0.070	0.037	0.041	0.226
	Cr		1	0.884[b]	0.953[a]	0.858[c]	0.865[c]	0.745	0.842[c]	−0.824[c]	0.849[c]
	p			0.008	0.001	0.014	0.012	0.055	0.018	0.023	0.016
	Cu			1	0.928[b]	0.558	0.747	0.880[b]	0.929[b]	−0.926[b]	0.829[c]
	p				0.003	0.193	0.054	0.009	0.003	0.003	0.021
	Ni				1	0.694	0.848[c]	0.884[b]	0.901[b]	−0.906[b]	0.932[b]
	p					0.083	0.016	0.008	0.006	0.005	0.002
	Pb					1	0.830[c]	0.318	0.458	−0.424	0.522
	p						0.021	0.487	0.302	0.343	0.229
	Zn						1	0.550	0.565	−0.567	0.667
	p							0.201	0.187	0.184	0.101

[a] $p < 0.001$.
[b] $0.001 < p < 0.01$.
[c] $0.01 < p < 0.05$.
n = 18 for the Laizhou Bay and n = 7 for the coastal Zhangzi Island.

that the deposition of fine grained materials and organic matter physically controls the abundance and distribution of metals in sediments [51]. The concentrations of Cu, Ni, Pb, and Zn appeared to be influenced by both the sediment grain size composition and the amount of organic matter; the concentration of Cr appeared to be more influenced by the sediment grain size composition than by the amount of organic matter; the concentration of Cd appeared to be influenced by neither the sediment grain size composition nor the amount of organic matter, perhaps because it is a typical anthropogenic element.

In the surface sediments of the Zhangzi Island, Cr was significantly correlated with the other studied metals except Cd, suggesting a wide origin of Cr in sediments of this area. The concentrations of Cr, Cu and Ni appeared to be influenced by both the sediment grain size composition and the amount of organic matter, and the concentration of Cd appeared to be influenced only by grain size composition. The concentrations of Pb and Zn were significantly correlated, whereas both of them had no significant correlation with grain size composition and the amount of organic matter, which indicated that they had a major common origin in sediments but not the same as the other metals. Zn might be released from the anti-corrosion paints used on ship hulls or from other anthropogenic sources [52,53], and Pb might be released with engine exhaust of ships [54,55,56].

Metal fractionation

The sequential extraction technique is proposed to provide information about the strength and ways of metals associating with sediments and thus predict the possible metal impact on biota in aquatic ecosystems [16,57]. The metals in acid soluble fraction (i.e., the exchangeable and bound to carbonate fractions) are mainly introduced by human activities and are considered to be weakly bound. This fraction may equilibrate with aqueous phase and thus become more rapidly bioavailable and cause environmental toxicity [58]. The reducible fraction (bound to Fe/Mn oxyhydroxides) and the oxidizable fraction (bound to organic matter) can be mobilized when environmental conditions become increasingly reducing or oxidizing, respectively [58]. The detrital fraction which is composed of metals present in the inert fraction, being of lattice origin or primary mineral phases, can be regarded as a measure of contribution by natural sources [59]. The percentages of heavy metal concentrations that were extracted in each step of the sequential extraction procedure used in this study were presented in Fig. 3, and the pearson correlation matrix for metal fractionations with grain size and TOC was shown in Table 4.

Each studied metal displayed the similar compositional characteristics between the surface sediments in the Laizhou Bay and the Zhangzi Island. On average, the residual fraction was the most dominant one for all the studied metals except Cd, indicating the

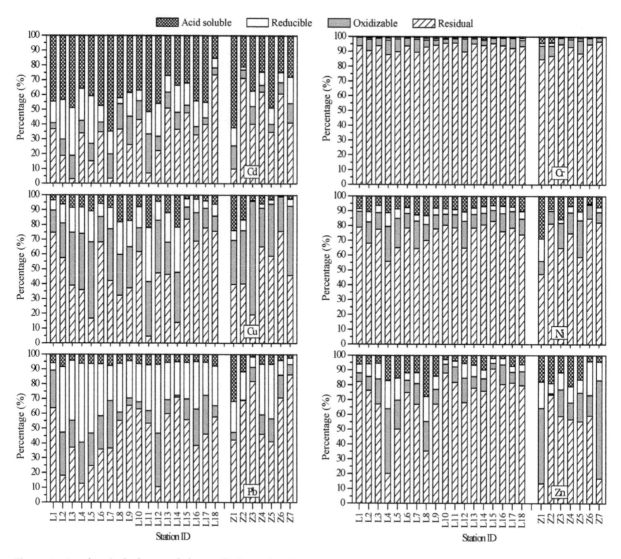

Figure 3. Geochemical phases of the studied metals. The distributions of studied metals in different geochemical phases of the surface sediments from the Laizhou Bay and the coastal Zhangzi Island.

paramount mineralogical origin of these metals; while differences among sampling sites were obvious, which might result from the combined effects of the physicochemical conditions of the sedimentary environment, the intensity of human activities and so on (Fig. 3).

Generally, except for Cd, the relative proportions of metals in the acid soluble fraction were very low, especially for Cr ($\approx$1% of total concentration). In the Laizhou Bay, the very low concentrations of Cr and Cu in the acid soluble fraction could still be partly from mineral sources, because significant positive correlation between this fraction and sand was observed (Table 4). Only Cd had observable contents of the acid soluble fraction with the mean values of 41.0% and 35.2% in the Laizhou Bay and the Zhangzi Island, respectively. The result was similar to the previous study carried out in the Bohai Bay by Gao and Chen [3]. The excessive input of Cd into water induced by phosphorus fertilizer has been widely reported [60,61]. In addition, the presence of Cd could also be a result of road traffic, which has been described as an important source of Cd emission [62].

Pb exhibited the highest proportion within the reducible fraction (5.0–53.2% in the Laizhou Bay; 4.4–36.7% in the

Zhangzi Island) among the six studied metals, which might be the result of the higher stability of Pb-oxides, and also could be attributed to the adsorption, flocculation and co-precipitation of heavy metals with the colloids of Fe and Mn oxyhydroxide [63]. The same result was also reported by other researchers [3,16,22,39]. In both of the studied areas, Pb in oxidizable fraction was significantly correlated with clay and silt, indicating this fraction might be mainly from terrestrial source (Table 4).

On average, the proportions of non-residual Cu, Pb and Zn were identified being the highest in the oxidizble fraction in the Laizhou Bay; in the Zhangzi Island, the proportions of non-residual Cu, Zn and Ni were the highest in the oxidizble fraction. In both of the two studied areas, Cu had the highest proportion (8.2–51.5% in the Laizhou Bay; 23.4–76.8% in the Zhangzi Island) among these metals. This result could be explained by the affinity of metals with organic matter, especially humic substances, which are both the components of natural organic matter and chemical actives in complexing metals [64,65]. The partitioning patterns of Zn and Cu were somewhat very similar with each other. The observed non-residual fractions of Zn and Cu might be the result of ZnO and Cu$_2$O released from the anti-corrosion

Table 4. Pearson correlation matrix for metal fractionations with grain size and TOC.

		Laizhou Bay				Coastal Zhangzi Island			
		F1	F2	F3	F4	F1	F2	F3	F4
Cd	%Clay	0.435	0.344	−0.516[c]	−0.235	−0.266	−0.288	−0.441	0.356
	p	0.071	0.163	0.028	0.347	0.564	0.531	0.322	0.433
	%Silt	0.381	0.452	−0.450	−0.275	−0.323	−0.277	−0.463	0.400
	p	0.119	0.060	0.061	0.270	0.480	0.548	0.295	0.374
	%Sand	−0.421	−0.424	0.498[c]	0.270	0.311	0.283	0.462	−0.392
	p	0.082	0.080	0.035	0.279	0.498	0.539	0.296	0.384
	%TOC	0.398	0.501[c]	−0.403	−0.323	−0.328	−0.272	−0.241	0.355
	p	0.102	0.034	0.097	0.191	0.473	0.556	0.603	0.434
Cr	%Clay	−0.673[b]	−0.511[c]	0.525[c]	−0.324	−0.484	−0.307	−0.184	0.338
	p	0.002	0.030	0.025	0.190	0.271	0.504	0.693	0.458
	%Silt	−0.560[c]	−0.343	0.600[b]	−0.436	−0.586	−0.385	−0.276	0.439
	p	0.016	0.164	0.009	0.070	0.167	0.394	0.550	0.324
	%Sand	0.634[b]	0.432	−0.593[b]	0.406	0.565	0.367	0.253	−0.416
	p	0.005	0.074	0.010	0.095	0.187	0.418	0.584	0.353
	%TOC	−0.566[c]	−0.297	0.671[b]	−0.511[c]	−0.544	−0.561	−0.399	0.520
	p	0.014	0.231	0.002	0.030	0.207	0.190	0.375	0.232
Cu	%Clay	−0.785[a]	−0.651[b]	−0.192	0.542[c]	−0.559	−0.701	−0.429	0.781[c]
	p	0.000	0.003	0.445	0.020	0.192	0.079	0.336	0.038
	%Silt	−0.695[a]	−0.535[c]	0.023	0.369	−0.598	−0.800[c]	−0.296	0.684
	p	0.001	0.022	0.927	0.132	0.156	0.031	0.519	0.090
	%Sand	0.764[a]	0.609[b]	0.070	−0.461	0.594	0.781[c]	0.337	−0.719
	p	0.000	0.007	0.782	0.054	0.160	0.038	0.460	0.069
	%TOC	−0.636[b]	−0.492[a]	0.060	0.319	−0.597	−0.853[c]	−0.070	0.472
	p	0.005	0.038	0.813	0.197	0.157	0.015	0.881	0.285
Ni	%Clay	−0.149	−0.131	0.290	−0.108	−0.485	−0.385	0.186	0.335
	p	0.556	0.604	0.244	0.671	0.270	0.394	0.689	0.463
	%Silt	−0.169	0.144	0.448	−0.266	−0.475	−0.414	0.237	0.316
	p	0.501	0.569	0.062	0.285	0.281	0.356	0.608	0.489
	%Sand	0.168	−0.029	−0.398	0.208	0.483	0.410	−0.226	−0.325
	p	0.506	0.908	0.102	0.407	0.272	0.361	0.626	0.477
	%TOC	0.025	0.106	0.445	−0.310	−0.400	−0.500	0.095	0.366
	p	0.923	0.676	0.064	0.210	0.374	0.254	0.840	0.420
Pb	%Clay	0.083	0.011	0.839[a]	−0.496[c]	−0.366	0.098	0.792[c]	−0.101
	p	0.744	0.965	0.000	0.036	0.420	0.834	0.034	0.830
	%Silt	−0.111	0.264	0.736[a]	−0.581[c]	−0.436	0.134	0.820[c]	−0.095
	p	0.661	0.289	0.000	0.012	0.329	0.774	0.024	0.839
	%Sand	0.031	−0.165	−0.813[a]	0.569[c]	0.421	−0.126	−0.822[c]	0.098
	p	0.903	0.514	0.000	0.014	0.347	0.788	0.023	0.835
	%TOC	0.271	0.258	0.841[a]	−0.660[b]	−0.470	−0.216	0.679	0.201
	p	0.277	0.301	0.000	0.003	0.287	0.642	0.093	0.666
Zn	%Clay	−0.410	−0.183	−0.112	0.249	−0.355	0.584	−0.415	0.392
	p	0.091	0.467	0.658	0.319	0.435	0.169	0.354	0.384
	%Silt	−0.189	0.119	0.216	−0.071	−0.236	0.493	−0.482	0.447
	p	0.453	0.639	0.390	0.780	0.611	0.261	0.273	0.315
	%Sand	0.294	0.009	−0.081	−0.066	0.271	−0.524	0.469	−0.437
	p	0.237	0.973	0.748	0.794	0.556	0.228	0.289	0.327

Table 4. Cont.

	Laizhou Bay				Coastal Zhangzi Island			
	F1	F2	F3	F4	F1	F2	F3	F4
%TOC	−0.204	0.035	−0.034	0.081	−0.482	0.613	−0.413	0.422
p	0.418	0.891	0.894	0.751	0.273	0.144	0.357	0.346

[a] $p < 0.001$.
[b] $0.001 < p < 0.01$.
[c] $0.01 < p < 0.05$.
$n = 18$ for the Laizhou Bay and $n = 7$ for the coastal Zhangzi Island; F1, F2, F3 and F4 represent acid soluble, reducible, oxidizable and residual fraction, respectively.

paints used on ship hulls during the maintenance of ships or from other anthropogenic sources [52,53,66]. The residual fractions of Cu in both the Laizhou Bay and the Zhangzi Island could be from rock weathering source because significant positive correlation between this fraction and clay was observed (Table 4).

The concentration of Cr showed a completely different pattern from the others. Among the non-residual fraction, Cr had the highest content in oxidizable fraction (2.8–9.7% in the Laizhou Bay; 2.6–9.1% in the Zhangzi Island). In the Laizhou Bay, Cr in oxidizable fraction was significantly correlated with clay, silt and TOC, indicating this fraction was mainly from terrestrial source and was influenced by the amount of organic matter (Table 4). Substantial amounts of Cr were found in the residual phases (87.8–95.7% in the Laizhou Bay; 84.8–96.8% in the Zhangzi Island). This suggested that Cr had the strongest associations with the crystalline sedimentary components.

Pearson correlation analysis can partly give information on the sources of the metals in the environment. To explore this topic further, risk indices, principal component analysis and factor analysis were used later in this article.

Risk assessment of heavy metals

1. Risk assessment according to SQGs. The higher boundary values of Class I sediment category of China [9] and the corresponding TEL and PEL concentrations were listed in Table 1 and marked in Fig. 2. All the concentrations of metals at all the sites in both the Laizhou Bay and the Zhangzi Island were below the values of the upper limit for Class I sediment except for Cr in the surface sediments of site L15. This indicated that the sedimentary environments of both the Laizhou Bay and the

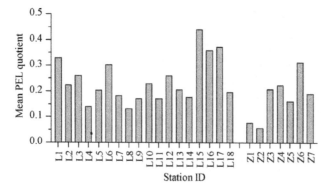

Figure 4. The spatial distribution of mean PEL quotient values.
The histogram shows spatial distribution of mean PEL quotient values in the surface sediments of the Laizhou Bay and the coastal Zhangzi Island.

Zhangzi Island were in good condition according to the marine sediment quality of China.

The data of this study also suggested that no site exceeded the TEL guideline for Cd or Zn in both the Laizhou Bay and the Zhangzi Island. In the case of other metals in the Laizhou Bay, 50%, 78%, 11% and 94% of sites were below the TEL guideline for Cr, Cu, Ni and Pb, respectively; site L15 was even above the PEL guideline for Ni. In the case of other metals in the Zhangzi Island, site Z6 was above the TEL guideline for Cr and Cu; sites Z4 and Z6 were above the TEL guideline for Ni; no site exceeded the TEL guideline for Pb. As shown in Fig. 4, in the surface sediments of the Laizhou Bay and the Zhangzi Island, the combination of the six studied metals of sites Z1 and Z2 might have an 8% probability of being toxic, and these metals might have a 21% probability of being toxic for all the other sites.

The result of the hierarchical cluster analysis of the sampling sites based on the data of metals in total concentrations and SQGs was shown in Fig. 5. Three main different clusters could be observed for the Laizhou Bay. Cluster 1 involved several sites (L1, L15, L16 and L17) near the Yellow River mouths and in the southwestern Laizhou Bay, respectively, indicating the sites were moderately to heavily contaminated according to the SQGs; Cluster 2 was made up of the sites L4, L7, L8, L9, L11 and L14 which were uncontaminated or less contaminated; Cluster 3 was made up of the rest sites which were moderately contaminated. Two main different clusters could be observed for the Zhangzi Island. Cluster 1 included two sites (Z1 and Z2) in the intertidal zone which were uncontaminated according to the SQGs; Cluster 2 was made up of the rest sites which were moderately contaminated.

2. Risk assessment according to contamination and ecological risk indices. According to Zhang and Liu [67], EF values between 0.5 and 1.5 indicate that the given metal is entirely derived from crustal materials or natural weathering processes, whereas EF values higher than 1.5 suggest that a significant portion of metal is delivered from non-crustal materials and the sources are more likely to be anthropogenic. The spatial distributions of calculated EFs for each of the studied metals were shown in Fig. 6. The mean EF values of Cd and Cr suggested their enrichments in most surface sediments of the Laizhou Bay and the Zhangzi Island. In the Laizhou Bay, the highest EF value of Cd was recorded at site L18 (5.7) near the estuary of the Xiaoqinghe River; in the Zhangzi Island, the highest EF value of Cd was found at site Z5 (9.6) 1.4 km away from the coast where scallops were farmed; sites Z2, Z4 and Z6 in the Zhangzi Island also had high EF values of Cd within the range of 5 to 10. This indicated that Cd in the surface sediments of these five sites was in moderately severe enrichment [12]. The highest EF values of Cr were recorded at site L15 (2.8) near the Guanglihe River estuary in the Laizhou Bay and at site Z4 (2.7) where sea cucumbers were farmed in the

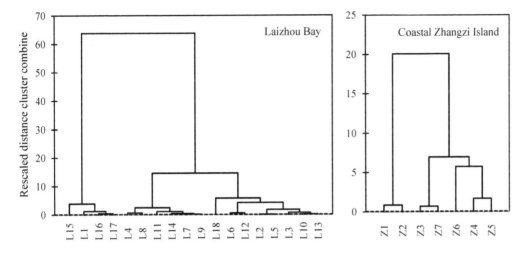

Figure 5. Dendrogram showing clustering of sampling sites in the two studied areas.

Zhangzi Island. According to the EF values, the contribution from anthropogenic sources was negligible for Zn at all sites in the Laizhou Bay and for Cu in all sites of the Laizhou Bay and the Zhangzi Island. Generally, Ni and Pb were slightly enriched in the surface sediments near the Yellow River Estuary and the southwestern Laizhou Bay and sites Z4–Z7 of the Zhangzi Island; Zn was also slightly enriched at Z7 which was near a small wharf.

The spatial distributions of calculated I_{geo} values for each of the studied metals were shown in Fig. 7. Most of the I_{geo} values of Cd were between 0 and 1 which showed that these sites were uncontaminated to moderately contaminated. I_{geo} values of Cd in the surface sediments of the sites near to the mouths of the Yellow River, the Guanglihe River and the Xiaoqinghe River in the

Laizhou Bay and all the three sites in the coastal waters of the Zhangzi Island were between 1 and 2, further indicating these sites were more affected by human activities than other sites. The I_{geo} values suggested that Cu and Zn at all sites in both the Laizhou Bay and the Zhangzi Island were in the uncontaminated level, and this was true for Pb except at site L15 where the I_{geo} value of Pb was a little higher than 0. The values of I_{geo} for Cr and Ni at most sites were <0 except at several ones that were near to the Yellow River mouths and in the southwestern Laizhou Bay; for all the sampling sites in the Zhangzi Island, the I_{geo} values indicated that their surface sediments were practically uncontaminated by Ni, and Cr presented the same situation like Ni except at site Z6 which had a condition of slight Cr pollution.

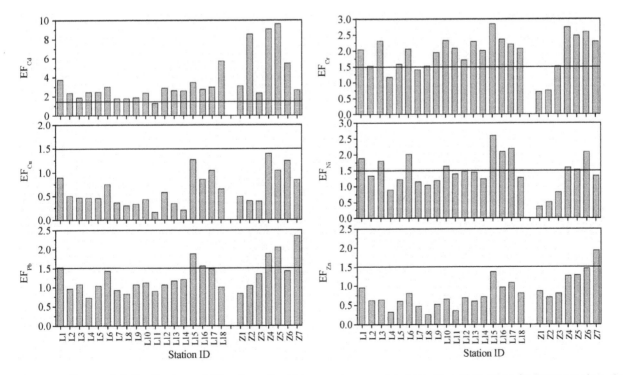

Figure 6. The spatial distributions of EF values. The histogram shows spatial distributions of EF values for heavy metals in the surface sediments of the Laizhou Bay and the coastal Zhangzi Island. The horizontal lines represent EF value of 1.5.

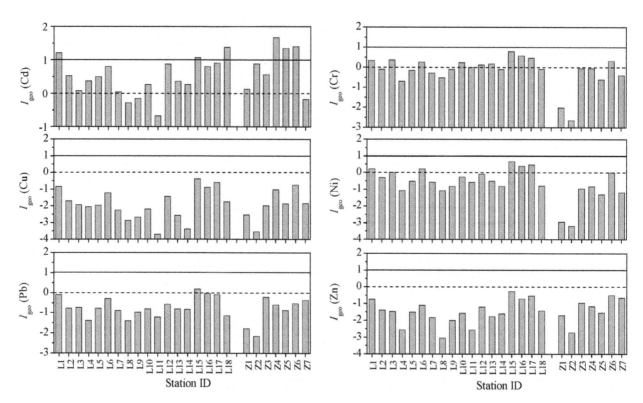

Figure 7. The spatial distributions of I_{geo} values. The histogram shows the spatial distributions of I_{geo} values for heavy metals in the surface sediments of the Laizhou Bay and the coastal Zhangzi Island. The horizontal dash and solid lines represent I_{geo} values of 0 and 1, respectively.

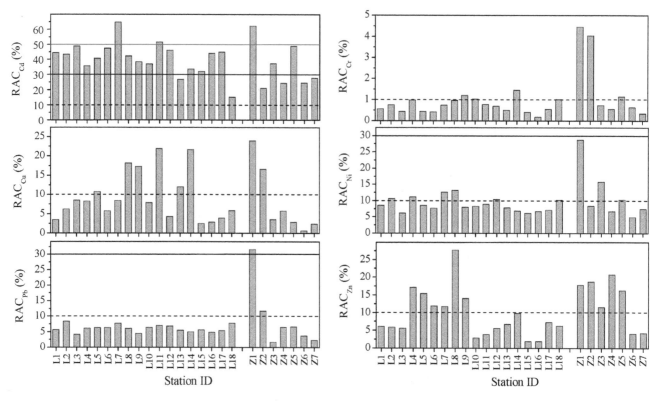

Figure 8. The spatial distributions of RAC values. The histogram shows the spatial distributions of RAC values (%) for heavy metals in the surface sediments of the Laizhou Bay and the coastal Zhangzi Island. The horizontal dash lines represent RAC values (%) of 1 or 10. The horizontal solid lines represent RAC values (%) of 30 or 50.

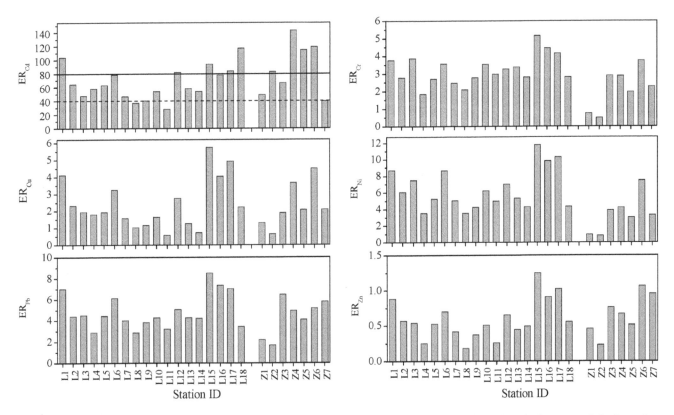

Figure 9. The spatial distributions of ER values. The histogram shows the spatial distributions of ER values for heavy metals in the surface sediments of the Laizhou Bay and the coastal Zhangzi Island. The horizontal dash and solid lines represent ER values of 40 and 80, respectively.

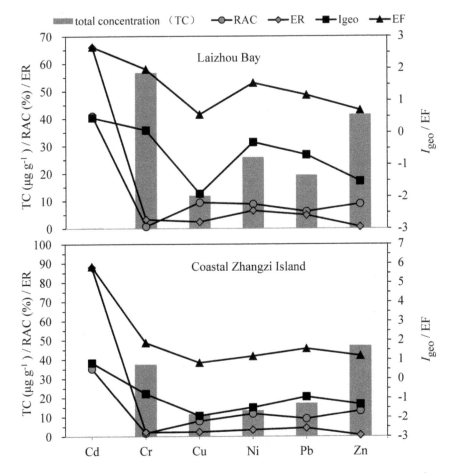

Figure 10. The mean values of EF, *I*geo, RAC, ER and total concentration for metals.

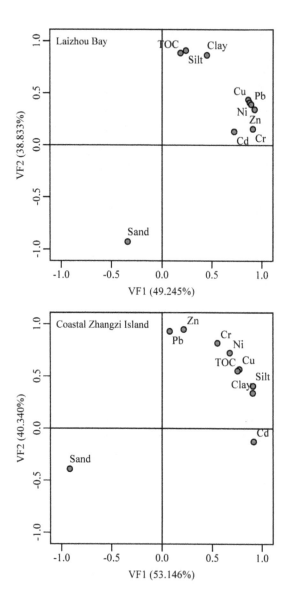

Figure 11. Loading plots of the principal components obtained for the data set.

Table 5. Loadings of experimental variables on significant principal components for the data sets of the Laizhou Bay and the coastal Zhangzi Island.

	Laizhou Bay		Coastal Zhangzi Island	
	PC1	**PC2**	**PC1**	**PC2**
Cd	**0.639**	0.359	**0.916**	−0.126
Cr	**0.800**	0.460	**0.553**	**0.818**
Cu	**0.944**	0.213	**0.774**	**0.567**
Ni	**0.937**	0.243	**0.678**	**0.725**
Pb	**0.937**	0.268	0.075	**0.930**
Zn	**0.933**	0.327	0.214	**0.949**
Clay	**0.897**	−0.379	**0.906**	0.339
Silt	**0.760**	−0.549	**0.909**	0.405
Sand	**−0.853**	0.498	**−0.918**	−0.391
TOC	**0.703**	**−0.566**	**0.755**	**0.552**
Eigenvalue	4.925	3.883	5.315	4.034
% Total variance	49.245	38.833	53.146	40.340
Cumulative % variance	49.245	88.078	53.146	93.486

Bold values indicate strong loadings.

the Laizhou Bay and Z7 in the Zhangzi Island, all the other sites at least suffered moderate risk from Cd ($40 < ER \leq 80$). Sites L1, L15, L17 and L18 in the Laizhou Bay and sites Z4–Z6 suffered considerable risk from Cd ($80 < ER \leq 160$). The sites which suffered considerable risk from Cd according to ER index were also moderately contaminated according to I_{geo} index.

The mean values of EF, I_{geo}, RAC and ER of the six studied metals and their total concentrations were summarized in Fig. 10. It showed clearly that Cd had the highest potential risk according to all the four indices. However, inconsistent conclusions for the other five metals according to these four indices could be drawn. In addition, the results of EF and I_{geo} were consistent with the trend of the total metal concentrations except for Cd; meanwhile the results of RAC and ER had no consistency with the trend of their corresponding total metal concentrations. The explanation for this might be as the following: i) the EF and I_{geo} indices are calculated mainly based on the concentrations of the total metals and enrichment levels. ii) The ER index is based on the toxic-response factor besides the total concentration. For example, though total concentration of Cd is usually pretty lower than Zn in sediments, but the higher value of toxic-response factor of Cd ($Tr = 30$) might make it much more toxic than Zn whose toxic-response factor is only 1. iii) The RAC index, which reflects the potential mobility of sedimentary metals, is based on the chemical form of a given metal which has no direct relationship with its total concentration. It is generally accepted that the fractionation can give more information on the bio-availability and bio-toxicity of a certain metal than the total concentration [3,16,38,68]. However, the results of the present study indicated the necessity of further verifying the prediction accuracy of the fractionation based methods.

It is generally known that the mechanisms of these SQS and indices are divergent and they are used to assess the environmental risk from different angles. Neither the total concentration based index nor the chemical fractionation based index alone could be sufficient in revealing the biogeochemical information of heavy metals in sediments. However, the trend of risk level according to

The spatial distributions of calculated RACs for each of the studied metals were shown in Fig. 8. It showed that most of the sites suffered high risks from Cd ($30\% < RAC \leq 50\%$). Risk from Cd in the surface sediments of sites L7 and L11 in the Laizhou Bay and site Z1 in the Zhangzi Island was very high ($RAC > 50\%$). Although the total concentration of Cd in the surface sediments of Z1 was very low, the proportion of acid soluble fraction of it was high; this might be because Z1 was near a seedling factory and was severely impacted by the discharge from that factory. The values of RAC at site Z1 were also higher than the other sites in the Zhangzi Island for Cu (24.0%), Ni (28.8%) and Pb (31.6%). Cr at all the sites had no to low risk; Cu, Ni and Zn at most sites had low to medium risk; Pb at all the sites had low risk except at sites Z1 and Z2 which had high risk and medium risk, respectively.

The spatial distributions of calculated ERs for each of the studied metals were shown in Fig. 9. According to ER index, potential ecological risk posed by Cr, Cu, Ni, Pb and Zn was very low at all the sites. However, the potential ecological risk posed by Cd was obvious at most of the sites. Except for sites L8 and L11 in the assessment results of them should be consistent or at least should not be opposite to each other; if not, misleading results of risk assessment may even cause misjudgment during the formulation and enforcement of public policies by government. Therefore, the merits and defects should be evaluated before the application of these assessment methods, and it is necessary to use multiple evaluation methods. Different metals may have different risk gradation ranges. For instance, the differences between metal fractionation characteristics of Cd and Cr are often substantial, but the risks of these two metals ranked by RAC indexes are within the same gradation range that this may not be true. We believe that a more comprehensive index to reveal the information of the concentration, the chemical fractionation and the toxic-response factor of heavy metals should be developed. Therefore, we suggest that more toxicity tests, benthic community analyses, bioaccumulation tests, and a combination of these should be carried out to make the risk assessment of sedimentary heavy metals more accurate and reliable.

Principal component analysis/factor analysis

PCA/FA was performed to identify interrelationships of the six studied heavy metals and the major constituents of the sediments (TOC, clay, silt and sand). Table 5 showed that there were two PCs for the surface sediments in both the Laizhou Bay and the Zhangzi Island. The loading plots of the VFs were presented in Fig. 11. These PCs were the ones with eigenvalues larger than 1, and altogether they accounted for 88.1% and 93.5% of the variance in the data of the Laizhou Bay and the Zhangzi Island, respectively.

In the Laizhou Bay, PC1 which explained 49.3% of the total variance was positively related to all the heavy metals and major constituents except sand. Sand was significantly negatively related to PC1. The high loading of clay, silt and TOC with PC1 highlighted the influence of fine grained minerals and organic matter on the distributions of heavy metals in the sediments of the Laizhou Bay, and revealed that these metals were mainly from terrestrial source especially via rivers [68]. PC2 which explained 38.8% of the total variance was only positively related to sand, indicating that sand could hardly capture metal ions.

In the Zhangzi Island, PC1 was positively related to Cd, Cr, Cu, Ni, Clay, Silt and TOC. This PC represented terrestrial sources. PC2 was also positively related to Cr, Cu, Ni and TOC, indicating they were from both terrestrial inputs and biogenic sources [68]. Zn and Pb were only positively related to PC2, indicating their different sources with the other studied metals. Zn and Pb might be released from ships and from biogenic sources. In addition, previous studies have shown that a large amount of Pb is supplied by the precipitation of aerosols in coastal environments [69]. So the precipitation of aerosols might be another important source of Pb in the surface sediments of the Zhangzi Island.

Conclusions

This study investigated the total concentrations and fractionation of heavy metals in the surface sediments from the Laizhou Bay and the Zhangzi Island. The relatively higher concentrations of metals in the Laizhou Bay were mainly distributed near to the new and old mouths of the Yellow River, the mouths of Guanglihe and Xiaoqinghe Rivers, and in the middle of the Bay. The relatively higher concentrations of metals in the Zhangzi Island were mainly distributed in and near the mariculture areas. In the Laizhou Bay, all the metals studied were mainly from terrestrial sources, and especially Cd, Cr and Ni had obvious anthropogenic sources; in the Zhangzi Island, both natural and anthropogenic sources contributed significantly to the metal contents.

The marine sediment quality of China showed that the sedimentary environment in both the Laizhou Bay and the Zhangzi Island were in good condition. TEL/PEL guidelines revealed that adverse biological effects might occur frequently in some areas in the Laizhou Bay and the Zhangzi Island especially from Cr and Ni. Based on the mean PEL quotient, surface sediments of sites Z1 and Z2 had an 8% probability of toxicity, and surface sediments of the rest sites of the two studied areas had a 21% probability of toxicity.

All the four risk assessment indices used in this study revealed an obvious pollution risk by Cd, especially in sites near the river mouths and in the southwestern Laizhou Bay and in the coastal waters of the Zhangzi Island. Nevertheless, contradictory conclusions could be obtained when different indices and SQGs are used. Significant negative correlations between RAC and the other indices and between RAC and the total metal concentration for Cd, Cr, Cu and Ni were found. We suggest that toxicity tests, bioaccumulation tests, and other related experiments should be further carried out in order to make the risk assessment methods more accurate and reliable in the analysis of sedimentary heavy metals.

Acknowledgments

We are thankful to the anonymous reviews and the Academic Editor whose pertinent comments have greatly improved the quality of this paper. We thank Yong Zhang, Peimiao Li, Fengxia Zhou and Jinfeng Zhang for participating in the sample collection.

Author Contributions

Conceived and designed the experiments: WZ XG. Performed the experiments: WZ. Analyzed the data: WZ XG. Contributed reagents/materials/analysis tools: XG. Wrote the paper: WZ XG.

References

1. Prica M, Dalmacija B, Rončević S, Krčmar D, Bečelić M (2008) A comparison of sediment quality results with acid volatile sulfide (AVS) and simultaneously extracted metals (SEM) ratio in Vojvodina (Serbia) sediments. Sci Total Environ 389: 235–244.

2. Yang YQ, Chen FR, Zhang L, Liu JS, Wu SJ, et al. (2012) Comprehensive assessment of heavy metal contamination in sediment of the Pearl River Estuary and adjacent shelf. Mar Pollut Bull 64: 1947–1955.

3. Gao XL, Chen CTA (2012) Heavy metal pollution status in surface sediments of the coastal Bohai Bay. Water Res 46: 1901–1911.

4. Gao XL, Zhou FX, Chen CTA (2014) Pollution status of the Bohai Sea, China: An overview of the environmental quality assessment related trace metals. Environ Int 62: 12–30.

5. Varol M (2011) Assessment of heavy metal contamination in sediments of the Tigris River (Turkey) using pollution indices and multivariate statistical techniques. J Hazard Mater 195: 355–364.

6. Gaur VK, Gupta SK, Pandey SD, Gopal K, Misra V (2005) Distribution of heavy metals in sediment and water of River Gomti. Environ Monit Assess 102: 419–433.

7. van Den Berg GA, Loch JPG, van Der Heijdt LM, Zwolsman JJG (1999) Mobilisation of heavy metals in contaminated sediments in the river Meuse, The Netherlands. Water Air Soil Poll 116: 567–586.

8. Long ER, MacDonald DD, Smith SC, Calder FD (1995) Incidence of adverse biological effects within ranges of chemical concentrations in marine and estuarine sediments. Environ Manage 19: 81–97.

9. SEPA (State Environmental Protection Administration of China) (2002) Marine Sediment Quality (GB 18668–2002). Standards Press of China, Beijing.

10. Caeiro S, Costa MH, Fernandes F, Silveira N, Coimbra A, et al. (2005) Assessing heavy metal contamination in Sado Estuary sediment: An index analysis approach. Ecol Indic 5: 151–169.

11. Porstner U (1989) Lecture Notes in Earth Sciences (Contaminated Sediments). Springer Verlag, Berlin. pp. 107–109.

12. Sakan SM, Dordević DS, Manojlović DD, Predrag PS (2009) Assessment of heavy metal pollutants accumulation in the Tisza river sediments. J Environ Manage 90: 3382–3390.

13. Zhao S, Feng CH, Yang YR, Niu JF, Shen ZY (2012) Risk assessment of sedimentary metals in the Yangtze Estuary: New evidence of the relationships between two typical index methods. J Hazard Mater 241–242: 164–172.

14. Li XY, Liu LJ, Wang YG, Luo GP, Chen X, et al. (2012) Integrated assessment of heavy metal contamination in sediments from a coastal industrial basin, NE China. PLoS ONE 7: e39690.

15. Long ER, MacDonald DD, Severn CG, Hong CB (2000) Classifying probabilities of acute toxicity in marine sediments with empirically derived sediment quality guideline. Environ Toxicol Chem 19: 2598–2601.

16. Sundaray SK, Nayak BB, Lin S, Bhatta D (2011) Geochemical speciation and risk assessment of heavy metals in the river estuarine sediments–a case study: Mahanadi basin, India. J Hazard Mater 186: 1837–1846.

17. Yu GB, Liu Y, Yu S, Wu SC, Leung AOW, et al. (2011) Inconsistency and comprehensiveness of risk assessments for heavy metals in urban surface sediments. Chemosphere 85: 1080–1087.

18. Ma SS, Xin FY, Cui Y, Qiao XY (2004) Assessment of main pollution matter volume into the sea from Yellow River and Xiaoqing River. Mar Fish Res 25: 47–51. (In Chinese with English abstract)

19. Wang HJ, Yang ZS, Saito Y, Liu JP, Sun XX (2006) Interannual and seasonal variation of the Huanghe (Yellow River) water discharge over the past 50 years: Connections to impacts from ENSO events and dams. Global Planet Change 50: 212–225.

20. Wang HJ, Yang ZS, Saito Y, Liu JP, Sun XX, et al. (2007) Stepwise decreases of the Huanghe (Yellow River) sediment load (1950–2005): Impacts of climate change and human activities. Global Planet Change 57: 331–354.

21. Rauret G, López-Sánche JF, Sahuquillo A, Rubio R, Davidson C, et al. (1999) Improvement of the BCR three step sequential extraction procedure prior to the certification of new sediment and soil reference materials. J Environ Monitor 1: 57–61.

22. Gao XL, Chen CTA, Wang G, Xue QZ, Tang C, et al. (2010) Environmental status of Daya Bay surface sediments inferred from a sequential extraction technique. Estuar Coast Shelf S 86: 369–378.

23. Rapin F, Tessier A, Campbell PGC, Carignan R (1986) Potential artifacts in the determination of metal partitioning in sediments by a sequential extraction procedure. Environ Sci Technol 20: 836–840.

24. Hjorth T (2004) Effects of freeze-drying on partitioning patterns of major elements and trace metals in lake sediments. Anal Chim Acta 526: 95–102.

25. Horowitz AJ, Elrick KA (1988) Interpretation of bed sediment traces metal data: methods for dealing with the grain size effect. In: Lichrenberg JJ, Winter JA, Weber CI, Fradkin L, editors. Chemical and Biological Characterization of Sludges, Sediments Dredge Spoils, and Drilling Muds. ASTM Special Technical Publication, Philadelphia 976. pp. 114–128.

26. Howari FM, Banat KM (2001) Assessment of Fe, Zn, Cd, Hg, and Pb in the Jordan and Yarmouk river sediments in relation to their physicochemical properties and sequential extraction characterization. Water Air Soil Poll 132: 43–59.

27. Gilliam FS, Richter DD (1988) Correlations between extractable Na, K, Mg, Ca, P & N from fresh and dried samples of two Aquults. Eur J Soil Sci 39: 209–214.

28. MacDonald DD, Scottcarr R, Calder FD, Long ER, Ingersoll CG (1996) Development and evaluation of sediment quality guidelines for Florida coastal waters. Ecotoxicology 5: 253–278.

29. Taylor SR, McLennan SM (1995) The geochemical evolution of the continental crust. Rev Geophys 33: 241–265.

30. Alexander CR, Smith RG, Calder FD, Schropp SJ, Windom HL (1993) The historical record of metal enrichments in two Florida estuaries. Estuaries 16: 627–637.

31. Müller G (1969) Index of geoaccumulation in sediments of the Rhine River. Geojournal 2: 108–118.

32. Perin G, Craboledda L, Cirillo M, Dotta L, Zanette ML, et al. (1985) Heavy metal speciation in the sediments of Northern Adriatic Sea: a new approach for environmental toxicity determination. In: Lekkas TD, editor. Heavy Metal in the Environment. CEP Consultant, Edinburgh 2. pp. 454–456.

33. Hakanson L (1980) An ecological risk index for aquatic pollution control. A sedimentological approach. Water Res 14: 975–1001.

34. Carr RS, Long ER, Windom HL, Chapman DC, Thursby G, et al. (1996) Sediment quality assessment studies of Tampa Bay, Florida. Environ Toxicol Chem 15: 1218–1231.

35. Gao XL, Li PM, Chen CTA (2013) Assessment of sediment quality in two important areas of mariculture in the Bohai Sea and the northern Yellow Sea based on acid-volatile sulfide and simultaneously extracted metal results. Mar Pollut Bull 72: 281–288.

36. Li GG, Hu BQ, Bi JQ, Leng QN, Xiao CQ, et al. (2013) Heavy metals distribution and contamination in surface sediments of the coastal Shandong Peninsula (Yellow Sea). Mar Pollut Bull 76:420–426.

37. Hu BQ, Li J, Zhao JT, Yang J, Bai FL, et al. (2013) Heavy metal in surface sediments of the Liaodong Bay, Bohai Sea: distribution, contamination and sources. Environ Monit Assess 185: 5071–5083.

38. Yu Y, Song JM, Li XG, Yuan HM, Li N, et al. (2013) Fractionation, sources and budgets of potential harmful elements in surface sediments of the East China Sea. Mar Pollut Bull 68: 157–167.

39. Gao XL, Li PM (2012) Concentration and fractionation of trace metals in surface sediments of intertidal Bohai Bay, China. Mar Pollut Bull 64: 1529–1536.

40. Hu NJ, Shi XF, Liu JH, Huang P, Yang G, et al. (2011) Distributions and impacts of heavy metals in the surface sediments of the Laizhou Bay. Adv Mar Sci 29: 63–72. (In Chinese with English abstract)

41. Luo XX, Zhang R, Yang JQ, Liu RH, Tang W, et al. (2010) Distribution and pollution assessment of heavy metals in surface sediment in Laizhou Bay. Ecol Environ Sci 19: 262–269. (In Chinese with English abstract)

42. Luo W, Lu YL, Wang TY, Hu WY, Jiao WT, et al. (2010) Ecological risk assessment of arsenic and metals in sediments of coastal areas of northern Bohai and Yellow Seas, China. Ambio 39: 367–375.

43. Zhang WG, Feng H, Chang JN, Qu JG, Xie HX, et al. (2009) Heavy metal contamination in surface sediments of Yangtze River intertidal zone: An assessment from different indexes. Environ Pollut 157: 1533–1543.

44. Durán I, Sánchez-Marín P, Beiras R (2012) Dependence of Cu, Pb and Zn remobilization on physicochemical properties of marine sediments. Mar Environ Res 77: 43–49.

45. Pekey H, Doğan G, 2013. Application of positive matrix factorisation for the source apportionment of heavy metals in sediments: A comparison with a previous factor analysis study. Microchem J 106: 233–237.

46. Hyun S, Lee C-H, Lee T, Choi J-W (2007) Anthropogenic contributions to heavy metal distributions in the surface sediments of Masan Bay, Korea. Mar Pollut Bull 54: 1059–1068.

47. Larrose A, Coynel A, Schäfer J, Blanc G, Massé L, et al. (2010) Assessing the current state of the Gironde Estuary by mapping priority contaminant distribution and risk potential in surface sediment. Appl Geochem 25: 1912–1923.

48. Lu XQ, Werner I, Young TM (2005) Geochemistry and bioavailability of metals in sediments from northern San Francisco Bay. Environ Int 31: 593–602.

49. Beck M, Böning P, Schückel U, Stiehl T, Schnetger B et al. (2013) Consistent assessment of trace metal contamination in surface sediments and suspended particulate matter: A case study from the Jade Bay in NW Germany. Mar Pollut Bull 70: 100–111.

50. Zhuang W, Gao XL (2013) Acid-volatile sulfide and simultaneously extracted metals in surface sediments of the southwestern coastal Laizhou Bay, Bohai Sea: Concentrations, spatial distributions and the indication of heavy metal pollution status. Mar Pollut Bull 76: 128–138.

51. Gomes F, Godoy JM, Godoy ML, Carvalho Z, Lopes R, et al. (2009) Metal concentrations, fluxes, inventories and chronologies in sediments from Sepetiba and Ribeira Bays: A comparative study. Mar Pollut Bull 59: 123–133.

52. Goh BPL, Chou LM (1997) Heavy metal levels in marine sediments of Singapore. Environ Monit Assess 44: 67–80.

53. Turner A (2010) Marine pollution from antifouling paint particles. Mar Pollut Bull 60: 159–171.

54. Guillen G, Ruckman M, Smith S, Broach L (1993) Marina impacts in Clean Lake and Galveston Bay. Special Report D7-001A prepared for the Texas Water Commission, Houston, TX.

55. Hinkey LM, Zaidi BR, Volson B, Rodriguez N (2005) Identifying sources and distributions of sediment contaminants at two US Virgin Islands marinas. Mar Pollut Bull 50: 1244–1250.

56. Hinkey LM, Zaidi BR (2007) Differences in SEM–AVS and ERM–ERL predictions of sediment impacts from metals in two US Virgin Islands marinas. Mar Pollut Bull 54: 180–185.

57. Tessier A, Campbell PGC, Bisson M (1979) Sequential extraction procedure for the speciation of particulate trace metals. Anal Chem 51: 844–851.

58. Karbassi AR, Shankar R (2005) Geochemistry of two sediment cores from the west coast of India. Int J Environ Sci Te 1: 307–316.

59. Salmonas W, Förstner U (1980) Trace metal analysis on polluted sediments, part-II: Evaluation of environmental impact. Environ Technol Lett 1: 506–517.

60. Lambert R, Grant C, Sauvé S (2007) Cadmium and zinc in soil solution extracts following the application of phosphate fertilizers. Sci Total Environ 378: 293–305.

61. Zhang H, Shan BQ (2008) Historical records of heavy metal accumulation in sediments and the relationship with agricultural intensification in the Yangtze-Huaihe region, China. Sci Total Environ 399: 113–120.

62. Alloway BJ (1990) Cadmium. In: Alloway BJ, editor. Heavy metals in soils. Blackie, Glasgow. pp. 100–124.

63. Rath P, Panda UC, Bhatta D, Sahu KC (2009) Use of sequential leaching, mineralogy, morphology and multivariate statistical technique for quantifying metal pollution in highly polluted aquatic sediments – a case study: Brahamani and Nandira rivers, India. J Hazard Mater 163: 632–644.

64. Fytianos K, Lourantou A (2004) Speciation of elements in sediment samples collected at lakes Volvi and Koronia, N. Greece. Environ Int 30: 11–17.

65. Ytreberg E, Karlsson J, Hoppe S, Eklund B, Ndungu K (2008) Effect of organic complexation on copper accumulation and toxicity to the estuarine Red Macroalga Ceramium tenuicorne: a test of the free ion activity model. Environ Sci Technol 45: 3145–3153.

66. Jain CK (2004) Metal fractionation study on bed sediments of River Yamuna, India. Water Res 38: 569–578.

67. Zhang J, Liu CL (2002) Riverine composition and estuarine geochemistry of particulate metals in China–weathering features, anthropogenic impact and chemical fluxes. Estuar Coast Shelf S 54: 1051–1070.

68. Duan LQ, Song JM, Xu YY, Li XG, Zhang Y (2010) The distribution, enrichment and source of potential harmful elements in surface sediments of Bohai Bay, North China. J Hazard Mater 183: 155–164.

69. Li XD, Wai OWH, Li YS, Coles BJ, Ramsey MH, et al. (2000) Heavy metal distribution in sediment profiles of the Pearl River estuary, South China. Appl Geochem 15: 567–581.

13

Transgenic Common Carp Do Not Have the Ability to Expand Populations

Hao Lian[1,2○]**, Wei Hu**[1○]**, Rong Huang**[1]**, Fukuan Du**[1]**, Lanjie Liao**[1]**, Zuoyan Zhu**[1]**, Yaping Wang**[1]*****

1 State Key Laboratory of Freshwater Ecology and Biotechnology, Institute of Hydrobiology, The Chinese Academy of Sciences, Wuhan, China, **2** University of Chinese Academy of Sciences, Beijing, China

Abstract

The ecological safety of transgenic organisms is an important issue of international public and political concern. The assessment of ecological risks is also crucial for realizing the beneficial industrial application of transgenic organisms. In this study, reproduction of common carp (*Cyprinus carpio*, CC) in isolated natural aquatic environments was analyzed. Using the method of paternity testing, a comparative analysis was conducted on the structure of an offspring population of "all-fish" growth hormone gene-transgenic common carp (af*gh*-CC) and of wild CC to evaluate their fertility and juvenile viability. Experimental results showed that in a natural aquatic environment, the ratio of comparative advantage in mating ability of af*gh*-CC over wild CC was 1:1, showing nearly identical mating competitiveness. Juvenile viability of af*gh*-CC was low, and the average daily survival rate was less than 98.00%. After a possible accidental escape or release of transgenic CC into natural aquatic environments they are unable to monopolize resources from eggs of natural CC populations, leading to the extinction of transgenic CC. Transgenic CC are also unlikely to form dominant populations in natural aquatic environments due to their low juvenile viability. Thus, it is expected that the proportion of af*gh*-CC in the natural environment would remain low or gradually decline, and ultimately disappear.

Editor: Justin David Brown, University of Georgia, United States of America

Funding: This work was financially supported by grants from the Development Plan of the State Key Fundamental Research of China (Grant No. 2009CB118701), and the National Natural Science Foundation of China (Grant No. 30771664). The funders had no role in study design, data collection and analysis, decision to publish, or preparation of the manuscript.

Competing Interests: The authors have declared that no competing interests exist.

* E-mail: Wangyp.ihb@gmail.com

○ These authors contributed equally to this work.

Introduction

Transgenic fish technology was first introduced in the 1980s and has been used for the genetic improvement of farmed fish [1,2]. In the past 20 years, important progress has been made in the research of transgenic fish breeding for the purposes of increasing growth rates. Several fast-growing transgenic fish strains have been bred, showing attractive commercial prospects [3–13]. However, there are currently no transgenic fish strains used in commercial production for human consumption.

Food security and the ecological safety of transgenic organisms are of great public concern and represent the last obstacles preventing transgenic fish from entering the market. Compared with food safety assessment, ecological security assessment poses a more difficult task. It is generally believed that fertility and viability are key fitness parameters for evaluating the ecological safety of transgenic fish [14,15]. In recent years, a number of transgenic fish viability-related traits have been reported, including characteristics of the appetite and feeding behavior of transgenic salmon and common carp [16–19], swimming ability [20–22], respiratory metabolism characteristics [23], viability under dissolved oxygen and ammonia nitrogen stress [24,25] and juvenile prey mortality [15,26,27]. Studies on transgenic fish fertility mainly involve the transgenic fish gonad index [5,7], the first sexual maturation time [28], sperm ejaculation volume, sperm motility and mating behavior [29]. Results of comparative studies of single factors of

viability and fertility showed that transgenic fish exhibited lower viability and fertility [15,20,22,27–29].

However, Muir et al. [30] obtained opposing results in a study on transgenic medaka (*Oryzias latipes*). By assessing a population genetics model based on the findings of fertility and viability of transgenic medaka, they proposed the "Trojan gene hypothesis". This hypothesis predicts that because of the significant fertility advantage of their transgenic medaka and the low viability of its juveniles, once transgenic medaka were released into a natural aquatic environment, transgenic and wild medaka populations would be extinct within 50 generations [30,31]. This hypothesis has resulted in a strong public response, but its universality has been questioned by the academic community [32].

We examined and analyzed the fertility and juvenile viability of af*gh*-CC and CC in an isolated pond. The study aimed to obtain fitness parameters of fertility and viability of the two populations in a natural state, thus further assessing the ecological risk for the environment into which af*gh*-CC could be released.

Materials and Methods

Experimental Fish

In this study, male parents were used from the same family population of two-year old transgenic fish. Transgenic males were heterozygotes and wild-type males were their full-sib controls, while female parents were three-year wild-type females, provided

Table 1. Primers used for PCR amplification.

Locus	Primer+	Primer-	H	PIC
MFW1	mGTCCAGACTGTCATCAGGAG	GAGGTGTACACTGAGTCACGC	0.880	0.867
MFW9	mGATCTGCAAGCATATCTGTCG	ATCTGAACCTGCAGCTCCTC	0.788	0.763
MFW11	mGCATTTGCCTTGATGGTTGTG	TCGTCTGGTTTAGAGTGCTGC	0.799	0.777
MFW15	mCTCCTGTTTTGTTTTGTGAAA	GTTCACAAGGTCATTTCCAGC	0.853	0.836
HLJ38	mCACAGAACGCATCAGTAA	TGTAAACCTTCAACCTCC	0.860	0.844
MFW19	mGAATCCTCCATCATGCAAAC	GCACAAACTCCACATTGTGCC	0.838	0.817
MFW18	mGTCCCTGGTAGTGAGTGAGT	GCGTTGACTTGTTTTATACTAG	0.708	0.665
MFW26	mCCCTGAGATAGAAACCACTG	CACCATGCTTGGATGCAAAAG	0.855	0.837
MFW29	mGTTGACCAAGAAACCAACATGC	GAAGCTTTGCTCTAATCCACG	0.726	0.728
PII- Pc	CATTTACAGTTCAGCCATGGCTAGA	AGCACCACCGACAACAGCACTAATG	–	–

NOTE:+"m" represents the M13 sequence (CACGACGTTGTAAAACGAC).

by the Zhan Dian Breeding Farm of Fisheries Institute of Sciences, Henan Province, China. Ethical approval for the work was obtained from Expert Committee of Biomedical Ethics, Institute of Hydrobiology of the Chinese Academy of Sciences. The Reference number obtained was 091110-1-303.

Artificial Reproduction and Reproduction Experiment in a Natural Ecosystem

Artificial insemination was conducted using the dry fertilization method. Fertilized eggs were hatched in culture dishes at an average water temperature of 18.5°C. The fertilization rate was calculated when fertilized eggs developed into the segmentation phase and myocomma appeared. The hatching rate was calculated when juveniles emerged from the membrane (0 days).

The reproduction experiment was conducted in an outdoor isolated -pond as a natural ecosystem. The pond had an area of approximately 1400 m² and a depth of 1.5–2.0 m. Six transgenic males (T1–T6), six wild-type males (N1–N6) and six wild-type females (W1–W6) constituted a natural reproductive population. The reproductive population was put into the experimental pond one month before the breeding season to perform natural spawning. Parents were removed after mating which was continued for eight days. Fertilized eggs and juveniles hatched and grew under the natural conditions of the pond. During the experiment, no human intervention, such as enriching oxygen, changing water or feeding were conducted to fully simulate the natural process of breeding. Forty-day-old juvenile offspring were used for paternity testing and PCR detection of transgenes.

DNA Sample Preparation and Transgene Detection

Genomic DNA of fin rays or fry were prepared using the phenol-chloroform extraction method. The PCR reaction (10 μL)

contained 1 μL genomic DNA (50 ng/μl), 1 μL 10×buffer (containing Mg²⁺), 0.5 μL Taq DNA polymerase (1 U/μL), 0.4 μL dNTPs (2.5 mmol/L each), 0.4 μL each of forward and reverse primers (10 μmol/L) and 6.3 μL ddH₂O. The thermocycling program consisted of pre-denaturation at 94°C for 3 min, 30 cycles of amplification (denaturation at 94°C for 30 s, annealing at 58°C for 30 s and extension at 72°C for 40 s), and a final extension at 72°C for 5 min. PCR products were detected on a 1% agarose gel by electrophoresis.

Microsatellite Marker Screening and Genotyping

Thirty candidate marker loci were selected from carp microsatellite MFW series [40], KOI series [41] and HLJ series [42]. The 18 parent fish described above were taken as the detection group for microsatellite marker screening. The PCR reaction system (25 μL) contained 1 μL genomic DNA (50 ng/μl), 2.5 μL 10×buffer (including Mg²⁺), 0.8 μL dNTPs (2.5 mmol/L each), 0.8 μL each of forward and reverse primers (10 μmol/L), 1 μL

Table 2. Fertilization, hatching rates and transgene segregation.

	Non-transgenic males			Transgenic males			P
Fertility (%)	87.05	94.15	97.27	94.15	89.66	87.31	>0.05
Hatchability (%)	73.38	92.68	89.55	94.15	84.48	78.17	>0.05
Segregation (%)	–	–	–	48	47	51	>0.05

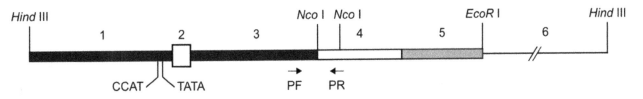

Figure 1. pCAgcGH structure diagram. 1: carp β-actin gene 5'-flanking sequence; 2: carp β-actin gene first exon; 3: carp β-actin gene first intron; 4: grass carp GH gene sequence; 5: grass carp GH gene 3'-flanking sequence; 6: plasmid pUC118. PF and PR indicate PCR primers of transgenes.

Taq DNA polymerase (1 U/μL) and 18.1 μL ddH$_2$O. The PCR reaction program consisted of pre-denaturation at 94°C for 3 min, 30 cycles of amplification (denaturation at 94°C for 30 s, annealing at 58°C for 30 s and extension at 72°C for 40 s), and a final extension at 72°C for 5 min. Genotyping of PCR products was detected using the LI-COR 4300 DNA gel electrophoresis system. The polymorphic information content (PIC) and heterozygosity (H) at each locus were analyzed using POPGEN (Version 1.32) software. Nine appropriate SSR loci were screened for paternity testing (Table 1). Marker genotyping of paternity testing samples was conducted using the same method.

Paternity Testing and Statistical Analysis

Paternity testing was conducted using Cervus 2.0 software. Allele frequencies (P), exclusion probabilities (PE), cumulative chance of exclusion (CCE), natural logarithm of likelihood ratio (LOD) values and delta values at each locus were calculated. The likelihood ratio of each parental candidate [43] was counted. Delta values of assumed parents were calculated by a simulation program to ensure assessment of parentage with high statistical confidence [44]. Significant differences between populations were detected by t-test or chi-square test.

Results

Genetics and Reproductive Biology Characteristics of Transgenic CC

Transgenic CC carried recombinant a*fgh*, which is the grass carp growth hormone gene driven by the CC β-actin promoter (pCAgcGH) (Figure 1). P0 transgenic males obtained by microinjection were hybridized with wild-type females to obtain F1 transgenic fish heterozygous groups [10]. Fast-growing F1 transgenic heterozygous males were selected to hybridize with wild-type females to screen for a fast-growing transgenic fish F2 family whose unit points were integrated. In this family group, transgenes showed Mendelian segregation (1:1) (transgenic fish:non-transgenic fish), where transgenic fish were transgenic heterozygotes. We conducted selfing of the F2 transgenic fish for the screening of the homozygous transgenic family; F2 transgenic males hybridized with wild-type females to establish F3 transgenic fish-segregated populations. Similarly, transgenic fish were transgenic heterozygotes. Transgenic fish and non-transgenic fish showed 1:1 segregation. Transgenes were passed between generations in a Mendelian way.

Experimental results of artificial fertility-hatching showed that reproductive biology characteristics of F3 transgenic fish were similar to those of the controls. Three F3 transgenic males and

three full-sib non-transgenic males were used to fertilise eggs of the same wild-type female, respectively. Average fertilization rates of transgenic fish and non-transgenic fish were 90.37±3.48% and 92.82±5.24%, respectively ($t=0.68$; $P=0.54$; $df=4$). Their average hatching rates were 85.60±8.05% and 85.20±10.36%, respectively ($t=-0.05$; $P=0.96$; $df=4$). Under laboratory conditions, no significant differences were found in the fertility between transgenic and non-transgenic fish (Table 2). Transgene detection by PCR was conducted on the offspring population of transgenic males. In offspring populations of three transgenic males, ratios of transgenic fish were 48% (48/100) ($\chi^2=0.08$; $P=0.78$), 47% (47/100) ($\chi^2=0.18$; $P=0.67$) and 51% (51/100) ($\chi^2=0.02$; $P=0.89$). The segregation of transgenes in the offspring population was consistent with a Mendelian segregation ratio of 1:1 (Table 2). These results indicate that transgenes passed steadily between generations of transgenic fish, that sperm carrying transgenes and their controls had the same level of fertilization ability, and they did not impact on the early development of embryos.

Composition of the Natural Fertility Population and their Offspring Population

From the segregated populations of two-year old F3 transgenic fish described above, six transgenic males (T1–T6) and six full-sib non-transgenic males (N1–N6), plus six wild-type females aged three years (W1–W6) were selected to constitute a fertility population including six females and 12 males. The weight distribution of transgenic males was 1.31–2.63 kg, and average weights of three small individuals (T1–T3) and three large individuals (T4–T6) were 1.49±0.16 kg and 2.55±0.09 kg, respectively. Non-transgenic males showed a smaller weight distribution (0.80–1.17 kg), and the average weight was 0.98±0.15 kg; the weight distribution of wild-type females was 3.50–4.84 kg, and the average weight was 4.22±0.53 kg (Table 3).

Six females and 12 males described above were released into an isolated pond before the reproductive season. In late March, experimental fish started spawning, and spawning was continued for eight days. After spawning, reproductive parents were removed from the pond, fertilized eggs and fry were naturally hatched and grown in the pond without human intervention. After 40 days, 1200 juveniles with an average weight of 0.28±0.03 g were randomly sampled from the offspring population for paternity testing.

Marking and typing were conducted on nine SSR loci. Complete SSR typing data were obtained from 1138 juvenile samples. The paternity testing results showed that these 1138 offspring were from 72 parental combinations (12×6), that is, six

Table 3. Number of offspring of all parental combinations.

Wild-type females (body weight, kg)	Non-transgenic males						Transgenic males					
	N1 (0.87)	N2 (0.87)	N3 (0.80)	N4 (1.05)	N5 (1.17)	N6 (1.10)	T1 (1.62)	T2 (1.31)	T3 (1.54)	T4 (2.46)	T5 (2.56)	T6 (2.63)
W1 (4.08)	9	15	16	20	44	19	12	12	9	5	16	8
W2 (3.89)	21	10	14	12	18	21	11	8	22	18	22	5
W3 (3.50)	9	20	8	22	33	17	19	6	13	10	9	9
W4 (4.14)	15	7	18	32	37	33	10	15	5	13	17	11
W5 (4.84)	35	27	18	10	34	13	8	19	23	31	8	13
W6 (4.84)	7	10	14	19	22	15	8	8	15	9	12	5

Table 4. Number of offspring form male parents.

	N1	N2	N3	N4	N5	N6	T1	T2	T3	T4	T5	T6
Transgenic offspring	–	–	–	–	–	–	9	8	12	9	13	3
Non-transgenic offspring	96	89	88	115	188	118	59	60	75	77	71	48
Sum	96	89	88	115	188	118	68	68	87	86	84	51

females successfully mated with 12 males and produced offspring populations ranging from 5 to 44 offspring (Table 3).

Juvenile Viability of af*gh*-CC Descendants

According to the results of paternity testing, six transgenic males in the testing samples produced 444 offspring (Table 3). Transgene PCR results showed 54 transgenic fish and 390 non-transgenic fish. In offspring populations of transgenic males, the transgenic fish ratio was 12.16%, strongly deviating from the theoretical value of 50% (Table 4). The juvenile viability of descendants of af*gh*-CC was significantly lower than that of their non-transgenic full-sib controls ($\chi^2 = 148.37$; $P = 0.00$). During the 40–48 day experiment, the relative viability of transgenic juveniles was 13.85%, and the average daily relative viability was 0.98.

In the offspring population of transgenic males at different sizes, ratios of transgenic fish showed no significant difference. Among 223 offspring of small transgenic males (T1–T3), there were 29 transgenic fish, and the ratio of transgenic fish was 13.00%. Among 221 offspring of large transgenic males (T4–T6), there were 25 transgenic fish, and the ratio of transgenic fish was 11.31% (Table 4). The size of transgenic males did not impact on their offspring viability ($\chi^2 = 0.30$; $P = 0.59$).

Effect of Body Size on Reproductive Success in af*gh*-CC Males

Offspring populations of transgenic males (T1–T6) at different sizes were compared. The numbers of offspring from small males (T1–T3) and large males (T4–T6) were 223 and 221, respectively

(Table 4), and ratios in offspring populations of transgenic males were 50.23% and 49.77%, respectively. A chi-square test showed that the number of offspring of T1–T3 and T4–T6 were not significantly different ($\chi^2 = 0.01$; $P = 0.95$) (Figure 2a).

Since no significant differences were found in the offspring viability, population size and composition of T1–T3 and T4–T6, the offspring population from T1–T6 was compared with that from non-transgenic males (N1–N6). The number of offspring from T1–T6 was 444, and that from N1–N6 was 694 (Table 4), while their ratios of offspring population were 39.01% and 60.09%, respectively. The number of offspring from T1–T6 was significantly lower than that from N1–N6 ($\chi^2 = 27.80$; $P = 0.00$) (Figure 2b).

Taking into account the high mortality of transgenic juveniles, the number of surviving offspring of transgenic males could not accurately reflect the number of fertilized eggs, and a direct comparison of the number of surviving offspring would underestimate the fertility of transgenic males. Therefore, twice the number of non-transgenic offspring of transgenic males was taken as the corrected value of the number of offspring from transgenic males; that is 780 (2×390) offspring. Results of the comparison after this correction showed that the ratio of offspring from T1–T6 in the population was 52.92%, and that from N1–N6 was 47.08% (Figure 2c). Based on this finding, the comparative advantage in mating ability of transgenic carp to wild-type carp was 1:1, with no significant difference between the two groups ($\chi^2 = 0.18$; $P = 0.67$).

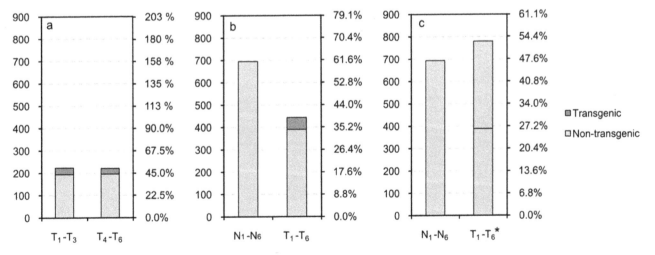

Figure 2. Number and ratio of each offspring population and the significance test between offspring populations. a: T1–T3 are the numbers of offspring of small transgenic males, T4–T6 are the numbers of offspring of large transgenic males; b: N1–N6 are the numbers of offspring of wild-type males, T1–T6 are the numbers of offspring of transgenic males; c: N1–N6 are the numbers of offspring of wild-type males, T1–T6* are the numbers of offspring of transgenic males after correction. Left vertical axis shows the number of individuals; right vertical axis shows the ratio of offspring populations.

Discussion

The "Trojan gene hypothesis" proposed by Muir et al. [30] assumes that transgenic males would be significantly larger than control males, thus holding a dominant position when competing for a mate. In contrast, the hypothesis states the viability of offspring from transgenic fish is significantly lower than that of wild-type fish. The combination of these two factors would lead to the extinction of the entire population, and transgenes show the so-called "Trojan effects" [30,31]. In this study of afgh-CC and wild CC, paternity testing of 1138 offspring showed that transgenic and wild-type males, regardless of their individual sizes, were successfully involved in breeding, including small males. This result differs substantially from the situation observed in medaka [31]. Carp follow the reproductive strategy of "group spawning" and have the characteristic of batch spawning (batch spawning and promiscuity of carp account for this similarity) [33–35]. The different reproductive strategies of carp and medaka are probably the main reason for the different results described above. These results suggest that the role of male size in competitive mating arenas and its effects on the structure of their offspring populations is different in fish employing different reproductive strategies. Large individuals do not result in an advantage in fertility. In a study of afgh-modified Atlantic salmon (Salmo salar), even the opposite phenomenon was found, with the fertility of large transgenic fish significantly decreased [36]. Based on findings from different transgenic fish species, it is hard to draw a general conclusion on the ecological risk assessment of transgenic fish. Targeted case analyses are still required for the ecological risk assessment of transgenic fish.

Male fertility is an embodiment of the competitive ability of mating and sperm. It is essentially the ability of males to fertilize eggs. It is difficult to accurately observe competitive breeding behavior under natural conditions. In this experiment, the number of surviving offspring of males was used to calculate the number of fertilized eggs, further calculating the relative fertility advantages of transgenic and wild-type males. As described in the results section of this article, because of the significant difference in the viability between transgenic and wild-type juveniles, the number of surviving transgenic juveniles could not truly reflect the number of fertilized eggs of transgenic males. A direct comparison of the number of surviving offspring between transgenic and wild-type males would underestimate the actual fertility of transgenic males. Artificial insemination-incubation experiments have shown that heterozygous transgenic males produce an equal number of

transgenic and wild-type sperms with the same fertilization capacity. Therefore, it would be more accurate to estimate the total number of fertilized eggs by using the number of wild-type surviving offspring of transgenic males. The fertility comparison indicated that the fertility ratio of transgenic males to wild-type males was 1:1, showing no significant differences. A larger body size of transgenic carp did not confer any advantage in breeding competitiveness.

Fast-growing transgenic fish generally exhibit lower juvenile viability [8,5,37,38]. The increased metabolism of fast-growing transgenic fish may be the main cause of high juvenile mortality [9]. Juvenile viability differs between different transgenic fish species. The daily viability of 3-day-old transgenic medaka reached 91.5–93.0% [37], while all juveniles of transgenic coho salmon died within a few weeks due to food shortage [38]. In the present study, the viability of 40–48-day-old transgenic carp was 13.8%, and the average daily viability was 98.0%. Because most transgenic juveniles die at an early stage of initial feeding, the daily viability in the early stage will be lower. According to the ecological risk assessment model of transgenic fish established by Muir et al. [30,39], if the relative fertility advantage was 1:1 and the daily survival rate was 98.00% or less, transgenic fish populations would survive. Transgenic carp differ from transgenic medaka in that they do not present "Trojan effects" leading to the extinction of the population.

Conclusions

Our results show that in a natural aquatic environment, fast-growing transgenic common carp have no advantages in fertility, and that their juvenile viability is low. We suggest that transgenic carp escaped or released into the natural aquatic environment would be unable to monopolize the eggs of natural CC populations, thus leading to population extinction. Moreover, transgenic CC are incapable of forming dominant populations in natural aquatic environments due to their low juvenile viability. Therefore, we predict that the ratio of transgenic CC in natural populations would remain at low levels or gradually decline, and ultimately disappear.

Author Contributions

Conceived and designed the experiments: YW WH ZZ. Performed the experiments: HL RH FD. Analyzed the data: WH LL YW. Contributed reagents/materials/analysis tools: YW WH. Wrote the paper: YW HL RH.

References

1. Zhu Z, Li G, He L, Chen S (1985) Novel gene transfer into the fertilized eggs of goldfish (Carassius auratus L. 1758). Z Angew Ichthyol 1: 31–34.
2. Zhu Z, Xu KS, Li GH, Xie YF (1986) The biological effect of human growth hormone when microinjected into the fertilized eggs of loach. Chin Sci Bull 31: 387–389.
3. Du SJ, Gong Z, Fletcher GL, Shears MA, King MJ, et al. (1992) Growth enhancement in transgenic Atlantic salmon by the use of an "all-fish" chimeric growth hormone gene constructs. Biotechnology 10: 176–181.
4. Tsai HJ, Tseng FS, Liao IC (1995) Electroporation of sperm to introduce foreign DNA into the genome of loach (Misgurnus anguillicaudatus). Can J Fish Aquat Sci 52: 776–787.
5. Rahman MA, Mak R, Ayad H, Smith A, Maclean N (1998) Expression of a novel piscine growth hormone gene results in growth enhancement in transgenic tilapia (Oreochromis niloticus). Transgenic Res 7: 357–370.
6. Rahman MA, Maclean N (1999) Growth performance of transgenic tilapia containing an exogenous piscine growth hormone gene. Aquaculture 173: 333–346.
7. Rahman MA, Ronyai A, Engidaw BZ, Jauncey K, Hwang GL, et al. (2001) Growth and nutritional trials of transgenic Nile tilapia containing an exogenous fish growth hormone gene. J Fish Biol 59: 62–78.

8. Cook JT, McNiven MA, Richardson GF, Sutterlin AM (2000a) Growth rate, body composition and feed digestibility/conversion of growth enhanced Atlantic salmon (Salmo salar). Aquaculture 188: 15–32.
9. Cook JT, McNiven MA, Sutterlin AM (2000b) Metabolic rate of presmolt growth enhanced transgenic Atlantic salmon (Salmo salar). Aquaculture 188: 33–45.
10. Wang YP, Hu W, Wu G, Sun Y, Chen S, et al. (2001) Genetic analysis of "all-fish" growth hormone gene transferred carp (Cyprinus carpio L.) and its F1 generation. Chin Sci Bull 46: 226–229.
11. Nam YK, Noh JK, Cho YS, Cho HJ, Cho KN, et al. (2001) Dramatically accelerated growth and extraordinary gigantism of transgenic mud loach (Misgurnus mizolepis). Transgenic Res 10: 353–362.
12. Kobayashi S, Morita T, Miwa M, Lu J, Endo M, et al. (2007) Transgenic Nile tilapia (Oreochromis niloticus) over-expressing growth hormone show reduced ammonia excretion. Aquaculture 270: 427–435.
13. Hallerman EM, McLean E, Fleming IA (2007) Effects of growth hormone transgenes on behavior and welfare of aquacultured fishes: a review identifying research needs. Appl Anim Behav Sci 104: 265–294.
14. Devlin RH, Donaldson EM (1992) Containment of genetically altered fish with emphasis on salmonids. In: Hew CL, Fletcher GL, editors. Singapore: World Scientific. 229–265.

15. Sundström LF, Lôhmus M, Johnsson J, Devlin RH (2004) Growth hormone transgenic salmon pay for growth potential with increased predation mortality. Proc R Soc Lond B 271: S350–S352.

16. Devlin RH, Johnsson JI, Smailus DE, Biagi CA, Jönsson E, et al. (1999) Increased ability to compete for food by growth hormone-transgenic coho salmon *Oncorhynchus kisutch* (Walbaum). Aquaculture Research 30: 479–482.

17. Fu C, Li D, Hu W, Wang Y, Zhu Z (2007) Growth and energy budget of F2 "all-fish" growth hormone gene transgenic common carp. J Fish Biol 70: 347–361.

18. Duan M, Zhang T, Hu W, Sundström LF, Wang Y, et al. (2009) Elevated ability to compete for limited food resources by 'all-fish' growth hormone transgenic common carp *Cyprinus carpio*. J Fish Biol 75: 1459–1472.

19. Duan M, Zhang T, Hu W, Li Z, Sundström F, et al. (2011) Behavioral alterations in GH transgenic common carp may explain enhanced competitive feeding ability. Aquaculture 317: 175–181.

20. Farrell AP, Bennett W, Devlin RH (1997) Growth-enhanced transgenic salmon can be inferior swimmers. Can J Zool 75: 335–337.

21. Lee CG, Devlin RH, Farrell AP (2003) Swimming performance, oxygen consumption and excess post-exercise oxygen consumption in adult transgenic and ocean-ranched coho salmon. J Fish Biol 62: 753–766.

22. Li D, Hu W, Wang Y, Zhu Z, Fu C (2009) Reduced swimming abilities in fast-growing transgenic common carp *Cyprinus carpio* associated with their morphological variations. J Fish Biol 74: 186–197.

23. Guan B, Hu W, Zhang T, Wang Y, Zhu Z (2008) Metabolism traits of 'all-fish' growth hormone transgenic common carp (*Cyprinus carpio* L.). Aquaculture 284: 217–223.

24. McKenzie DJ, Martínez R, Morales A, Acosta J, Morales R, et al. (2003) Effects of growth hormone transgenesis on metabolic rate, exercise performance and hypoxia tolerance in tilapia hybrids. J Fish Biol 63: 398–409.

25. Guan B, Hu W, Zhang TL, Duan M, Li DL, et al. (2010) Acute and chronic un-ionized ammonia toxicity to 'all-fish' growth hormone transgenic common carp (*Cyprinus carpio* L.). Chin Sci Bull 55: 4032–4036.

26. Zhang T, Hu W, Duan M (2006) The *gh*-transgenic common carp (*Cyprinus carpio*): growth performance, viability and predator avoidance. The Seventh International Congress on the Biology of Fish, St John's, Newfoundland, Canada, 18–22.

27. Duan M, Zhang T, Hu W, Guan B, Wang Y, et al. (2010) Increased mortality of "all-fish" growth hormone transgenic common carp (*Cyprinus carpio* L.) under a short-term risk of predation. J Appl Ichthyol 26: 908–912.

28. Bessey C, Devlin RH, Liley NR, Biagi CA (2004) Reproductive Performance of Growth-Enhanced Transgenic Coho Salmon. Trans Am Fish Soc 133: 1205–1220.

29. Fitzpatrick JL, Akbarashandiz H, Sakhrani D, Biagi CA, Pitcher TE, et al. (2011) Cultured growth hormone transgenic salmon are reproductively out-competed by wild-reared salmon in semi-natural mating arenas. Aquaculture 312: 185–191.

30. Muir WM, Howard RD (1999) Possible ecological risks of transgenic organism release when transgenes affect mating success: Sexual selection and the Trojan gene hypothesis. Proc Natl Acad Sci USA 96: 13853–13856.

31. Howard RD, Andrew DeWoody J, Muir WM (2004) Transgenic male mating advantage provides opportunity for Trojan gene effect in a fish. Proc Natl Acad Sci USA 10: 2934–2938.

32. Maclean N, Laight RJ (2000) Transgenic fish: An evaluation of benefits and risks. Fish and Fisheries 1: 146–172.

33. Li DS (1961) Inland fish farming. In: Xin Hailian fishery technical college (Agriculture), editor. 11.

34. Yang ZJ (1981) The artificial propagation of carps in three seasons. Freshwater fishery 1: 31–32.

35. Li MD (1989) Fish Ecology. Tianjin Technology Translated Press Company Press, 266 p.

36. Gage MJG, Stockly P, Parker GA (1995) Effects of alternative male mating strategies on characteristics of sperm production in Atlantic salmon (*Salmo salar*). Philos Trans R Soc London Ser B 350: 391–399.

37. Muir WM, Howard RD (2002) Assessment of possible ecological risks and hazards of transgenic fish with implications for other sexually reproducing organisms. Transgenic Res 11: 101–114.

38. Devlin RH, D'Andrade M, Uh M, Biagi CA (2004) Population effects of growth hormone transgenic coho salmon depend on food availability and genotype by environment interactions. Proc Natl Acad Sci USA 101: 9303–9308.

39. Howard RD, Andrew DeWoody J, Muir WM (2004) Transgenic male mating advantage provides opportunity for Trojan gene effect in a fish. Proc Natl Acad Sci USA 10: 2934–2938.

40. Crooijmans RPMA, Bierbooms VAF, Komen J, Van der Poel JJ, Groenen MAM (1997) Microsatellite markers in common carp (*Cyprinus carpio* L.). Anim Genet 28: 129–134.

41. David L, Jinggui F, Palanisamy R, Hillel J, Lavi U (2001) Polymorphism in ornamental and common carp strains (*Cyprinus carpio* L.) as revealed by AFLP analysis and a new set of microsatellite markers. Mol Gen Genomics 266: 353–362.

42. Quan YC, Sun XW, Liang LQ (2005) Genetic diversity of four breeding populations of common carps revealed by SSR. Zoological research/Dongwuxue Yanjiu 26: 595–602.

43. Bekkevold D, Hansen MM, Loeschcke V (2002) Male reproductive competition in spawning aggregations of cod (*Gadus morhua*, L.). Mol Ecol 11: 91–102.

44. Slate J, Marshall T, Pemberton J (2000) A retrospective assessment of the accuracy of the paternity inference program CERVUS. Mol Ecol 9: 801–808.

14

Integrated Assessment of Heavy Metal Contamination in Sediments from a Coastal Industrial Basin, NE China

Xiaoyu Li[1,2], Lijuan Liu[1], Yugang Wang[1], Geping Luo[1], Xi Chen[1], Xiaoliang Yang[3], Bin Gao[4], Xingyuan He[2]*

1 State Key Laboratory of Desert and Oasis Ecology, Xinjiang Institute of Ecology and Geography, Chinese Academy of Sciences, Xinjiang, China, 2 State Key Laboratory of Forest and Soil Ecology, Institute of Applied Ecology, Chinese Academy of Sciences, Liaoning, China, 3 College of Environmental Science and Forestry, State University of New York, Syracuse, New York, United States of America, 4 College of Resources Science and Technology, Beijing Normal University, Beijing, China

Abstract

The purpose of this study is to investigate the current status of metal pollution of the sediments from urban-stream, estuary and Jinzhou Bay of the coastal industrial city, NE China. Forty surface sediment samples from river, estuary and bay and one sediment core from Jinzhou bay were collected and analyzed for heavy metal concentrations of Cu, Zn, Pb, Cd, Ni and Mn. The data reveals that there was a remarkable change in the contents of heavy metals among the sampling sediments, and all the mean values of heavy metal concentration were higher than the national guideline values of marine sediment quality of China (GB 18668-2002). This is one of the most polluted of the world's impacted coastal systems. Both the correlation analyses and geostatistical analyses showed that Cu, Zn, Pb and Cd have a very similar spatial pattern and come from the industrial activities, and the concentration of Mn mainly caused by natural factors. The estuary is the most polluted area with extremely high potential ecological risk; however the contamination decreased with distance seaward of the river estuary. This study clearly highlights the urgent need to make great efforts to control the industrial emission and the exceptionally severe heavy metal pollution in the coastal area, and the immediate measures should be carried out to minimize the rate of contamination, and extent of future pollution problems.

Editor: Jack Anthony Gilbert, Argonne National Laboratory, United States of America

Funding: This study was financed by National Natural Science Foundation of China (No. 40830746 and 40971272). The funders had no role in study design, data collection and analysis, decision to publish, or preparation of the manuscript.

Competing Interests: The authors have declared that no competing interests exist.

* E-mail: hexy@iae.ac.cn

Introduction

Coastal and estuarine areas are among the most important places for human inhabitants [1]; however, with rapid urbanization and industrialization, heavy metals are continuously carried to the estuarine and coastal sediments from upstream of tributaries [2–5]. Heavy metal contamination in sediment could affect the water quality and bioaccumulation of metals in aquatic organisms, resulting in potential long-term implication on human health and ecosystem [6–7]. In most circumstances, the major part of the anthropogenic metal load in the sea and seabed sediments has a terrestrial source, from mining and industrial developments along major rivers and estuaries [8–10]. The hot spots of heavy metal concentration are often near industrial plants [11]. Heavy metal emissions have been declining in some industrialized countries over the last few decades [12,13], however, anthropogenic sources have been increasing with rapid industrialization and urbanization in developing countries [14,15].Heavy metal contaminations in sediment could affect the water quality, the bioassimilation and bioaccumulation of metals in aquatic organisms, resulting in potential long-term affects on human health and ecosystem [16–19]. Quantification of the land-derived metal fluxes to the sea is therefore a key factor to ascertain at which extent those inputs can influence the natural biogeochemical processes of the elements in the marine [20,21]. The spatial distribution of heavy metals in marine sediments is of major importance in determining the pollution history of aquatic systems [22,23], and is basic information for identifying the possible sources of contamination and to delineate the areas where its concentration exceeds the threshold values and the strategies of site remediation [24]. Therefore, understanding the mechanisms of accumulation and geochemical distribution of heavy metals in sediments is crucial for the management of coastal environment.

China's rapid growth of the economy since 1979 under the reform policies has been accompanied by considerable environmental side effects [25]. China is one of the largest coastal countries in the world. Booming coastal urban areas are increasingly dumping huge industrial and domestic waste at sea [26]. The elevated metal discharges put strong pressure on China's costal and estuarine area. The average annual input of metals by major rivers was approximately 30,000 t between 2002 and 2008 [27]. Chinese government indicates that 29,720 km^2 of offshore areas of China are heavily polluted [28]. "Hot spots" of metal contamination can be found along the coast of China [29], from the north to the south, especially in the industry-developed estuaries, such as the Liaodong Bay [30] and Yangtze River catchment [31] and Xiamen Bay [32]. In 2002, China enforced Marine Sediment Quality (GB 18668-2002) to protect marine environment (CSBTS, 2002). Therefore, Marine Sediment Qual-

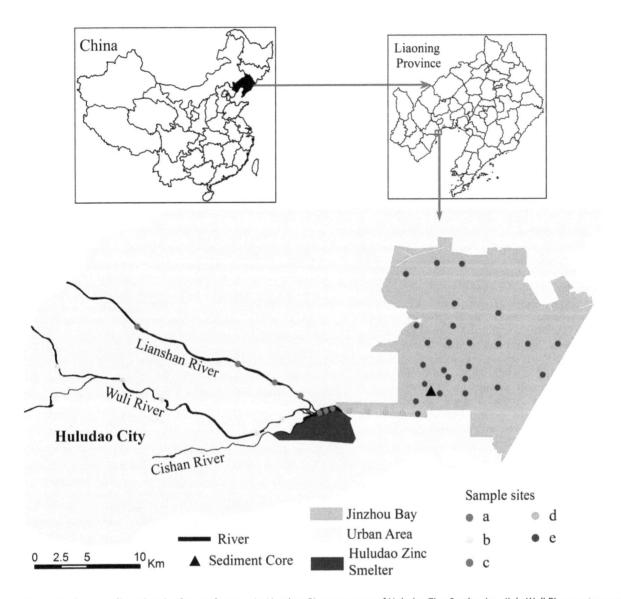

Figure 1. The sampling sites in the study area. (a, Lianshan River upstream of Huludao Zinc Smelter (n = 4); b, Wuli River upstream of Huludao Zinc Smelter (n = 5); c, Converged river Downstream of Huludao Zinc Smelter (n = 3); d, Estuary (n = 4); e, Jinzhou Bay (n = 25)).

ity (GB 18668-2002) is used as a general measure of marine sediment contamination in China.

Jinzhou Bay, surrounded by highly industrialized regions, is considered as one of the most contaminated coastal areas in China [33]. China produces the largest amount of zinc (Zn) in the world, which was 1.95 million tons in 2000 and will grow to 14.9 million tons in 2010 [26,34]. And the largest zinc smelting plant in Asia was located at the coast of Jinzhou bay. From 1951 to 1980, the amount of Zn, Cu, Pb and Cd discharged from Huludao Zinc Smelter to Jinzhou bay reached 33745, 3689, 3525 and 1433 t respectively [35]. Although several heavy metal contamination studies have conducted in Jinzhou Bay area recently, these studies were focused on coastal urban soils [36], river sediments [37] and seawater [38] separately. Few researches take the coastal stream, estuary and bay as a whole unit to assess the heavy metal contamination of coastal industrial area spatially and temporally. Thus, it is necessary to understand the process of heavy metal contamination and to evaluate the potential ecological risks of heavy metals in the coastal stream, estuary and bay integratively.

In recent decades different metal assessment indices applied to sediment environments have been developed. Caeiro et al [9] classified them in three types: contamination indices, background enrichment indices and ecological risk indices. The geo-accumulation index (I_{geo}) [39] and the potential ecological risk index (RI) [40] are the most popular methods used to evaluate the ecological risk posed by heavy metals in sediments [41–44]. RI method considers the toxic-response of a given substance and the total risk index, and can exhibit the actual pollution condition of seriously polluted sediment [45,46].

Over the last few decades the study of the sediment cores has shown to be an excellent tool for establishing the effects of anthropogenic and natural processes on depositional environments [44,47]. Sediment cores can be used to study the pollution history of aquatic ecosystem [48,49]. Within an individual sediment core, differences in pollutant concentrations at different depths reflect how heavy metal input and accumulation changes over time [50,51].

The purpose of this study is (1) to quantify and explain the spatial distribution of heavy metal contaminants in modern sediments of Jinzhou bay, NE China; (2) to investigate the natural and anthropogenic processes controlling sediment chemistry; and (3) to identify the potential ecological risks of such heavy metals.

Materials and Methods

Study Area

This study was carried out in Jinzhou Bay and its coastal city, Huludao City, in Liaoning Province, northeast of China (Fig. 1). Jinzhou Bay is one of the important bays in the northwest of Liaodong Bay at the northwestern bank of China's Bohai Sea. It is a semi-closed shallow area with an average depth of 3.5 m and an approximate area of 120 km^2. Huludao city is located at southwestern coast of Jinzhou bay. The city is an important non-ferrous smelting and chemical industry area in northeast China. More than forty different mineral resources have been discovered in the Huludao region, including gold, zinc, molybdenum, lime and manganese. The economy is dominated by some of China's most important industrial enterprises, such as Asia's biggest zinc manufacturing operation, the Huludao Zinc Smelter (HZS), the Jinxi Oil Refinery and Jinhua Chemical Engineering, and Huludao's Massive Shipyard. The Wuli River, Lianshan River and the Cishan River are three main rivers in the city, flowing into Jinzhou Bay. The water, soil and sediment in the city and Jinzhou bay were heavily polluted by industrial activities. Land reclamation from sea by landfill of soils and solid wastes further increase the level of pollution of the sedimentary environments in this area. These anthropogenic activities have created great threat to the public health and the regional biological and geochemical conditions.

No specific permits were required for the described field studies. The studying area is not privately-owned or protected in any way and the field studies did not involve endangered or protected species.

Sampling and Analysis

Twelve samples of river sediment were collected from the two major rivers (Lianshan river and Wuli river) and and four samples were collected from their estuary of Huludao City, using a stainless steel shovel. Twenty-five of surface sediments (0–5 cm) and one sediment core were collected in Jinzhou Bay using a stainless gravity corer (40 cm length and 5 cm diameter). The sediment core was sectioned at 2 cm intervals, and each fraction (subsamples) was sliced into 50 ml polyethylene centrifuge tubes with the help of PVC spatula. All the samples were collected in October 2009 in one week.

The samples were oven-dried at 45°C for 3 days, and sieved through a 2-mm plastic sieve to remove large debris, gravel-size materials, plant roots and other waste materials, and stored in closed plastic bags until analysis. Soil was digested with a mixture 5:2:3 of HNO_3–$HClO_4$–HF. The digested solutions were analyzed via an inductively coupled plasma-atomic emission spectroscopy (ICP-AES; Perkin Elmer Optima 3300 DV). All of the soil samples were analyzed for total concentrations of Cu, Zn, Pb, Ni, Mn and Cd.

Statistical Analysis

Statistical methods were applied to process the analytical data in terms of its distribution and correlation among the studied parameters. The commercial statistics software package SPSS version 17.0 for Windows was used for statistical analyses in present study. Basic statistical parameters such as mean, median, standard deviation (SD), coefficient of variation (CV), skewness and kurtosis were computed. To identify the relationship among heavy metals in sediments and their possible sources, Pearson's correlation coefficient analysis were performed.

Geostatistical Methods

Semivariogram is a basic tool of geostatistics and also the mathematical expectation of the square of regional variable $z(x_i)$ and $z(x+h_i)$ increment, namely the variance of regional variable. Its general form is:

$$\gamma(h) = \frac{1}{2N(h)} \sum_{i=1}^{N(h)} [z(x_i) - z(x_i + h)]^2$$

where $r(h)$ is semivariogram; h is step length, namely the spatial interval of sampling points used for the classification to decrease the individual number of spatial distance of various sampling point assemblages; $N(h)$ is the logarithm of sampling point when the spacing is h; $z(x_i)$ and $z(x_i+h)$ are the values when the variable Z is at the x_i and x_i+h positions respectively. The residual sums of squares (RSS), the determining coefficient (R^2) and F test were used to evaluate the accuracy of the interpolated results.

Kriging, as a geostatistical interpolation method, uses the semivariogram to quantify the spatial variability of regionalized variables, and provides parameters for spatial interpolation. The maps of spatial distribution of heavy metal concentrations were generated by Kriging interpolation with the support of the statistical module of ArcGIS-Geostatistical Analyst.

Potential Ecological Risk

To assess the effect of multiple metal pollutions in the sediments from the river, estuary and Jinzhou bay, potential ecological Risk Index (RI) was used, which was originally developed by Håkanson [40] and is widely used in ecological risk assessments of heavy metals in sediments. According to this methodology, the potential ecological risk index (RI) is defined as

$$RI = \sum Er^i \qquad (1)$$

$$Er^i = Tr^i \cdot C_f^i \qquad (2)$$

$$C_f^i = C_o^i / C_n^i \qquad (3)$$

where RI is calculated as the sum of all risk factors for heavy metals in sediments; Er^i is the monomial potential ecological risk factor; TR^i is the toxic-response factor for a given substance (e.g., Cu = Pb = Ni = 5, Zn = 1, Cd = 30); C_f^i, C_o^i and C_n^i are the contamination factor, the concentration of metals in the sediment and the background reference level, respectively. The background values of Cu, Zn, Pb and Cd are defined as the maximum values of the first category standard of national guideline values of marine sediment quality of China (GB18668-2002) and Ni is defined as the average value of Ni in residual fraction determined, they were 35 mg/kg for Cu, 150 mg/kg for Zn, 60 mg/kg for Pb, 0.5 mg/kg for Cd, and 9 mg/kg for Ni. The concentration of Mn in the sediments showed very weak relationship with the industrial activities, so it was not included in the calculation process of RI.

Still according to Hakanson [40] the following terminology is indicated to be used for the RI value:

RI <150, low ecological risk for the sediment;
150≤ RI <300, moderate ecological risk for the sediment;
300≤ Ri <600, considerable ecological risk for sediment;
RI ≥600, very high ecological risk for the sediment.

Results and Discussion

Heavy Metal in the Sediments

Descriptive statistics of heavy metal concentrations of sediments present in rivers of Huludao city, estuary and Jinzhou bay (Fig. 1) are presented in Table 1, 2, 3. As confirmed by the skewness values (Table 1, 3), the concentrations of elements (except Mn) are characterized by large variability, with positively skewed frequency distributions. This is common for heavy metals, because they usually have low concentrations in the environment, so that the presence of a point source of contamination may cause a sharp increase in local concentration, exceeding the thresholds [24].

Heavy metal in river sediments. Among the concentrations of heavy metal of river sediments (Table 1), the low values are from the sediments sampled before the river flowing by the Huludao Zinc Smelter (HZS), and the highest heavy metal concentrations come from the two sediment samples collected after the river flows by the HZS and accepted the wastewater discharged from HZS. Although the Wuli river is the least contaminated river in Huludao City; however the concentration of Cd exceeded the national guideline values of marine sediment quality of China (GB 18668-2002) by 22 times. In the upstream of HZS the sediment contamination of Lianshan river was higher than that of Wuli river. These data showed many other industrial operations at upper reaches of HZS also contributed great heavy metals to the river sediments. According to historical data, wastes from Jinxi Chemical Factory and Jinxi Petroleum Chemical Factory were discharged into Wuli River directly for nearly 40 years till 2000 and caused heavy contamination of river sediments. At the same time, wastewater from other small industrial plants and residents, and nonpoint pollution from soils with runoff or atmospheric deposition contributed additional pollution sources [52,53]. This is be confirmed by the previous studies about contamination of urban soils [36,54] and river sediments [37] in Huludao city.

The mean concentrations of Cu, Zn, Pb and Cd in the sediment of converged river downstream of HZS adjacent to the estuary exceeding the national guideline values of marine sediment quality of China (GB 18668-2002) by 42, 42, 40 and 1006 times respectively. These indicated that the Huludao Zinc Smelter (HZS) is the largest source of heavy metals in river sediment adjacent to the estuary. The annual amount of discharged wastewater from HZS is estimated to be more than 8 million tons directly to the Wuli river [35]. This situation lasted for more than 50 years till the water reuse was realized around year 2000 [28].

Heavy metal in the estuary sediments. Four sediment samples were collected along the estuary, and the distance from HZS to the sampling sites increases from 400 m with an 800 m-interval to the Jinzhou Bay (Fig. 1). All the investigated heavy metals clearly showed the same distribution trend. The maximum concentrations of heavy metals declined as the distance increased from HZS (Table 2). The mean concentration of Cu, Zn, Pb and Cd at estuary were exceeded the national guideline values of marine sediment quality of China (GB 18668-2002) by 24, 31, 11 and 332 times respectively.

Table 1. Heavy metal concentrations (mg/kg) of River sediments.

		Minimum	Maximum	Mean	Median	SD	CV%	Skewness	Kurtosis	national guideline values
	a	67.00	186.50	116.50	106.25	50.22	43.11	1.15	2.24	
Cu	b	32.15	85.50	50.61	42.40	24.55	48.51	1.45	1.73	35.00
	c	795.00	2535.00	1533.33	1270.00	899.39	58.66	1.20	–	
	a	471.61	965.31	633.85	549.24	224.08	35.35	1.83	3.51	
Zn	b	153.52	473.93	256.78	199.85	146.71	57.13	1.84	3.50	150.00
	c	1825.96	11010.02	6546.57	6803.73	4597.42	70.23	-.251	–	
	a	62.64	185.10	112.28	100.69	52.34	46.62	1.18	1.76	
Pb	b	40.15	98.47	57.60	45.89	27.47	47.69	1.90	3.66	60.00
	c	417.58	6090.90	2431.09	784.81	3174.79	130.59	1.70	–	
	a	35.69	109.99	57.41	41.98	35.23	61.36	1.94	3.79	
Ni	b	28.28	35.94	31.49	30.885	3.31	10.51	0.92	0.52	–
	c	40.77	87.93	62.58	59.06	23.77	37.98	0.65	–	
	a	520.89	1283.46	812.91	723.65	327.98	40.34	1.47	2.75	
Mn	b	520.21	903.43	710.96	710.11	205.78	28.94	0.004	−5.81	–
	c	337.93	1358.55	915.18	1049.07	523.31	57.18	−1.08	–	
	a	25.53	98.78	53.18	44.21	32.61	61.32	1.29	1.30	
Cd	b	8.04	17.75	11.12	9.35	4.49	40.38	1.81	3.29	0.50
	c	136.73	1019.10	503.51	354.71	459.62	91.28	1.30	–	

[a]Lianshan River upstream of Huludao Zinc Smelter (n = 4);
[b]Wuli River upstream of Huludao Zinc Smelter (n = 5);
[c]Converged river Downstream of Huludao Zinc Smelter (n = 3). The locations of sampling sites are on Fig. 1.

Table 2. Heavy metal concentrations (mg/kg) of Estuary sediments.

Distance from HZS to the sampling sites	Cu	Zn	Pb	Ni	Mn	Cd
400 m	1510.00	9304.24	1414.14	59.95	744.13	269.34
1200 m	805.00	3898.09	561.21	64.10	677.46	181.90
2000 m	675.00	2750.57	486.43	48.14	514.07	118.08
2800 m	505.00	3221.59	421.28	53.85	682.69	98.29

The locations of sampling sites are on Fig. 1.

Contaminant distributions in the Hudson River estuary were identified two types of trends: increasing trend down-estuary dominated by down-estuary sources such as wastewater effluent, and decreasing trend toward bay dominated by upriver sources, where they are removed and diluted downstream along with the sediment transport [55]. The result of this study showed obvious decreasing trend toward bay. This confirmed that the wastewater discharged from HZS and other industrial plants were the main sources of heavy metals in sediment.

Heavy metal in Jinzhou Bay sediments. Twenty-five of surface sediments from Jinzhou Bay were tested and the minimum, maximum and mean values are all located in Table 3. There were remarkable changes in the concentrations of heavy metals in the sediments of Jinzhou bay (Fig. 1). The heavy metal concentration levels are comparable to those with previous studies [56–58], especially for Cu, Pb and Zn, Mean concentrations of Cu, Zn, Pb and Cd were as high as 2.1, 4.6, 2.1 and 53.6 times of the national guideline values of marine sediment quality of China (GB 18668-2002). According to the research results from Institute of Marine Environmental protection, State oceanic Administration of China in 1984, the background value of Cu, Zn, Pb and Cd in Jinzhou bay was 10.0, 48.59, 9.0 and 0.29 mg/kg respectively [35]. This clearly demonstrates an anthropogenic contribution and reveals a serious pollution of sediments in Jinzhou bay. For all metals, total concentrations had a great degree of variability, shown by the large coefficients of variation (CV) from 20.49% of Mn to 92.51% of Pb. The elevated coefficients of variations reflected the inhomogeneous distribution of concentrations of discharged heavy metals. Large standard deviations were found in all heavy metals levels. The results of the K–S test (P<0.05) showed that the concentrations of measured metals were all normally distributed.

The comparison of contaminant concentrations observed in this study with those reported for other regions (Table 4) indicates that levels and ranges of variation of our data are similar to those reported from sites with high anthropogenic impact. It was found that the concentrations of Cd measured in this study were greatly higher than other studies except that of Algeciras Bay was a little close to this study. The levels of Zn and Pd in this study were only lower than that of Izmit Bay and Gulf of Naples respectively. The contents of Cu were relatively higher than other study (except Izmit Bay, Bay of Bengal and Hong Kong). All these show that Jinzhou bay was a highly polluted area in the world.

Correlation between Heavy Metals

Correlation analyses have been widely applied in environmental studies. They provided an effective way to reveal the relationships between multiple variables in order to understand the factors as well as sources of chemical components [50,65]. Heavy metals in environment usually have complicated relationships among them. The high correlations between heavy metals may reflect that the accumulation concentrations of these heavy metals came from similar pollution sources [66,67]. Results of Pearson's correlation coefficients and their significance levels (P<0.01) of correlation analysis were shown in Table 5. The concentrations of Cu, Zn, Pb, Ni and Cd showed strong positive relationship (P<0.01) with each other. This shows that Cu, Zn, Pb, Ni and Cd come from the same source. However, the concentration of Mn showed very weak correlations with the concentrations of the other metals, except Ni. This indicates that Mn have different sources than Cu, Zn, Pb and Cd. Han et al [68] also found the similar result about Mn by multivariate analysis.

Spatial Distribution of Heavy Metals in the Sediments of Jinzhou Bay

Geostatistics is increasingly used to model the spatial variability of contaminant concentrations and map them using generalized least-squares regression, known as kriging [69–71]. The probabil-

Table 3. Heavy metal concentrations (mg/kg) of Jinzhou Bay sediments (n = 25).

	Minimum	Maximum	Mean	Median	SD	CV%	Skewness	Kurtosis	national guideline values
Cu	24.45	327.50	74.11	51.50	68.54	92.48	2.706	7.964	35.00
Zn	168.06	2506.33	689.39	550.58	568.50	82.46	2.183	4.656	150.00
Pb	29.17	523.45	123.98	89.29	114.70	92.51	2.398	5.934	60.00
Ni	26.29	85.99	43.47	41.41	11.92	27.42	1.835	8.867	–
Mn	445.57	1123.03	750.64	774.00	153.82	20.49	0.311	0.137	–
Cd	7.91	105.31	26.81	20.74	22.23	82.92	2.388	6.155	0.50

The locations of sampling sites are on Fig. 1.

Table 4. Mean concentrations (mg/kg) of heavy metals found in Jinzhou bay compared to the reported average concentrations for other world impacted coastal systems.

Area	Cu	Zn	Pb	Ni	Mn	Cd	Reference
Izmit Bay, Turkey	89.4	754	94.9	52.1	–	6.3	[56]
Ribeira Bay, Brazil	24.6	109	22.9	47	466	0.207	[59]
Sepetiba Bay, Brazil	31.9	567	40	22.3	595	3.22	[59]
Mejillones Bay, Chile	–	29.7	–	20.6	93.8	21.9	[58]
Algeciras Bay, Spain	17	73	24	65	534	0.3	[60]
Taranto Gulf, Italy	47.4	102.3	57.8	53.3	893	–	[11]
Tivoli South Bay, USA	17.6	92.8	26.3	–	–	–	[61]
Bay of Bengal, India	677.7	60.39	25.66	34.03	366.66	5.24	[57]
Gulf of Mannar, India	57	73	16	24	305	0.16	[62]
Gulf of Naples, Italy	27.2	602	221	6.93	1550	0.57	[63]
Hong Kong, China	118.68	147.73	53.56	24.72	523.99	0.33	[64]
This study	74.11	689.39	123.98	43.47	750.64	26.81	

Note: "–" = no data.

Table 5. Correlations between heavy metal concentrations.

	Cu	Zn	Pb	Ni	Mn	Cd
Cu	1.000					
Zn	0.919**	1.000				
Pb	0.870**	0.824**	1.000			
Ni	0.499**	0.524**	0.472**	1.000		
Mn	0.115	0.270	0.238	0.515**	1.000	
Cd	0.906**	0.885**	0.972**	0.488**	0.285	1.000

Levels of significance:
*$P<0.05$;
**$P<0.01$.

ity map produced based on kriging interpolation and kriging standard deviation integrates information about the location of the pollutant source and transport process into the spatial mapping of contaminants [71,72]. There are a lot of studies of the performance of the spatial interpolation methods, but the results are not clear-cut [73]. Some of them found that the kriging method performed better than inverse distance weighting (IDW) [74]; while others showed that kriging was no better than alternative methods [75]. For example, Kazemi and Hosseini [76] compared the ordinary kriging (OK) and other three spatial interpolation methods for estimating heavy metals in sediments of Caspian Sea, they found that the OK realization smoothed out spatial variability and extreme measured values between the range of observed minimum and maximum values for all of the contaminants.

The spatial distribution of metal concentrations is a useful tool to assess the possible sources of enrichment and to identify hot-spot area with high metal concentrations [48,49]. Semivariogram calculation was conducted and experimental semivariogram of sediment heavy metal concentrations could be fitted with the Gaussian model for Cu, Zn, Pb, Cd, Ni and Mn. The theoretical variation function and experimental variation function exhibits a better fitting (Table 6). The values of R were significant at the 0.01 level by F test, which shows that the semivariogram models well reflect the spatial structural characteristics of sediemnt heavy metals.

The estimated maps of Cu, Zn, Pb, Cd, Ni and Mn clearly identified that the river, where the Huludao Zinc Smelter (HZS) is located at, is the most important source of heavy metals except Mn (Fig. 2). Among these metals, Cu, Zn, Pb and Cd showed a very similar spatial pattern, with contamination hotspots located at the

estuary area, and their concentrations decreased sharply with the distance farther away from the estuary, indicating that they were from the same sources. The concentration of Ni also showed a similar pattern with the concentrations of Cu, Zn, Pb and Cd, but it changed not as sharply as the latter one. However, the concentration of Mn showed a completely different pattern with the others, indicating the industrial activities are not the source of Mn in the Jinzhou bay, and it may be related to geological factor. The mineral constituents of Jinzhou bay consist mainly of hornblende, epidote and magnetite. The percentage of hornblende, which is rich in Mn element, varied from 24.78% up to 64.70% [77]. This confirms that the concentration of Mn comes from the geological sources. The feature of point sources lies in that the inputs of heavy metals occur over a finite period of time and may have been effectively retained in the sediments near the sources, rather than re-suspended and distributed uniformly throughout the region [66]. Distinct from point sources, metals from non-point sources are more uniformly distributed throughout the area [50]. The results of this study closely correspond to this association.

The Spatial pattern of heavy metal in Jinzhou bay also provided a refinement and reconfirmation of the results in the statistical analysis, in which strong associations were found among Cu, Zn, Pb, Cd and Ni and very weak relations were found between Mn and the other heavy metals except Ni.

Assessment of Potential Ecological Risk, RI

Almost all the RI values of sampling sites were higher than 600 except the two samples from Wuli river, indicating that the

Table 6. Parameters and F-test of fitted semivariogram models (Gaussian model) for heavy metals in sediment.

	Nugget (C_0)	Sill (C_0+C)	C/(C_0+C)	Range	R^2	RSS	F test
Cu	5200	111500	0.953	0.0744	0.833	1.41E+09	24.94**
Zn	170000	4450000	0.962	0.0831	0.868	1.38E+12	32.88**
Pb	7900	106900	0.926	0.0883	0.837	7.23E+08	25.67**
Ni	10	7460	0.999	0.0277	0.513	3.81E+07	5.27**
Mn	7370	38310	0.808	0.0710	0.946	2.71 E+07	87.59**
Cd	220	5550	0.960	0.0935	0.921	8.49 E+05	58.29**
RI	890000	22880000	0.961	0.1004	0.927	1.12E+13	63.49**

**Significance at $\alpha=0.01$ level of F test.

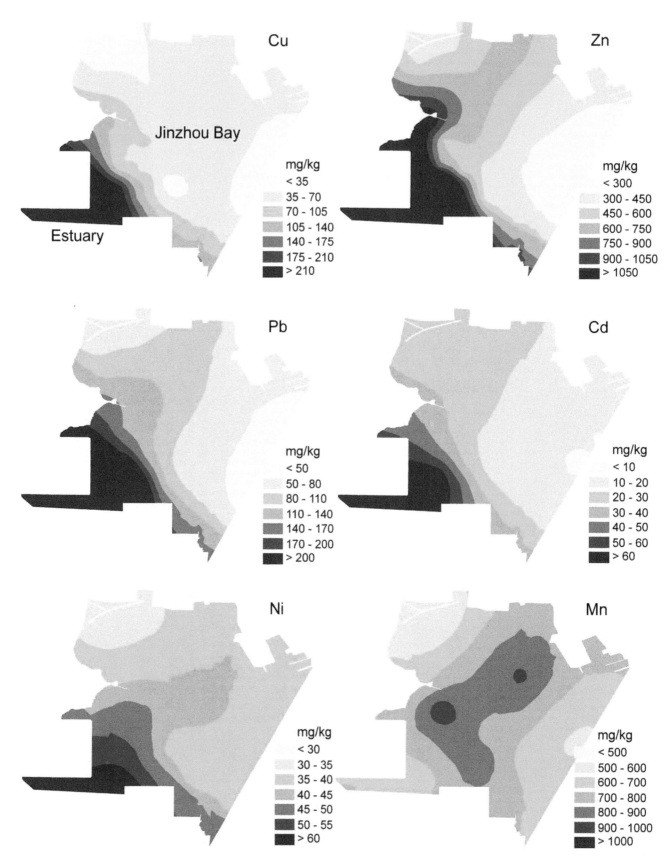

Figure 2. Estimated concentration maps for Cu, Zn, Pb, Cd, Ni and Mn (mg/kg).

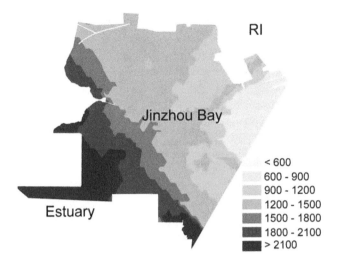

RI

Jinzhou Bay

Estuary

	< 600
	600 - 900
	900 - 1200
	1200 - 1500
	1500 - 1800
	1800 - 2100
	> 2100

Figure 3. The spatial distribution pattern of RI of sediments in Jinzhou Bay.

sediments in the rivers of Huludao city and their estuary and Jinzhou bay exhibited very high ecological risk of heavy metals (Fig. 3, Table 7). The RI of sediments in the estuary was as high as 34.6 times of the line value for very high ecological risk level, suggesting that the sediments in the estuary were extremely polluted by heavy metals because of industrial discharge. Cd showed the highest potential ecological risk in the heavy metals, which contributed more than 95% of RI in the sampled sediments.

Temporal Distribution of Heavy Metals in the Sediments of Jinzhou Bay

Sediment cores can be used to study the pollution history of aquatic ecosystem [44,48]. Vertical distribution (0–36 cm) of heavy metals in Jinzhou Bay indicate that the concentration of Cu, Zn, Pb and Cd show similar vertical patterns (Fig. 4). The values of Cu, Zn, Pb and Cd increased sharply from the surface to its highest concentration at the depth of 8 cm and then decreased rapidly at a depth of 20 cm. The values of Cu, Zn, Pb and Cd varied slightly from a depth of 21–36 cm. The concentration of Ni

Table 7. The heavy metal potential ecological risk indexes in sediments.

Sediments	Er^i						Pollution degree
	Cu	Zn	Pb	Ni	Cd	RI	
Lianshan River	16.64	4.23	9.36	31.89	3190.804	3252.96	very high
Wuli River	7.23	1.71	4.80	17.50	667.43	698.69	very high
Estuary	162.50	41.10	135.54	34.54	20414.20	20787.87	very high
Jinzhou Bay	10.59	4.60	10.33	24.15	1608.68	1658.34	very high

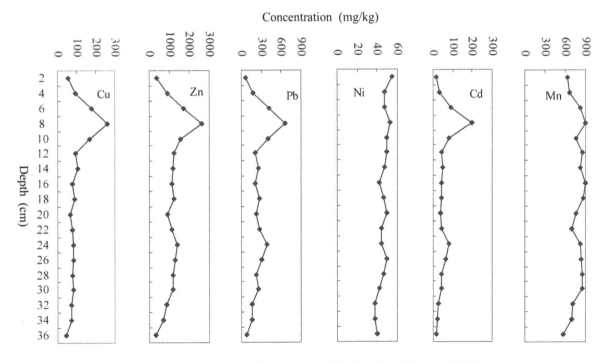

Figure 4. Vertical profiles of heavy metals for sediment core of Jinzhou bay, NE coast of China.

in the sediment core decreased gradually from the surface with small fluctuations while the Mn remained relatively consent throughout the core.

According to the sedimentation rate of about 1.0 cm/yr [78], the bottom of the sediment core (at the depth of 36 cm) was polluted about in 1973, 36 years after the set up of Huludao Zinc Smelter (HZS) in 1937. The heavy metals concentrations were already greatly higher than the national guideline values of marine sediment quality of China (GB 18668-2002), especially the concentration of Cd as high as 31 times of the latter. Related to annual Zn yields of HZS between 1973 and 2010 with the concentrations of Cu, Zn, Pb and Cd in sediment core, they showed similar temporal patterns, especially around the year 2000, the peak of Zn yield was followed the highest heavy metal concentrations at the depth of 8 cm of sediment core. This is conclusive evidence that Zn smelting operation was the dominant pollution source of aquatic environment in Jinzhou Bay. With the disposal and reuse of heavy metal wastewater from HZS around 2000, although the Zn yields increased year by year, the sediment pollution was alleviated gradually since 2000. The concentration of Ni in the sediment core decreased gradually from the surface with small fluctuations, the value varied from 54.36 mg/kg at the surface to 37.09 mg/kg at the depth of 32 cm. This also clearly indicated that the content of Ni in the sediment was come from the industrial discharges. Vertical profile of Mn shows some fluctuations in its concentration, and no obvious correlation with the depth of sediment core was observed. This indicates that the concentration of Mn is not under the control of human factors, as the result showed by the spatial pattern of Mn.

Conclusion

This study investigated the concentrations of heavy metals in the sediments from urban-stream, estuary and Jinzhou bay of the coastal industrial city, NE China. The results showed the impact of anthropogenic agents on abundances of heavy metals in sediments. The sediments are found to be extremely contaminated due to many years of random dumping of hazardous waste and free discharge of effluents by industries like Huludao Zinc Smelter (the largest zinc smelting plant in Asia), the Jinxi oil refinery and Jinhua chemical engineering, Huludao's massive shipyard and several arms factories. The potential ecological risk of sediments in lower river reaches, estuary and Jinzhou bay is at very high level, and Cd contributed more than 95% of RI in the sampled sediments. The estuary is the most polluted area, and its RI value was as high as 34.6 times of the line value for very high ecological risk level. The closer the distance to the estuary is, the higher RI values of sediments in Jinzhou bay are. The results of this research updated the information for effective environmental management in the industrial region. This study clearly highlights the urgent need to make great efforts to control the industrial discharges in the coastal area, and the immediate measures should be carried out to minimize the contaminations, and to prevent future pollution problems.

Acknowledgments

Many thanks to the three anonymous reviews and the Academic Editor whose pertinent comments have greatly improved the quality of this paper and to Dr. V. Achal and Dr. X. Pan for lingual edit.

Author Contributions

Conceived and designed the experiments: XL XH. Performed the experiments: XL LL BG. Analyzed the data: XL YW GL. Contributed reagents/materials/analysis tools: YW XC. Wrote the paper: XL XC XY.

References

1. McKinley AC, Dafforn KA, Taylor MD, Johnston EL (2011) High Levels of Sediment Contamination Have Little Influence on Estuarine Beach Fish Communities. PLoS ONE 6(10): e26353. doi: 10.1371/journal.pone.0026353.

2. Morton B, Blackmore G (2001) South China Sea. Marine Pollution Bulletin 42: 1236–1263.

3. Jha SK, Chavan SB, Pandit GG, Sadasivan S (2003) Geochronology of Pb and Hg pollution in a coastal marine environment using global fallout [137]Cs. Journal of Environmental Radioactivity 69: 145–157.

4. Muniz P, Danula E, Yannicelli B, Garcia-Alonso J, Medina G, et al. (2004) Assessment of contamination by heavy metals and petroleum hydrocarbons in sediments of Montevideo Harbour (Uruguay). Environment International 29: 1019–1028.

5. Xia P, Meng X, Yin P, Cao Z, Wang X (2011) Eighty-year sedimentary record of heavy metal inputs in the intertidal sediments from the Nanliu River estuary, Beibu Gulf of South China Sea. Environmental pollution 159: 92–99.

6. Fernandes C, Fontainhas-Fernandes A, Peixoto F, Salgado MA (2007) Bioaccumulation of heavy metals in Liza saliens from the Esomriz-Paramos coastal lagoon, Portugal. Ecotoxicology and Environmental Safety 66: 426–431.

7. Abdel-Baki AS, Dkhil MA, Al-Quraishy S (2011) Bioaccumulation of some heavy metals in tilapia fish relevant to their concentration in water and sediment of Wadi Hanifah, Saudi Arabia. African Journal of Biotechnology 10: 2541–2547.

8. Ridgway J, Breward N, Langston WJ, Lister R, Rees JG, et al. (2003) Distinguishing between natural and anthropogenic sources of metals entering the Irish Sea. Applied Geochemistry 18: 283–309.

9. Caeiro S, Costa MH, Fernandes F, Silveira N, Coimbra A, et al. (2005) Assessing heavy metal contamination in Sado Estuary sediment: An index analysis approach. Ecological Indicators 5: 151–169.

10. Sundaray SK, Nayak BB, Lin S, Bhatta D (2011) Geochemical speciation and risk assessment of heavy metals in the river estuarine sediments – a case study: Mahanadi basin, India. Journal of Hazardous Materials 186: 1837–1846.

11. Buccolieri A, Buccolieri G, Cardellicchio N, Dell'Atti A, Leo AD, et al. (2006) Heavy metals in marine sediments of Taranto Gulf (Ionian Sea, Southern Italy). Marine Chemistry 99: 227–235.

12. Voet E, Guinée JB, Udo de Haes H (2000) Heavy Metals: A Problem Solved? Methods and Models to Evaluate Policy Strategies for Heavy Metals. Kluwer, Dordrecht, the Netherlands.

13. Hjortenkrans D, Bergback B, Haggerud A (2006) New metal emission patterns in road traffic environments. Environmental Monitoring and Assessment 117: 85–98.

14. Govil PK, Sorlie JE, Murthy NN, Sujatha D, Reddy GLN, et al. (2008) Soil contamination of heavy metals in the Katedan Industrial Development Area, Hyderabad, India. Environmental Monitoring and Assessment 140: 313–323.

15. Wu SH, Zhou SL, Li XG (2011) Determining the anthropogenic contribution of heavy metal accumulations around a typical industrial town: Xushe, China. Journal of Geochemical Exploration 110: 92–97.

16. Snodgrass JW, Casey RE, Joseph D, Simon JA (2008). Microcosm investigations of stormwater pond sediment toxicity to embryonic and larval amphibians: variation in sensitivity among species. Environmental Pollution 154: 291–297.

17. Besser J, Brumbaugh W, Allert A, Poulton B, Schmitt C, et al (2009) Ecological impacts of lead mining on Ozark streams: toxicity of sediment and pore water. Ecotoxicology and Environmental Safety 72: 516–526.

18. Ip CCM, Li XD, Zhang G, Wai OWH, Li YS (2007) Trace metal distribution in sediments of the Pearl River. Estuary and the surrounding coastal area, South China. Environmental Pollution 147: 311–323.

19. Suthar S, Arvind KN, Chabukdhara M, Gupta SK (2009) Assessment of metals in water and sediments of Hindon River, India: Impact of industrial and urban discharges. Journal of Hazardous Materials 178: 1088–1095.

20. IGBP (1995) Global Change. In: Pernetta JC, Milliman JD, editors. Land-ocean interactions in the coastal zone: implementation plan. Report No. 33. Stockholm: ICSU.

21. Cobelo-Garcia A, Prego R, Labandeira A (2004) Land inputs of trace metals, major elements, particulate organic carbon and suspended solids to an industrial coastal bay of the NE Atlantic. Water Research 38: 1753–1764.

22. Birch GF, Taylor SE, Matthai C (2001) Small-scale spatial and temporal variance in the concentration of heavy metals in aquatic sediments: a review and some new concepts. Environmental Pollution 113: 357–372.

23. Rubio B, Pye K, Rae JE, Rey D (2001) Sedimentological characteristics, heavy metal distribution and magnetic properties in subtidal sediments, Ria de Pontevedra, NW Spain. Sedimentology 48: 1277–1296.

24. Sollitto D, Romic M, Castrignanò A, Romic D, Bakic H (2010) Assessing heavy metal contamination in soils of the Zagreb region (Northwest Croatia) using multivariate geostatistics. CATENA 80: 182–194.

25. Liu JG (2010) China's Road to Sustainability. Scinece 328: 50.

26. Pan K, Wang WX (2012) Trace metal contamination in estuarine and coastal environments in China. Science of the Total Environment 421–422: 3–16.

27. NBSC (2010) National Bureau of Statistics of China. China Statistical Yearbook (2001–2009). Beijing.

28. NBO National Bureau of Oceanography of China (2010) Bulletin of Marine Environmental Quality, 2008–2009.

29. Yang ZF, Wang Y, Shen ZY, Niu JF, Tang ZW (2009) Distribution and speciation of heavy metals in sediments from the mainstream, tributaries, and lakes of the Yangtze River catchment of Wuhan, China. Journal of Hazardous Materials 166: 1186–1194.

30. Fang TH, Li JY, Feng HM, Chen HY (2009) Distribution and contamination of trace metals in surface sediments of the East China Sea. Marine Environmental Research 68: 178–187.

31. Müller B, Berg M, Yao ZP, Zhang XF, Wang D, et al. (2008) How polluted is the Yangtze river? Water quality downstream from the Three Gorges Dam. Science of the Total Environment 402: 232–247.

32. Chen C, Lu Y, Hong J, Ye M, Wang Y, Lu H (2010) Metal and metalloid contaminant availability in Yundang Lagoon sediments, Xiamen Bay, China, after 20 years continuous rehabilitation. Journal of Hazardous Materials 175: 1048–1055.

33. Zhang YF, Wang LJ, Huo CL, Guan DM (2008) Assessment on heavy metals pollution in surface sediments in Jinzhou Bay. Marine Environmental Science 2: 178–181.

34. Research and Markets website. Business monitor international. China Metals report. Available: http://www.researchandmarkets.com/research/703645/china_metals_repor. Accessed 2012 May 30.

35. Institute of Marine Environmental protection, State oceanic Administration. (1984) Studies on the contamination and protection of Jinzhou Bay.

36. Lu CA, Zhang JF, Jiang HM, Yang JC, Zhang JT, et al. (2010) Assessment of soil contamination with Cd, Pb and Zn and source identification in the area around the Huludao Zinc Plan. Journal of Hazardous Materials 182: 743–748.

37. Zheng N, Wang QC, Liang ZZ, Zheng DM (2008) Characterization of heavy metal concentrations in the sediments of three freshwater rivers in Huludao City, Northeast China. Environmental Pollution 154: 135–142.

38. Wang J, Liu RH, Yu P, Tang AK, Xu LQ, et al. (2012) Study on the Pollution Characteristics of Heavy Metals in Seawater of Jinzhou Bay. Procedia Environmental Sciences 13: 507–1516.

39. Porstner U (1989) Lecture Notes in Earth Sciences (Contaminated Sediments). Springer Verlag, Berlin, 107–109.

40. Håkanson L (1980) An ecological risk index for aquatic pollution control: A sedimentological approach. Water Research 14: 975–1001.

41. Selvaraj K, Mohan Ram V, Piotr S (2004) Evaluation of Metal Contamination in Coastal Sediments of the Bay of Bengal, India: Geochemical and Statistical Approaches. Marine Pollution Bulletin 49: 174–185.

42. Verca P, Dolence T (2005) Geochemical Estimation of Copper Contamination in the Healing Mud from Makirina Bay, Central Adriatic. Environment International 31: 53–61.

43. Yi YJ, Yang ZF, Zhang SH (2011) Ecological risk assessment of heavy metals in sediment and human health risk assessment of heavy metals in fishes in the middle and lower reaches of the Yangtze River basin. Environmental Pollution 159: 2575–2585.

44. Harikumar PS, Nasir UP (2010) Ecotoxicological impact assessment of heavy metals in core sediments of a tropical estuary. Ecotoxicology and Environmental Safety 73: 1742–1747.

45. Huang YL, Zhu WB, Le MH, Lu XX (2011) Temporal and spatial variations of heavy metals in urban riverine sediment: An example of Shenzhen River, Pearl River Delta, China. Quaternary International doi: 10.1016/j.quaint. 2011.05.026.

46. Uluturhan E, Kontas A, Can E (2011) Sediment concentrations of heavy metals in the Homa Lagoon (Eastern Aegean Sea): Assessment of contamination and ecological risks. Marine Pollution Bulletin 62: 1989–1997.

47. Rosales-Hoz L, Cundy AB, Bahena-Manjarrez JL (2003) Heavy metals in sediment cores from a tropical estuary affected by anthropogenic discharges: Coatzacoalcos estuary. Estuarine, Coastal and Shelf Science 58: 117–126.

48. Karbassi AR, Nabi-Bidhendi GHR, Bayati I (2005) Environmental geochemistry of heavy metals in a sediment core off Bushehr, Persian Gulf. Iranian Journal of Environmental Health Science & Engineering 2: 255–260.

49. Viguri JR, Irabien MJ, Yusta I, Soto J, Gomez J, et al. (2007) Physico-chemical and toxicological characterization of the historic estuarine sediments. A multidisciplinary approach. Environment International 33: 436–444.

50. Shine JP, Ika RV, Ford TE (1995) Multivariate statistical examination of spatial and temporal patterns of heavy-metal contamination in New-Bedford Harbor marine-sediments. Environmental Science and Technology 29: 1781–1788.

51. White HK, Xu L, Lima ANL, Egliton TI, Reddy CM (2005) Abundance, composition and vertical transport of PAHs in marsh sediments. Environmental Science and Technology 39: 8273–8280.

52. Berthelsen BO, Steinnes E, Solberg W (1995) Heavy metal concentrations in plants in relation to atmospheric heavy metal deposition. Journal of Environmental Quality 24: 1018–1026.

53. Gray CW, McLaren RG, Roberts AHC (2003) Atmospheric accessions of heavy metals to some New Zealand pastoral soils. Science of the Total Environment 305: 105–115.

54. Li LL, Yi YL, Wang YS, Zhang DG (2006) Spatial distribution of soil heavy metals and pollution evaluation in Huludao City. Chinese Journal of Soil Science 37: 495–499.

55. Feng H, Cochran JK, Lwiza H, Brownawell BJ, Hirschberg DJ (1998) Distribution of heavy metal and PCB contaminants in the sediments of an urban estuary: The Hudson River. Marine Environmental Research. 45: 69–88.

56. Pekey H (2006) Heavy metal pollution assessment in sediments of the Izmit Bay, Turkey. Environmental Monitoring and Assessment 123: 219–231.

57. Raju K, Vijayaraghavan K, Seshachalam S, Muthumanickam J (2011) Impact of anthropogenic input on physicochemical parameters and trace metals in marine surface sediments of Bay of Bengal off Chennai, India. Environmental Monitoring and Assessment 177: 95–114.

58. Valdés J, Vargas G, Sifeddine A, Ortlieb L, Guiņez M (2005). Distribution and enrichment evaluation of heavy metals in Mejillones Bay Northern Chile: Geochemical and statistical approach. Marine Pollution Bulletin 50: 1558–1568.

59. Gomes F, Godoy J, Godoy M, Carvalho Z, Lopes R, et al. (2009) Metal concentrations, fluxes, inventories and chronologies in sediments from Sepetiba and Ribeira Bays: A comparative study. Marine Pollution Bulletin 59: 123–133.

60. Alba MD, Galindo-Riaño MD, Casanueva-Marenco MJ, García-Vargas M, Kosore CM (2011) Assessment of the metal pollution, potential toxicity and speciation of sediment from Algeciras Bay (South of Spain) using chemometric tools. Journal of Hazardous Materials 190: 177–187.

61. Benoit G, Wang EX, Nieder WC, Levandowsky M, Breslin VT (1999) Sources and history of heavy metal contamination and sediment deposition in Tivoli South Bay, Hudson River, New York. Estuaries 22: 167–178.

62. Jonathan M P, Stephen-Pichaimani V, Srinivasalu S, RajeshwaraRao N, Mohan SP (2007) Enrichment of trace metals in surface sediments from the northern part of Point Calimere, SE coast of India. Environmental Geology 55: 1811–1819.

63. Romano E, Ausili A, Zharova N, Magno MC, Pavoni B, et al. (2004) Marine sediment contamination of an industrial site at Port of Bagnoli, Gulf of Naples, Southern Italy. Marine Pollution Bulletin 49: 487–495.

64. Zhou F, Guo HC, Hao ZJ (2007) Spatial distribution of heavy metals in Hong Kong's marine sediments and their human impacts: a GIS-based chemometric approach. Marine Pollution Bulletin 54: 1372–84.

65. Al-Khashman OA, Shawabkeh RA (2006) Metals distribution in soils around the cement factory in southern Jordan. Environmental Pollution 140: 387–394.

66. Facchinelli A, Sacchi E, Mallen L (2001) Multivariate statistical and GIS-based approach to identify heavy metal sources in soils. Environmental Pollution 114: 313–324.

67. Manta DS, Angelone M, Bellanca A, Neri R, Sprovieri M (2002) Heavy metals in urban soils: a case study from the city of Palermo (Sicily), Italy. The Science of the Total Environment 300: 229–243.

68. Han Y, Du P, Cao J, Posmentier ES (2006) Multivariate analysis of heavy metal contamination in urban dusts of Xi'an, Central China. Science of the Total Environment 355: 176–186.

69. Carlon C, Critto A, Marcomini A, Nathanail P (2001) Risk based characterisation of contaminated industrial site using multivariate and geostatistical tools. Environmental Pollution 111: 417–427.

70. Romic M, Romic D (2003) Heavy metals distribution in agricultural topsoils in urban area. Environmental Pollution 43: 795–805.

71. McGrath D, Zhang CS, Carton OT (2004) Geostatistical analyses and hazard assessment on soil lead in Silvermines area, Ireland. Environmental Pollution 127: 239–248.

72. Saito H, Goovaerts P (2001) Accounting for source location and transport direction into geostatistical prediction of contaminants. Environmental Science & Technology 35: 4823–4829.

73. Xie YF, Chen TB, Lei M, Yang J, Guo QJ, et al. (2011) Spatial distribution of soil heavy metal pollution estimated by different interpolation methods: Accuracy and uncertainty analysis. Chemosphere 82: 468–476.

74. Yasrebi J, Saffari M, Fathi H, Karimian N, Moazallahi M, et al. (2009) Evaluation and comparison of ordinary kriging and inverse distance weighting methods for prediction of spatial variability of some soil chemical parameters. Research Journal of Biological Sciences 4: 93–102.

75. Gotway CA, Ferguson RB, Hergert GW, Peterson TA (1996) Comparison of kriging and inverse-distance methods for mapping soil parameters. Soil Science Society of America Journal 60: 1237–1247.

76. Kazemi SM, Hosseini SM (2011) Comparison of spatial interpolation methods for estimating heavy metals in sediments of Caspian Sea. Expert Systems with Applications 38: 1632–1649.

77. Compiling Council of Chinese Embayment (1997) Chinese Embayment (Part 2). Beijing: China Ocean Press.

78. Ma JR, Shao MH (1994) Variation in heavy metal pollution of offshore sedimentary cores in Jinzhou Bay. China Environmental Science 14: 22–29.

15

Efficiency of Household Reactive Case Detection for Malaria in Rural Southern Zambia: Simulations Based on Cross-Sectional Surveys from Two Epidemiological Settings

Kelly M. Searle[1], **Timothy Shields**[2], **Harry Hamapumbu**[3], **Tamaki Kobayashi**[1], **Sungano Mharakurwa**[2,3], **Philip E. Thuma**[3], **David L. Smith**[1], **Gregory Glass**[2], **William J. Moss**[1,2]*

1 Department of Epidemiology, Bloomberg School of Public Health, Johns Hopkins University, Baltimore, Maryland, United States of America, 2 W. Harry Feinstone Department of Molecular Microbiology and Immunology, Bloomberg School of Public Health, Johns Hopkins University, Baltimore, Maryland, United States of America, 3 Macha Research Trust, Choma, Zambia

Abstract

Background: Case detection and treatment are critical to malaria control and elimination as infected individuals who do not seek medical care can serve as persistent reservoirs for transmission.

Methods: Household malaria surveys were conducted in two study areas within Southern Province, Zambia in 2007 and 2008. Cross-sectional surveys were conducted approximately five times throughout the year in each of the two study areas. During study visits, adults and caretakers of children were administered a questionnaire and a blood sample was obtained for a rapid diagnostic test (RDT) for malaria. These data were used to estimate the proportions of individuals with malaria potentially identified through passive case detection at health care facilities and those potentially identified through reactive case finding. Simulations were performed to extrapolate data from sampled to non-sampled households. Radii of increasing size surrounding households with an index case were examined to determine the proportion of households with an infected individual that would be identified through reactive case detection.

Results: In the 2007 high transmission setting, with a parasite prevalence of 23%, screening neighboring households within 500 meters of an index case could have identified 89% of all households with an RDT positive resident and 90% of all RDT positive individuals. In the 2008 low transmission setting, with a parasite prevalence of 8%, screening neighboring households within 500 meters of a household with an index case could have identified 77% of all households with an RDT positive resident and 76% of all RDT positive individuals.

Conclusions: Testing and treating individuals residing within a defined radius from an index case has the potential to be an effective strategy to identify and treat a large proportion of infected individuals who do not seek medical care, although the efficiency of this strategy is likely to decrease with declining parasite prevalence.

Editor: Joshua Yukich, Tulane University School of Public Health and Tropical Medicine, United States of America

Funding: This work was supported by the Johns Hopkins Malaria Research Institute, the Bloomberg Family Foundation, and the Division of Microbiology and Infectious Diseases, National Institute of Allergy and Infectious Diseases, National Institutes of Health. The funders had no role in study design, data collection and analysis, decision to publish, or preparation of the manuscript.

Competing Interests: The authors have declared that no competing interests exist.

* E-mail: wmoss@jhsph.edu

Introduction

In the past decade, international support and funding for malaria control increased dramatically and targets were set to reduce the burden of malaria by 75% by 2015 and eliminate malaria in 8–10 countries by 2015 [1]. This renewed commitment to malaria elimination has been made possible with increased coverage of four key interventions: long-lasting insecticide-treated nets (ITNs), indoor residual spraying (IRS), case identification with rapid diagnostic tests (RDT) and treatment with artemisinin-combination therapy (ACT), and intermittent preventive treatment for pregnant women and infants. Programs that achieved high coverage with these interventions showed dramatic decreases in the number of malaria cases, hospital admissions and deaths [1–4] and 11 African countries demonstrated large (>50%) and sustained decreases in the burden of malaria [1].

Case detection and treatment are critical to malaria elimination as infectious individuals serve as reservoirs for transmission [5]. Several case detection strategies have been developed and implemented. Passive case detection, involving identification of symptomatic patients seeking care at health facilities based on RDT or microscopy, requires the least resources. This strategy, however, does not identify asymptomatic (those with no symptoms), minimally symptomatic (those with mild symptoms or the perception that symptoms do not require medical treatment), or symptomatic, infected individuals who do not seek medical care, as these individuals do not present to health care facilities. The proportion of all infected persons who are asymptomatic, minimally symptomatic or do not seek medical care can be substantial and as high as 96% [6–8], suggesting that a majority of infectious cases could be missed with passive case detection.

Reactive case detection [9] extends this strategy based on the assumption that malaria cases are spatially clustered and that cases identified at health centers (index cases) represent foci of infection within households and surrounding neighborhoods. With reactive case detection, residents of households of index cases, and possibly of neighboring households, are screened using RDT and offered treatment if infected. In a study of reactive case detection in rural southern Zambia, the prevalence of malaria was found to be significantly higher among residents of households of index cases than among residents of randomly selected households in the study area [10]. Importantly, both passive and reactive case detection strategies based on standard diagnostic tests (RDT and microscopy) fail to identify individuals with low-level parasitemia below the limits of detection of these tests.

Little data exist, however, on the appropriate radius from the index household that should be screened with reactive case detection using RDT, and the efficiency and cost-effectiveness of this radius likely varies in different epidemiological settings. Using serial cross-sectional household surveys and model simulations in two settings with different levels of malaria transmission in southern Zambia, we sought to quantify the efficiency of screening individuals within households and the neighbors of index cases who present for treatment at health care facilities, and to estimate the radii necessary to achieve different levels of treatment coverage.

Methods

Study Site

The study was conducted in two epidemiological settings within the catchment area of Macha Hospital in Choma District, Southern Province, Zambia between April 2007 and December 2008. Households sampled in 2008 were selected from a different geographic area than those sampled in 2007 (Figure 1). Macha Hospital is approximately 70-kilometers from the town of Choma and lies on a plateau 1,100-meters above sea level. The single rainy season lasts from December through April, followed by a cool season from April until August, and a hot dry season through November. The primary malaria vector in this region is *Anopheles arabiensis*, and transmission peaks during the rainy season (December-April) [11]. The catchment area is populated by villagers living in small, scattered homesteads. Southern Province, Zambia was reported to have hyperendemic *P. falciparum* transmission [12]. However, the prevalence of malaria has declined over the past decade [13]. ACTs were introduced as first-line anti-malarial therapy in Zambia in 2002 [14] and into the study area in 2004, and insecticide treated bed nets (ITNs) were widely distributed in the study area in 2007 [14].

Study Population

The development of the sampling frame and enumeration of households were reported elsewhere [14]. Briefly, satellite images were used to construct a sampling frame from which households were selected by simple random sampling for enrollment into prospective longitudinal and cross-sectional surveys of malaria parasitaemia. Households enrolled in the longitudinal cohort were repeatedly surveyed every two months, whereas households enrolled in the cross-sectional cohort were surveyed once. The household survey was conducted from April through December in 2007 and from February through December in 2008 [14]. This analysis was restricted to households enrolled in the cross-sectional surveys and the first study visit of households enrolled in the longitudinal surveys.

The study was approved by the University of Zambia Research Ethics Committee and the Institutional Review Board at the Johns Hopkins Bloomberg School of Public Health. Informed consent was obtained from all participants. During each study visit, a questionnaire was administered to consenting participants over 18 years of age and to the guardians of participants younger than 18 years of age. Data collected included demographic information, current signs and symptoms of malaria, history of recent malaria and antimalarial treatment, reported health seeking behavior, knowledge of malaria transmission and prevention, and the use of ITNs. Participant's temperature was measured using a Braun Thermoscan® ear thermometer. A blood sample was collected by finger prick for malaria rapid diagnostic testing (RDT). The RDT (ICT Diagnostics, Cape Town, South Africa) detected *P. falciparum* histidine-rich protein 2 and was shown to detect 82% of test samples with wild-type *P. falciparum* at a concentration of 200 parasites/μL and 98% of test samples with a concentration of 2000 parasites/μL, with false positives in 0.6% of negative samples [15]. Participants who were RDT positive were offered treatment with artemether-lumefantrine (Coartem®).

Spatial Risk Map

A spatial risk map was previously developed using ecological and survey data [14]. Logistic regression was used to identify environmental factors associated with the odds of a household having an RDT positive resident. Each household in the study area was assigned a malaria risk according to its location on the spatial risk map ranging from. 065 to. 797, referred to as the ecological risk.

Sample Survey Data

Data from 2007 and 2008 were analyzed to compare characteristics under the different transmission settings represented by each year. For each year, differences between RDT positive and RDT negative individuals were compared using Fisher's exact test for dichotomous variables and two-sample t test for continuous variables. The Wilcoxon-ranksum test was used to compare mean ages between households.

Generation of Passively Detected Index Cases

Individuals were classified as likely to be passively detected (index cases) if they were RDT positive, had malaria specific symptoms and displayed care-seeking behavior. Care seeking behavior was determined if the individual reported visiting a health post or clinic for their most recent febrile illness. Malaria specific symptoms consisted of having a fever with either a headache or chills in the prior two weeks. An alternative algorithm was developed for individuals receiving antimalarial medication at the time of the survey. Individuals currently taking antimalarials from a health care facility, and who thus displayed care-seeking behavior, were classified as likely to be passively detected index cases.

RDT positive individuals likely to be detected and missed through passive case detection based on the algorithm were compared based on care seeking behavior, symptoms and ecological risk using Fisher's exact test and two-sample t test.

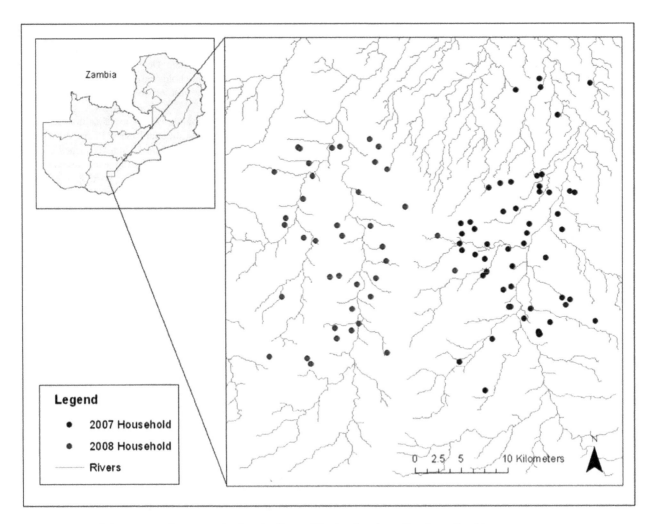

Figure 1. Map of the 2007 and 2008 study sites in Choma District, Southern Province, Zambia.

Classification of Households and Individuals Detected through Reactive case Detection

All households with one or more RDT positive resident were classified as positive households. Positive households were further classified as "identified" or "missed" based on whether or not at least one RDT positive resident was likely to be passively detected (i.e. was classified as seeking care for malaria-like illness). Individuals likely to be detected through reactive case detection were those who were RDT positive but were asymptomatic or minimally symptomatic, did not display care seeking behavior, or both, but resided in a household likely to be identified.

Positive households were compared on the basis of being identified or missed. The variables used for analysis included: mean age of household residents, number of residents in the household, number of RDT positive residents in the household, number of symptomatic and asymptomatic RDT positive residents in the household, and the household ecological risk.

Spatial Analysis of Sample Survey Data

Positive households were mapped using ArcGIS version 10.0 (Environmental Systems Research Institute [ESRI], Redlands, California). The identified and missed households were added as data layers geo-referenced to Universal Transverse Mercator (UTM), Southern Hemisphere, Zone 35, WGS1984. Identified and missed households were uniquely coded and distances between identified and missed households were determined. These distances were used to determine radii around identified households that would potentially need to be traversed to identify missed households.

Population Level Simulation

Simulations were performed using predictive models to extrapolate from sampled to non-sampled households based on household level data from those surveyed in 2007 and 2008. Individual and household survey data were selected from the dataset to create a household level dataset with covariates of interest for the development of predictive models. From this aggregated household level dataset, predictive models were determined for each covariate of interest to locate houses that would potentially be identified, according to the passive case detection and household identification algorithms, in order to fit the optimal chained equations to be used in the simulation. Dichotomous covariates of interest were predicted using logistic regression and continuous covariates of interest were predicted using linear regression. Each of the following variables was predicted at the household level: RDT status, antimalarial treatment status, number of RDT positive residents, at least one symptomatic resident, at least one care seeking resident, at least one symptomatic and care seeking RDT positive resident, and residents treated for malaria who sought care. For the predictive models, geographic coordinates, ecological risk, mean age of household residents, and number of household residents were used as initial predictive covariates.

Logistic regression models were evaluated using the Hosmer & Lemeshow goodness-of-fit test and the area under the receiver-operating curve (AUC). Linear regression models were evaluated using the R^2. The predictive models were initially built using the 2007 household level data. The AUC measurements for all dichotomous models were greater than 0.70 and p-values for the Hosmer & Lemeshow goodness-of-fit test were greater than 0.05 (Table S1 in File S1). The R^2 values for continuous models were greater than 0.50. The models were validated using the 2008 household data to ensure that the same model was fit under both transmission settings. Using the 2008 household data, the AUC measurements for all dichotomous models were greater than 0.65 and p-values for the Hosmer & Lemeshow goodness-of-fit test were greater than 0.05 (Table S2 in File S1). The R^2 values for continuous models were greater than 0.50. Households with ecological risk of less than 0.196 were not included in the simulation and were assumed to be negative (i.e. no RDT positive residents).

The simulation was performed using a multiple imputation by chained equations (MICE) method in STATA version 12.0 (StataCorps, College Station, TX), also referred to as fully conditional specification or sequential regression multivariate imputation [16–18]. In this analysis, all sampled household values were observed while non-sampled households had full data only for geographic coordinates and ecological risk of malaria. All other values for non-sampled households were missing. With MICE, initially all missing values are temporarily filled by a simple random sample of the observed values [16]. The first variable imputed in the chain is regressed on the variables specified by the model to predict that variable as well as the observed values for this variable. The subsequent variables imputed in the chain are regressed on the variables specified by their prediction model as well as their observed values, with the addition of variables previously imputed in the chain that are in their prediction model. MICE enabled the incorporation of multiple predictive covariates to simulate the population represented by the sampled households and allowed the use of outcome values imputed for a household to be used in the prediction of outcomes imputed in each subsequent chain.

All non-sampled households had covariates for ecological risk and longitude and latitude coordinates (i.e. X and Y coordinates). Numbers of persons per household and household mean age for non-sampled households were predicted first in the chain using predicted mean matching based on covariates from the survey sample data. Household RDT status (having at least one RDT positive individual in a household) and household antimalarial medication status (having at least one person receiving antimalarial medication in the household) were predicted next in the chain from the ecological risk, household spatial coordinates, persons per household (household sample level and imputed), and household mean age (household sample level and imputed).

The number of RDT positives per household was imputed next in the chain from the ecological risk, household coordinates, persons per household (household sample level and imputed), household mean age (household sample level and imputed), and household RDT status (household sample level and imputed). Restrictions were placed on this predicted outcome to ensure that the number of RDT positive residents per household did not exceed the number of persons per household and, if the household

was predicted to be positive, the number of RDT positive residents was at least one. Additionally, if the household was predicted to be RDT negative, no RDT positive residents resided in the household.

RDT positive residents with symptoms and care seeking behavior were imputed next in the chain from the ecological risk, household coordinates, persons per household (household sample level and imputed), household mean age (household sample level and imputed), household RDT status (household sample level and imputed), number of positives per household (household sample level and imputed) and household antimalarial medication status (household sample level and imputed). A person who received antimalarial medication in a household and visited a healthcare facility to obtain the medication was imputed last in the chain from the ecological risk, household coordinates, persons per household (household sample level and imputed), household mean age (household sample level and imputed), household RDT status (household sample level and imputed), and antimalarial medication status of the household (household sample level and imputed).

The simulated data were assessed to ensure that the simulated household population (persons per household), simulated mean household age and simulated household level malaria prevalence did not differ significantly from the sampled data. Since only household level data were used in the predictive models, if a simulated RDT positive household was classified as likely to be identified, all simulated RDT positive residents of that household also were classified as likely to be identified.

The prediction models used to perform the imputation were evaluated with the simulated data for each year to ensure that the models fit the simulated data. The same methods for evaluating the models in the sampled data were used to evaluate the models in the simulated data (Tables S3 and S4 in File S1).

Spatial Analysis of Population Level Simulated Data

Simulated RDT positive households were plotted on the map of the study area and differentiated as identified or missed. The identified and missed households were added as data layers using the projected Universal Transverse Mercator (UTM), 1983 Southern Hemisphere, Zone 35 coordinate system. Distance-based buffers were set surrounding identified households (index households). These buffers represented varying distances surrounding identified households that would be screened for malaria to detect and treat RDT positive individuals. Buffers of different distances were evaluated to determine the buffer size needed to identify maximum proportions of missed positive households under each transmission setting (2007 and 2008). In addition to using the distances provided from the sample data, buffer distances ranging from 500 to 3,000 meters surrounding an index household were evaluated. The buffers were dissolved to ensure that a household could only be counted once in the event that a missed household was located within the buffer of more than one identified household. Each buffer layer was then spatially joined to the missed household data layer. The sum of all missed households (as well as residents likely to be RDT positive in missed households) within each buffer layer, the proportions of missed RDT positive households and residents within each buffer of identified RDT positive households (relative to all RDT positive missed household), and the proportions of all RDT positive households within each buffer were calculated.

In addition, negative households from the simulation and those assumed to be negative by having an ecological risk less than 0.196 were added as new data layers to the map. The sum of all negative households within each buffer layer, and the proportions of negative households that potentially would be screened within each buffer (relative to all households screened in each buffer) were calculated. These proportions were then compared to proportions of positive households screened within each buffer to determine the impact of reactive case finding in each transmission setting.

Statistical analyses were performed using STATA 12.0 (StataCorps, College Station, TX). Spatial analyses were performed using ArcGIS 10.0 (Environmental Systems Research Institute [ESRI], Redlands, California).

Results

Characteristics of Sampled Households

The 2007 study area represented a setting of moderate transmission with a parasite prevalence of 23% by RDT, whereas the 2008 data represented a setting of low transmission (recently transitioned from moderate transmission) with a parasite prevalence of 8% by RDT. Two demographic characteristics differed significantly across the two study sites and years, care seeking behavior (43% in 2007 vs. 56% in 2008; p = .001) and reported malaria symptoms (37% in 2007 vs. 24% in 2008; p<.001).

Households Sampled in the 2007 Study Area

In 2007, RDT positive individuals were younger than RDT negative individuals (mean age 13.4 years vs. 23.4 years, p<.001) and were more likely to report symptoms consistent with malaria during the previous two weeks than RDT negative individuals (53.0% vs. 32.9%, p = .004). RDT positive and negative participants did not differ significantly on other demographic characteristics analyzed. Only 13 of 66 (19.7%) RDT positive participants would likely have been identified in 2007 using passive case detection using the algorithm. Of the remaining RDT positive participants, 11 (16.7%) were symptomatic with no care seeking behavior, 13 (19.7%) were asymptomatic with care seeking behavior, and 29 (43.9%) were asymptomatic with no care seeking behavior. With reactive case detection of household members residing with an index case, an additional 20 RDT positive malaria cases would likely have been detected, resulting in identification of half (33 of 66) of all RDT positive individuals among the sampled households. Of the RDT positive persons likely to have been identified, 73% were symptomatic. In contrast, only 33.3% of the RDT positive persons missed were symptomatic. No significant differences were observed for care seeking behavior or ecological risk of RDT positive persons identified and missed in 2007.

For all individuals residing in sampled households, those identified or residing in an identified household were more likely to be RDT positive (34% vs.17.6% p = .003), have malaria specific symptoms (58.8% vs. 26.6% p<.001) and have care seeking behavior (55.7% vs. 36.7% p = .002) than those not identified or residing in missed households. There were no significant differences observed for ecological risk. Thirty-five of 48 households (73%) had either an RDT positive individual or an individual receiving antimalarial drugs from a health care facility. Of these households, 41% would likely have been identified using the algorithm through passive case detection.

Households Sampled in the 2008 Study Area

In 2008, RDT positive and negative participants differed in numbers of persons per household, with RDT positive persons residing in larger households (82.3% in households with 5 or more persons per household vs. 60.9% in households with less than 5 persons per household; p = .03). RDT positive and negative individuals did not differ significantly by other demographic characteristics. Four of 34 (12%) RDT positive cases would likely

have been identified through passive case detection using the algorithm. Nine additional cases would likely have been identified through reactive case detection within the household, resulting in detection of 13 of 34 (38%) of all RDT positive individuals within the study area. Of the RDT positive individuals, there were no differences in symptoms, care seeking behavior or ecological risk, between those identified and missed using reactive case finding.

For individuals residing in sampled households, those identified or residing in an identified household were more likely to be RDT positive (22.81% vs.6.03%, p<.001) and to reside in an area of slightly lower ecological risk (0.310 vs. 0.388, p<.001) than those not identified or residing in missed households. There were no differences between symptoms and care seeking behavior among individuals identified and missed. Twenty-two of 75 households (29%) had either an RDT positive individual or a person taking antimalarial medication received from a health care facility. Of these households, 41% would likely have been identified using the algorithm, through passive case finding.

Simulated Data from Households Surveyed in 2007

Extrapolation from the households surveyed in 2007 to the non-sampled households resulted in data estimated for 7,980 households with 47,058 individual residents. Household level characteristics of the simulated households did not differ significantly from the sampled households with the exception of the household level of care seeking behavior (i.e. an individual in the house displays care seeking behavior): 70.8% in the sampled households and 86.82% (p = .004) in the non-sampled households (Table 1).

The simulation resulted in 5942 of 7,980 (74.4%) households having an RDT positive resident, with 2,397 (40.3%) of these households likely to have been identified through passive case detection (i.e. index households with a symptomatic, RDT positive individual who would seek care), and 3,545 (59.7%) households likely not to have been identified through passive case detection because the infected individuals were asymptomatic, did not seek care, or both (Table 1).

Spatial Analysis of Simulated Data from the 2007 Survey Sample

Of the non-identified households, 2,873 (81%) were located within a 500 meter radius of an index household and 3,362 (94.8%) were located within a one kilometer radius of an index household. When the radius surrounding index households was expanded to two kilometers, 3,519 (99.3%) of the non-identified households were within this range. All non-identified households were within a three kilometer radius of an identified household (Table 2). Testing and treating individuals residing within 500 meters of an index household identified 81% of households missed through passive case finding and 79% of all RDT positive individuals who would not have been identified and treated in a health care facility (Table 2). Of all households in the 500 meter radius, 62% were positive households, with a total of 53% of all households screened (Table 2, Figure 2). When combined with the RDT positive index households and residents, this strategy of screening all households within 500 meters of an index household would result in identifying 89% of all households with an RDT positive resident and 90% of all RDT positive individuals. If reactive case detection were increased from 500 meters to one kilometer from all index households, 95% of all households with an RDT positive resident and 94% of all RDT positive individuals would be identified, with 62% of all households screened (Table 2, Figures 2 and 3).

Table 1. Characteristics of sampled and simulated households: 2007 and 2008.

	2007			2008		
	Sampled Households	Simulated Households	p-value	Sampled Households	Simulated Households	p-value
Number of households	48	7,980		75	7,961	
Number of individuals screened	284	47,058		403	42,620	
Residents per household (mean, SD)	5.93 (3.13)	5.90 (3.07)	.927	5.37 (2.69)	5.35 (2.65)	.949
Mean age (mean, SD)	25.26 (15.77)	25.04 (15.37)	.942	26.74 (16.32)	26.76 (16.13)	.975
Households with an RDT positive individual (%)	66.7	74.4	.245	24.0	16.8	.119
Households with an individual taking antimalarials (%)	10.4	21.1	.076	9.3	22.1	.007
Households with an individual with care seeking behavior (%)	70.8	86.8	.004	85.3	82.4	.647
Households with an individual with malaria-like symptoms (%)	75.0	70.0	.529	64.0	70.0	.258
Households with an individual with malaria-like symptoms and care seeking behavior (%)	37.5	31.1	.349	34.7	34.1	.903
Households with an individual taking antimalarials with care seeking behavior (%)	22.9	24.7	.868	6.7	4.5	.067
Total households identified through reactive case detection (%)	34.3	42.5	.393	40.9	49.9	.521

Simulated Data from Households Surveyed in 2008

Extrapolation from the households surveyed in 2008 to the non-sampled households resulted in data estimated for 7,961 households with 42,620 individual residents. The household level characteristics of the simulated households did not differ significantly from the sampled households with the exception of the proportion of households with a resident taking antimalarial medication (9.3% in the sampled households vs. 22.1% in the non-sampled households; p = .007) (Table 1). The simulation resulted in 1,340 of 7,961 (16.8%) households with an RDT positive resident, with 470 (35.1%) of these households potentially

identified through passive case detection (i.e. index households with a symptomatic, RDT positive individual who would seek care), and 870 (66.7%) households with RDT positive residents likely not to have been identified through passive case detection, either because the infected individual was asymptomatic, lacked care seeking behavior, or both (Table 1).

Spatial Analysis of Simulated Data from the 2008 Survey Sample

Of the non-identified households, 476 (54.7%) were located within a 500-meter radius of an index household and 685 (78.7%)

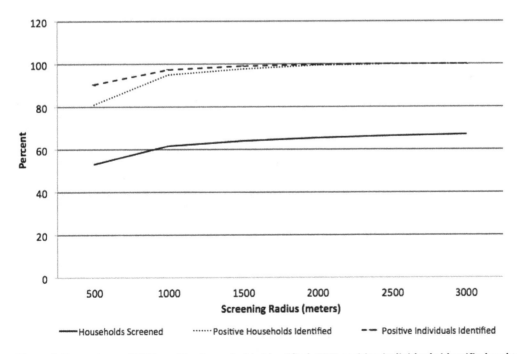

Figure 2. Percentage of RDT positive households identified, RDT positive individuals identified and total households screened by screening radii surrounding index households: 2007.

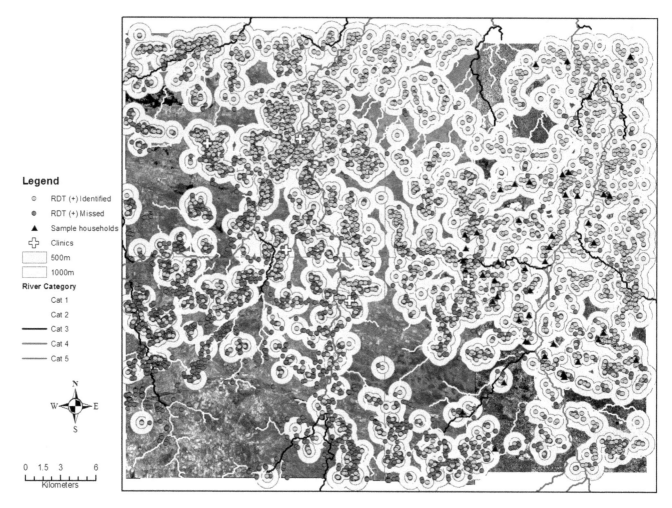

Figure 3. Map of screening radii surrounding RDT positive identified and missed households: 2007.

were located within a one kilometer radius of an index household (Table 2). When the radius surrounding the index households was expanded to two kilometers, 828 (95.2%) of the positive households were identified and 854 (98.2%) were identified within a radius of 3 kilometers (Table 2, Figure 4). Testing and treating individuals within 500 meters of an index household identified 54.7% of households missed through passive case finding, accounting for over 54.4% of RDT positive individuals who would not have been identified in a health care facility. Of all households within 500 meters, 11% were positive households, with a total of 48% of all households screened (Table 2, Figure 4). When combined with the RDT positive index households and residents, screening all households within 500 meters of an index household would result in identifying 77% of all households with an RDT positive resident and 76% of all RDT positive individuals. If the screening radius was increased from 500 meters to 1 kilometer, combined with the RDT positive index households and residents, 89% of all households with an RDT positive resident would be identified and 89% of all RDT positive individuals, while screening a total of 69% of all households (Table 2, Figure 5).

Discussion

In areas where malaria transmission has recently declined following implementation of effective control measures, additional strategies are needed to identify and treat infected individuals who

do not seek medical care to eliminate gametocyte reservoirs, interrupt transmission and achieve elimination [9]. Extrapolating from data collected in two settings in southern Zambia with different levels of malaria transmission, we demonstrated that reactive case detection within a 500 meter radius from the household of an index case would identify more than three quarters of infected individuals, although the proportion detected was lower as parasite prevalence declined. We are unaware of other published studies that assessed the simulated efficiency of reactive case detection. Testing and treating individuals residing in neighboring households of an index case could be useful in interrupting transmission in regions of declining malaria burden, although cost effectiveness studies are needed to determine the incremental costs associated with expanding the screening radius.

The maps generated by this analysis provide insight into the clustering of RDT positive households under different transmission settings. In addition, the maps show the distances surrounding index households to be screened to maximize the number of infected individuals identified within these foci. In foci where a large proportion of positive households would have been identified by passive case detection, screening and treating household members of an index case would have been sufficient to identify a high proportion of infected individuals. In foci where few households would have been identified passively, screening and treating contacts in the index household and surrounding

Table 2. Proportions of positive and negative households, missed households, missed individuals, total households, and total individuals identified at various screening radii: 2007 and 2008.

Buffer (m)	Positive households	Negative households	Total households screened (%)	Missed positive households identified through reactive case detection (%)	Total positive households identified through reactive case detection (%)	Total positive individuals	Missed individuals identified through reactive case detection (%)	Total positive individuals identified through reactive case detection (%)
2007								
500	2873	1778	53.2	81.0	89.1	5730	79.3	90.5
1000	3362	2036	61.7	94.8	97.0	6826	94.4	97.4
1500	3466	2142	64.1	97.8	98.7	7059	97.7	98.9
2000	3519	2195	65.3	99.3	99.6	7178	99.3	99.7
2500	3541	2251	66.2	99.9	99.9	7219	99.9	99.9
3000	3545	2316	67.0	100.00	100.0	7228	100.0	100.0
2008								
500	476	3684	47.5	54.7	77.3	721	54.4	75.8
1000	685	5331	68.8	78.7	89.3	1050	79.2	89.0
1500	795	6060	78.3	91.4	95.7	1221	92.1	95.8
2000	828	6410	82.7	95.2	97.6	1269	95.7	97.7
2500	843	6598	85.0	96.9	98.4	1289	97.2	98.5
3000	854	6712	86.5	98.2	99.1	1307	98.6	99.2

A positive household refers to a household with an RDT positive resident.
A negative household refers to a household in which all residents are RDT negative.

households would be needed to identify a high proportion of infected individuals.

This analysis showed that reactive case finding has the potential to identify individuals who would have been otherwise missed, simply by screening household members of RDT positive cases that present to the clinic. However, for both transmission settings, the benefits of screening household members was likely insufficient to eliminate the reservoir. Screening within 500 meters of the index households would have a significant impact on identifying and treating a large proportion of the asymptomatic reservoir in both moderate and low transmission settings.

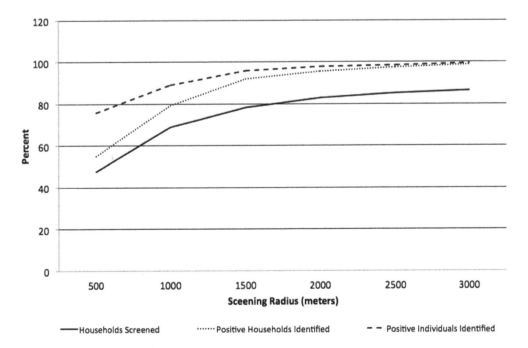

Figure 4. Percentage of RDT positive households identified, RDT positive individuals identified and total households screened by screening radii surrounding index households: 2008.

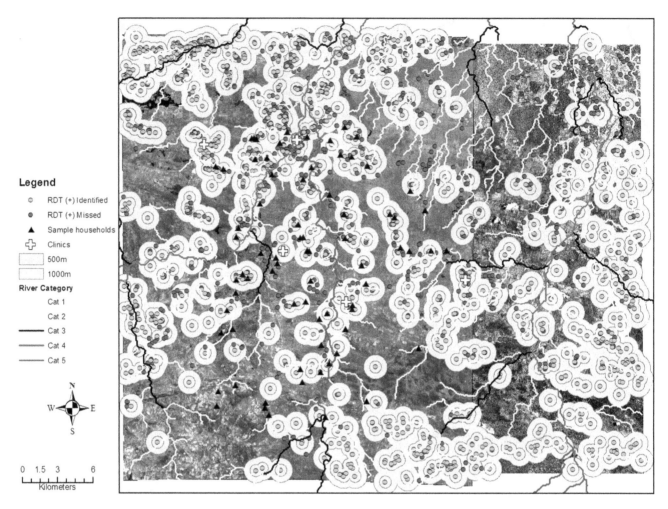

Figure 5. Map of screening radii surrounding RDT positive identified and missed households: 2008.

These analyses were based on the results of RDTs to identify infected individuals. However, RDTs are insufficiently sensitive to identify individuals with low-level parasitemia [19,20], who may account for up to 25% of transmissions to mosquitos [20]. Therefore, our results underestimate the human malaria reservoir. However, reactive case detection as a malaria control and elimination strategy is likely to rely on RDTs for screening, as is currently being done in southern Zambia, and not more sensitive nucleic acid detection tests until low-cost, field friendly assays become available. An alternative strategy to eliminate the infectious reservoir, including those with low-level parasitemia, is to administer ACT and primaquine to all household members of the index case (or within a defined radius) without diagnostic testing. Future analyses may consider the likelihood that undetected infectious individuals become gametocyte carriers and that sufficient mosquitoes feed on them, acquire infection, and become infectious to continue the transmission cycle.

The models were based on several assumptions: the data represent one transmission season; the population was homogeneous with regard to access to care; reinfection did not occur; and complete coverage is achieved of all individuals in all households within the screening radii of identified households shortly after an index case is identified.

The model assumes that these data represent one transmission season; however, the survey sample data was collected cross-sectionally across several months (April–December 2007 and February–December 2008). By making this assumption, any seasonal or temporal trends in malaria incidence were not captured in this analysis. Using data from serial cross-sectional surveys to simulate a closed population without a temporal dimension assumed that spatial clustering of malaria is static and stable over transmission seasons. In support of this assumption, malaria clusters were shown to be fairly stable over time, specifically clusters of asymptomatic parasitemia [21]. However, the spatial clustering of infected individuals is likely seasonal as the force of infection changes, resulting in different efficiencies for reactive case detection within defined radii. Future studies should explore the impact of seasonal malaria transmission on optimal reactive case detection strategies.

The assumption that all persons have equal access to care and treatment may be justified by the multiple health care facilities within the study area and the relatively homogeneous socio-economic status of residents. The assumption regarding reinfection is made likely by data from the longitudinal cohort: 17 of 330 individuals were re-infected in 2007, accounting for 5.2% of the total sample, and only 1 of 435 individuals was re-infected in 2008, accounting for 0.2% of the sample. However, this may be an underestimate due to the effects of repeated treatment within the longitudinal cohort [13]. The potential impact of reactive case detection on onward malaria transmission during this time frame could not be evaluated using this static model. Therefore, the efficiency of reactive case detection in the field may be quite different than the results presented here.

Assumptions were made in extrapolating from sampled to non-sampled households. The simulation was performed based on data from a small but random sample of the entire population. The model fit the data well and was accurate in predicting the data. However, the model was not formally validated externally. While the simulated data did not differ from the sample data, the simulated data may not fully account for heterogeneity between sampled and non-sampled households. The models assumed 100% coverage of all households and residents located within the screening radii of index households. Therefore, the results represent a best-case scenario of the efficiency of reactive case detection. In practice, coverage would not be 100% and the logistics and operational costs, specifically the resources needed to screen all households surrounding index households could impair the feasibility of reactive case detection.

Conclusions

Identifying and appropriately treating infected individuals, including those who do not seek medical care, is essential to achieve malaria elimination. Reactive case detection may be an additional, important strategy to achieve this goal, although the efficiency may vary in different transmission settings, and cost and effort will likely increase as the transmission level decreases. Testing and treating individuals residing within a defined radius of an index case has the potential to be an effective strategy to identify and treat a large proportion of asymptomatic, minimally symptomatic, and symptomatic individuals who do not seek care in regions with a declining burden of malaria. While this analysis based on the use of RDTs is unable to determine whether reactive case detection can eliminate the human malaria reservoir, including infected individuals who are RDT negative, it can provide insight into the potential impact that may be observed using currently available strategies under different epidemiological conditions.

Acknowledgments

We thank the residents of Choma District who participated in the study. This work was supported by the Johns Hopkins Malaria Research Institute, the Bloomberg Family Foundation, and the Division of Microbiology and Infectious Diseases, National Institutes of Allergy and Infectious Diseases, National Institutes of Health.

Author Contributions

Conceived and designed the experiments: KMS TS GG WJM. Performed the experiments: KMS TS HH TK SM PET DLS GG WJM. Analyzed the data: KMS TS DLS GG WJM. Wrote the paper: KMS TS DLS GG WJM.

References

1. WHO (2010) World malaria report 2010. Geneva, Switzerland: World Health Organization.
2. Aregawi MW, Ali AS, Al-Mafazy AW, Molteni F, Katikiti S, et al. (2011) Reductions in malaria and anaemia case and death burden at hospitals following scale-up of malaria control in Zanzibar, 1999–2008. Malar J 10.
3. Otten M, Aregawi M, Were W, Karema C, Medin A, et al. (2009) Initial evidence of reduction of malaria cases and deaths in Rwanda and Ethiopia due to rapid scale-up of malaria prevention and treatment. Malar J 8.
4. Barnes KI, Chanda P, Ab Barnabas G (2009) Impact of the large-scale deployment of artemether/lumefantrine on the malaria disease burden in Africa: case studies of South Africa, Zambia and Ethiopia. Malar J 8 Suppl 1: S8.
5. Macauley C (2005) Aggressive active case detection: a malaria control strategy based on the Brazilian model. Soc Sci Med 60: 563–573.
6. Mabunda S, Aponte JJ, Tiago A, Alonso P (2009) A country-wide malaria survey in Mozambique. II. Malaria attributable proportion of fever and establishment of malaria case definition in children across different epidemiological settings. Malar J 8: 74.
7. Owusu-Agyei S, Smith T, Beck HP, Amenga-Etego L, Felger I (2002) Molecular epidemiology of Plasmodium falciparum infections among asymptomatic inhabitants of a holoendemic malarious area in northern Ghana. Trop Med Int Health 7: 421–428.
8. Harris I, Sharrock WW, Bain LM, Gray KA, Bobogare A, et al. (2010) A large proportion of asymptomatic Plasmodium infections with low and sub-microscopic parasite densities in the low transmission setting of Temotu Province, Solomon Islands: challenges for malaria diagnostics in an elimination setting. Malar J 9: 254.
9. Moonen B, Cohen JM, Snow RW, Slutsker L, Drakeley C, et al. (2010) Operational strategies to achieve and maintain malaria elimination. Lancet 376: 1592–1603.
10. Stresman GH, Kamanga A, Moono P, Hamapumbu H, Mharakurwa S, et al. (2010) A method of active case detection to target reservoirs of asymptomatic malaria and gametocyte carriers in a rural area in Southern Province, Zambia. Malar J 9: 265.
11. Kent RJ, Thuma PE, Mharakurwa S, Norris DE (2007) Seasonality, blood feeding behavior, and transmission of Plasmodium falciparum by Anopheles arabiensis after an extended drought in southern Zambia. Am J Trop Med Hyg 76: 267–274.
12. Larkin GL, Thuma PE (1991) Congenital malaria in a hyperendemic area. Am J Trop Med Hyg 45: 587–592.
13. Sutcliffe CG, Kobayashi T, Hamapumbu H, Shields T, Mharakurwa S, et al. (2012) Reduced risk of malaria parasitemia following household screening and treatment: a cross-sectional and longitudinal cohort study. PLoS One 7: e31396.
14. Moss WJ, Hamapumbu H, Kobayashi T, Shields T, Kamanga A, et al. (2011) Use of remote sensing to identify spatial risk factors for malaria in a region of declining transmission: a cross-sectional and longitudinal community survey. Malar J 10: 163.
15. WHO (2009) Malaria rapid diagnostic test performance: results of WHO product testing of malaria RDTs: round 1 (2008). France: World Health Organization.
16. White IR, Royston P, Wood AM (2011) Multiple imputation using chained equations: Issues and guidance for practice. Stat Med 30: 377–399.
17. van Buuren S (2007) Multiple imputation of discrete and continuous data by fully conditional specification. Stat Methods Med Res 16: 219–242.
18. Raghunathan TL, J.; Van Hoewyk, J.; Solenberger, P. (2001) A multivariate technique for multiply imputing missing values using a sequence of regression models. Survey Methodology 27: 85–95.
19. Baltzell KA, Shakely D, Hsiang M, Kemere J, Ali AS, et al. (2013) Prevalence of PCR detectable malaria infection among febrile patients with a negative Plasmodium falciparum specific rapid diagnostic test in Zanzibar. Am J Trop Med Hyg 88: 289–291.
20. Okell LC, Bousema T, Griffin JT, Ouedraogo AL, Ghani AC, et al. (2012) Factors determining the occurrence of submicroscopic malaria infections and their relevance for control. Nat Commun 3: 1237.
21. Bejon P, Williams TN, Liljander A, Noor AM, Wambua J, et al. (2010) Stable and unstable malaria hotspots in longitudinal cohort studies in Kenya. PLoS Med 7: e1000304.

16

Scientific Foundations for an IUCN Red List of Ecosystems

David A. Keith[1,2]*, Jon Paul Rodríguez[3,4,5,6], Kathryn M. Rodríguez-Clark[3], Emily Nicholson[7], Kaisu Aapala[8], Alfonso Alonso[9], Marianne Asmussen[3,5], Steven Bachman[10], Alberto Basset[11], Edmund G. Barrow[12], John S. Benson[13], Melanie J. Bishop[14], Ronald Bonifacio[15], Thomas M. Brooks[6,16], Mark A. Burgman[17], Patrick Comer[18], Francisco A. Comín[19], Franz Essl[20,21], Don Faber-Langendoen[16], Peter G. Fairweather[22], Robert J. Holdaway[23], Michael Jennings[24], Richard T. Kingsford[1], Rebecca E. Lester[25], Ralph Mac Nally[26], Michael A. McCarthy[7], Justin Moat[10], María A. Oliveira-Miranda[4], Phil Pisanu[15], Brigitte Poulin[27], Tracey J. Regan[7], Uwe Riecken[28], Mark D. Spalding[29], Sergio Zambrano-Martínez[3]

1 Australian Wetlands Rivers and Landscapes Centre, University of New South Wales, Sydney, New South Wales, Australia, 2 New South Wales Office of Environment and Heritage, Hurstville, New South Wales, Australia, 3 Centro de Ecología, Instituto Venezolano de Investigaciones Científicas, Caracas, Venezuela, 4 Provita, Caracas, Venezuela, 5 EcoHealth Alliance, New York, New York, United States of America, 6 IUCN Commission on Ecosystem Management and IUCN Species Survival Commission, Gland, Switzerland, 7 Centre of Excellence for Environmental Decisions, University of Melbourne, Victoria, Australia, 8 Finnish Environment Institute, Helsinki, Finland, 9 Smithsonian Conservation Biology Institute, National Zoological Park, Washington, D.C., United States of America, 10 Royal Botanic Gardens, Kew, England, 11 Department of Biological and Environmental Science, Ecotekne Center, University of Salento, Lecce, Italy, 12 IUCN Global Ecosystem Management Programme, Nairobi, Kenya, 13 Royal Botanic Gardens Trust, Sydney, New South Wales, Australia, 14 Department of Biological Sciences, Macquarie University, New South Wales, Australia, 15 Science Resource Centre, Department of Environment and Natural Resources, Adelaide, South Australia, Australia, 16 NatureServe, Arlington, Virginia, United States of America, 17 Australian Centre of Excellence for Risk Assessment, University of Melbourne, Victoria, Australia, 18 NatureServe, Boulder, Colorado, United States of America, 19 Pyrenean Institute of Ecology, Zaragoza. Spain, 20 Environment Agency Austria, Vienna, Austria, 21 Department of Conservation Biology, Vegetation and Landscape Ecology, University of Vienna, Vienna, Austria, 22 School of Biological Sciences, Flinders University, Adelaide, South Australia, Australia, 23 Landcare Research, Lincoln, New Zealand, 24 Department of Geography, University of Idaho, Moscow, Idaho, United States of America, 25 School of Life and Environmental Sciences, Deakin University, Warnambool, Victoria, Australia, 26 Australian Centre for Biodiversity, School of Biological Sciences Monash University, Victoria, Australia, 27 Tour du Valat Research Center, Arles, France, 28 German Federal Agency for Nature Conservation, Bonn, Germany, 29 The Nature Conservancy and Conservation Science Group, Department of Zoology, University of Cambridge, Cambridge, England

Abstract

An understanding of risks to biodiversity is needed for planning action to slow current rates of decline and secure ecosystem services for future human use. Although the IUCN Red List criteria provide an effective assessment protocol for species, a standard global assessment of risks to higher levels of biodiversity is currently limited. In 2008, IUCN initiated development of risk assessment criteria to support a global Red List of ecosystems. We present a new conceptual model for ecosystem risk assessment founded on a synthesis of relevant ecological theories. To support the model, we review key elements of ecosystem definition and introduce the concept of ecosystem collapse, an analogue of species extinction. The model identifies four distributional and functional symptoms of ecosystem risk as a basis for assessment criteria: A) rates of decline in ecosystem distribution; B) restricted distributions with continuing declines or threats; C) rates of environmental (abiotic) degradation; and D) rates of disruption to biotic processes. A fifth criterion, E) quantitative estimates of the risk of ecosystem collapse, enables integrated assessment of multiple processes and provides a conceptual anchor for the other criteria. We present the theoretical rationale for the construction and interpretation of each criterion. The assessment protocol and threat categories mirror those of the IUCN Red List of species. A trial of the protocol on terrestrial, subterranean, freshwater and marine ecosystems from around the world shows that its concepts are workable and its outcomes are robust, that required data are available, and that results are consistent with assessments carried out by local experts and authorities. The new protocol provides a consistent, practical and theoretically grounded framework for establishing a systematic Red List of the world's ecosystems. This will complement the Red List of species and strengthen global capacity to report on and monitor the status of biodiversity

Editor: Matteo Convertino, University of Florida, United States of America

Funding: The authors gratefully acknowledge funding support from the MAVA Foundation, Gordon and Betty Moore Foundation, Smithsonian Institution, EcoHealth Alliance, Provita, the Fulbright Program, Tour du Valat, the Australian Centre of Excellence for Environmental Decisions and Centre de Suive Ecologique. The funders had no role in study design, data collection and analysis, decision to publish, or preparation of the manuscript.

Competing Interests: Provita is a non-governmental conservation organization based in Venezuela, focused on the conservation of threatened species and ecosystems (www.provitaonline.org or www.provita.org.ve). Provita has been involved with the IUCN Red List of Ecosystems effort for several years, and has therefore sponsored some of the authors' activities (which they acknowledge).

* E-mail: david.keith@unsw.edu.au

Introduction

The world's biodiversity continues to diminish as human populations and activities expand [1,2,3,4]. A sound understanding of risks to biodiversity is needed to plan actions to slow rates of decline, secure future ecosystem services for human use and foster investment in ecosystem management [5]. By identifying species most at risk of extinction, the IUCN Red List criteria [6] inform governments and society about the current status of biodiversity [7] and trends in extinction risks [8], and also provide data with which to formulate priorities and management strategies for conservation [9].

Despite the strengths and widespread acceptance of the IUCN Red List of Threatened Species [10], the need for biodiversity assessments that address higher levels of biological organisation has long been recognised [11,12]. This need is reflected in the emergence of recent national and regional listings of ecosystems, communities and habitats [13], and recent resolutions by the World Conservation Congress to develop quantitative criteria for assessing ecosystems [14]. Opportunities to meet the need for ecosystem risk assessment are supported by emerging theories on ecosystem dynamics and function [15,16,17], methods for handling uncertainty [18,19], ecosystem-specific measures of ecological change [20,21,22] and developing temporal data sets on ecosystem distribution and processes [23,24].

The scientific challenges in building a unified risk assessment framework for ecosystems are likely greater than those faced during development of Red List criteria for species [25]. Foremost among these challenges is balancing the need for specificity (to support consistent, quantitative evaluation of risk) with the need for generality (to support application of common theoretical concepts across the wide variety of ecosystems). To achieve this trade-off, and to address other scientific challenges outlined below, we first construct a framework comprising generic concepts and models derived from relevant ecological theories, and second, propose requirements or 'standards' for translating the concepts into practical assessments, illustrated by examples. Our intent is to outline the concepts in enough detail that applications will be consistent in a very broad range of contexts. We also aim to avoid prescriptive or arbitrarily exact definitions that would exclude or misclassify many cases or prove to be unworkable in the variety of contexts in which ecosystem assessment is required. Although we recognise that this approach carries some risk of inconsistent application between ecosystems defined in different regions or environments, we believe this trade-off is necessary to achieve the generality and flexibility required of a globally applicable risk assessment protocol.

Early development of Red List criteria for ecosystems drew from analogies with species criteria and existing protocols designed for regional applications [12,13]. Existing risk assessment protocols were primarily focussed on terrestrial plant communities and were national or regional in scope (e.g. [26,27,28]). Their assessment of declines in ecological function was mostly qualitative and they applied different treatments of common risk factors such as rates of decline and restricted distribution [13]. The reasons for differences between existing protocols were difficult to understand because their documentation provides limited theoretical rationale for their construction [13]. Our aim here is to develop a generic assessment method based on an explicit conceptual model for ecosystem risk. The intended scope of assessments spans terrestrial, subterranean, aquatic continental and marine realms, and transitional environments at their interfaces. The scope also includes semi-natural and cultural environments [29]. We first elucidate the goals and key concepts that underpin our approach to risk assessment. We then describe the conceptual model for assessing risks of ecosystem collapse, and justify the construction of risk assessment criteria with reference to relevant ecological theory. Finally, we trial the criteria on contrasting ecosystems from around the world to evaluate their applicability and performance relative to existing assessments, and to identify challenges for future research.

Goals and Key Concepts of Risk Assessment

Goals of a Red List of Ecosystems

Ideally, a Red List may be expected to identify ecosystems at risk of losing biodiversity, ecological functions and/or ecosystems services, since all three are inter-related and important objects for conservation [30]. However, an approach that simultaneously seeks to assess risks to all three is fraught with complexities in the relationships among them (we elaborate on these in the next section). Ecological changes that promote some ecosystem services may be detrimental to biodiversity or vice versa, leading to logical conflicts if a single assessment were to conflate biodiversity, functions and services. Therefore, to provide essential conceptual clarity for a simple and widely applicable risk assessment process, we have chosen to focus on risks to biodiversity as the primary goal for a Red List of Ecosystems, since this underpins many ecosystem functions [30,31]. Under this approach, changes in functions and services may contribute to assessments of risk if they threaten the persistence of characteristic ecosystem biota, but not if they are unlikely to generate a biotic response.

Complex relationships among biodiversity, ecosystem functions, and services. There is growing empirical and theoretical evidence that ecosystem functions and services are linked with biodiversity [30,32,33,34,35,36,37]. However, several complexities in these relationships preclude presuming that one can serve as a proxy for the others or that they can be conflated into a single objective for risk analysis. Firstly, functional roles of many species are only detectable at particular spatial and temporal scales [16,37]. Some ecosystem services may be initially insensitive to biotic loss because multiple species may perform similar functions in a replaceable manner (functional redundancy); some species may contribute little to overall function; or some functions may depend on abiotic components of ecosystems [34]. Conversely, small declines in species' abundance can seriously disrupt or cease the supply of critical ecosystem services before any characteristic biota is actually lost [38]. The subset of biota that sustain functions and services is therefore uncertain, scale-dependent and temporally variable within any ecosystem. Consequently the relationship between biodiversity and many ecosystem services is poorly defined [30].

Secondly, the identification and valuation of ecosystem services depend on social, cultural and economic factors, and may vary locally [39]. Thus risks to ecosystem services may not always be concordant with risks to biodiversity; some processes that promote services may increase risks to biodiversity.

Thirdly, whether particular directional changes in ecosystem function or the abiotic environment are 'good' or 'bad' for conservation often involves local value judgements [16]. In contrast, the loss of characteristic biota is unambiguously negative for conservation goals [40], and therefore provides a clear and simple objective for risk assessment.

Units of Assessment

Our purpose here is to develop a robust and generic risk assessment method that can be applied to any internally consistent classification of ecosystems. A generic risk assessment protocol requires clearly defined assessment units, yet it also requires flexibility to assess risks across contrasting ecosystems that vary greatly in biological and environmental characteristics, as well as scales of organisation, and for which varying levels of knowledge are available. Therefore we first propose an operational definition of ecosystems to guide delineation of assessment units that will be informative about the conservation status of higher levels of biodiversity. Second, we identify the potential sensitivities of risk assessment to scale of the assessment units and suggest a suitable level of ecosystem classification for global biodiversity assessment. Finally, we outline a number of requirements for ecosystem description that are necessary to translate the operational definition into a practical assessment unit.

Operational definition of ecosystems. In Appendix S1 we define terms used to describe ecosystems and other concepts required for risk assessment. We use the term 'ecosystem types' for units of assessment that represent complexes of organisms and their associated physical environment within an area (after [41]). Although many authors have proposed revised definitions of an ecosystem, most encapsulate four essential elements implicit in Tansley's original concept [42]: i) a biotic complex or assemblage of species; ii) an associated abiotic environment or complex; iii) the interactions within and between those complexes; and iv) a physical space in which these operate. Thus, ecosystems are defined by a degree of uniqueness in composition and processes (involving the biota and the environment) and a spatial boundary. For our purposes, we regard other terms applied in conservation assessments, such as 'ecological communities', 'habitats', 'biotopes' and (largely in the terrestrial context) 'vegetation types', as operational synonyms of 'ecosystem types' [13].

The influence of scale. The unique features that define individual ecosystem types are scale-dependent. The four key elements of an ecosystem type may be organised on spatial, temporal and thematic scales [43]. Spatially, ecosystems vary in extent and grain size from water droplets to oceans [44], with boundaries delimited physically or functionally [45]. Temporally, ecosystems may develop, persist and change over time frames that vary from hours to millenia. They appear stable at some temporal scales, while undergoing trends or fluctuations at others [44]. Thematic scale refers to similarity of features within and between ecosystems, their degree of uniqueness in composition and processes, which may be depicted hierarchically [46].

The outcomes of ecosystem assessments are also likely to depend on spatial, temporal and thematic scales [13,43]. Nonetheless, the applicability of the ecosystem concept across terrestrial, subterranean, freshwater and marine environments at any scale [47] offers important flexibility and generality for risk assessment. The diversity of conservation planning needs will likely require ecosystem risk assessments at multiple scales from global to local.

We do not consider ecological classifications in detail here, although we recognise that a global Red List will require a global classification of ecosystem types [12,14]. To provide initial guidance, we suggest that a classification comprising a few hundred ecosystem types on each continent and in each ocean basin will be a practical thematic scale for global assessment. These globally recognisable ecosystem types should be finer units than ecoregions and biomes [48,49], and should encompass variation that may be recognisable as distinct communities at regional and local scales. For example, a classification of approximately 500 assessment units has been adopted for an assessment of terrestrial ecosystems across the Americas [14]. These units correspond to the Macrogroup level of vegetation classification (see [50,51]). Similar classifications may prove suitable for global assessments of freshwater and marine ecosystems. We anticipate that sub-global ecosystem assessments will be most useful when based on established national or regional classifications that are cross-referenced to global assessment units and justified as suitable proxies for ecological assemblages (see examples in Appendix S2).

Describing Ecosystem Types

Since no universally accepted global taxonomy of ecosystems yet exists, lucid description of the assessment unit of interest is an important first step for a repeatable assessment process. Following from our operational definition of an ecosystem, we suggest that a description should address the four elements that define the *identity* of the ecosystem type (Table 1): the characteristic native biota; abiotic environment, key processes and interactions; and spatial distribution [41,45]. For each of these elements, a description should: i) justify conformity of an ecosystem type with the operational definition; and ii) elucidate the scale of the assessment unit, its salient and unique features, and its distinctions and relationships with other units. Essential supporting information includes reference to the classification and more detailed descriptions from which the assessment unit was derived, as well as cross-referencing to the IUCN habitat classification to elucidate context and facilitate comparisons. A description should furthermore establish reference states and appropriate proxies of defining features that will be used to diagnose loss of biodiversity from the ecosystem (we address this in the section on Ecosystem Collapse). Detailed case studies (Appendix S2) illustrate the translation of our operational ecosystem definition into workable assessment units, using a variety of existing ecosystem classification schemes across a wide range of terrestrial, freshwater, marine and subterranean ecosystems.

Characteristic native biota. The concept of 'characteristic native biota' (Appendix S1) is central to risk assessment in ecosystems and therefore to their description (Table 1): we define this as a subset of all native biota that either distinguishes an ecosystem from others (diagnostic components) or plays a non-trivial role in ecosystem function and persistence of other biota (functional components). Conversely, characteristic biota exclude uncommon or vagrant species that contribute little to function and may be more common in other ecosystems. The diagnostic components of an ecosystem exhibit a high abundance or frequency within it, relative to other ecosystems [52], and therefore demonstrate a level of compositional uniqueness within the domain of an assessment (i.e. global, regional, national).

The functional components of characteristic biota include species that drive ecosystem dynamics as ecosystem engineers, trophic or structural dominants, or functionally unique elements (see examples, Appendix S2). These essential components of ecosystem identity play key roles in ecosystem organisation by providing conditions or resources essential for species to complete their life cycles or by helping to maintain niche diversity or other mechanisms of coexistence. Typically they are common within the ecosystem [53], although sometimes they may be more common in other ecosystems. Examples include predators that structure animal communities in many ecosystems, tree species that create differential microclimates in their canopies or at ground level, reef-building corals and oysters that promote niche diversity for cohabiting fish and macro-invertebrates, nurse plants and those that provide sites for predator avoidance, flammable plants that promote recurring fires, etc.

Thus, characteristic native biota may be described using taxonomic or functional traits. To be useful for risk assessment, descriptions need not include exhaustive species inventories. However, they should demonstrate a level a compositional uniqueness and identify functionally important elements salient

Table 1. Description template for ecosystem types.

Elements of operational definition	Components of ecosystem description
1. Characteristic assemblage of biota	Identify defining biotic features
	a) List diagnostic native species and describe their relative dominance and uniqueness
	b) List functional component of characteristic biota and identify their roles
	c) Describe limits of variability in the ecosystem biota
	d) Exemplar photographs
2. Associated physical environment	Identify defining abiotic features (e.g. climate, terrain, water chemistry, depth, turbidity, ocean currents, substrate, etc.)
	a) Text descriptions and citations for characteristic states or values of abiotic variables
	b) Graphical descriptions of abiotic variables
	c) Exemplar photographs
3. Processes & interactions between components	Describe key ecosystem drivers and threatening processes
– among biota	a) Text descriptions and citations
– between biota & environment	b) Diagrammatic process models
	c) Exemplar photographs
4. Spatial extent	Describe distribution and extent
	a) Maps
	b) Estimates of area
	c) Time series, projections (past, present, future)
5. Classification context	Cross-references to relevant ecological classifications
	a) Source classification
	b) IUCN habitat classification
	c) Ecoregional classifications
6. Reference state(s)	Describe ecosystem-specific point of collapse
	a) Proxy variable
	b) Bounded threshold of collapse

See Appendix S2 for examples.

to the assessment of each ecosystem type (see Appendix S2 for examples).

Abiotic characteristics. Abiotic features are the second essential element of the ecosystem concept. Descriptions should similarly identify salient abiotic features that influence the distribution or function of an ecosystem type, define its natural range of variability and differentiate it from other systems (Table 1). For terrestrial ecosystems, salient abiotic features may include substrates, soils and landforms, as well as ranges of key climatic variables, while those of freshwater and marine ecosystems may include key aspects of water regimes, tides, currents, climatic factors and physical and chemical properties of the water column (see Appendix S2 for examples).

Characteristic processes and interactions. Characteristic ecological processes are a third element important to include in ecosystem description for risk assessment (Table 1). A qualitative understanding of the processes that govern ecosystem dynamics is essential for assessing risks related to functional declines. Again, to be practical this element of ecosystem description should not require extensive knowledge of interaction networks or fluxes of matter and energy: many ecosystems lack direct studies of ecological processes. However, generic mechanisms of ecosystem dynamics can often be inferred from related systems. For example, pelagic marine systems are invariably dominated by trophic interactions in which elements of the main trophic levels are

known, even if most particular predator-prey relationships are not. Similarly, the tree/grass dynamic in savannas throughout the world is influenced by fire regimes, herbivores and rainfall, although their relative roles may vary between savanna types. In many cases, a broad understanding of ecosystem processes may be a sufficient basis for assigning an ecosystem to a risk category, especially if key threats to ecosystem persistence can be identified. The basic requirements for assessments based on ecological processes are to identify the major drivers of change, deduce reference states and infer measureable symptoms of ecosystem transformation (see next section).

Simple diagrammatic process models [54] are a useful means of summarising understanding of salient ecosystem processes for risk assessment (see examples in Appendix S2). These models may be structured to describe transitions among alternative states of an ecosystem (e.g. [55,56]) or to show cause-effect dependencies between components and processes within the system (e.g. [57]). More complex models may identify variables and thresholds that define alternative states, pathways of transition between them and conditions or processes that drive the transitions (e.g. [58,59]). Detailed simulation models can predict the relative dominance of alternative states, given estimates of environmental drivers, although these have been developed for relatively few ecosystems [60,61].

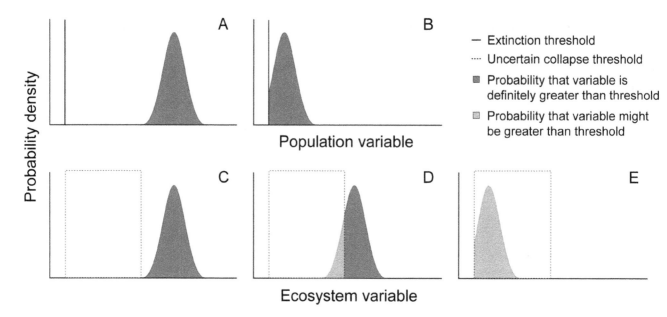

Figure 1. Probability density functions for the population and ecosystem variables that measure proximity to the thresholds that define species extinction (A, B) and ecosystem collapse (C, D). The probability density functions represent uncertainty in the measurement of the variables. For species, the population threshold that defines extinction is known with certainty (e.g. zero abundance of a species, defined by the vertical line in A and B). In A, the estimated population is definitely greater than the extinction threshold, so there is no doubt that the species is extant. Alternatively, the probability that the abundance is above the threshold (the area under the curve) might be less than one (B), in which case the species could be extinct or extant. The shaded area is the probability that the species remains extant. For ecosystems, the x-axis could represent spatial distribution, number of species, water quality, etc. In contrast to species, uncertainty about the definition of ecosystem collapse leads to a range of possible values for this threshold (dashed box in C and D). The ecosystem variable is above this upper bound in some cases (C), so there is no doubt that the ecosystem persists. Alternatively, probable values for the ecosystem variable might intersect the uncertain threshold (D), in which case the ecosystem may be collapsed or not. In this case, there is some probability that the ecosystem parameter is above the upper bound of the threshold (shaded dark grey), which places a lower bound on the probability that the ecosystem persists (i.e. that it has not collapsed). There is an additional probability (pale grey) that the ecosystem parameter is above the threshold that depends on the amount of uncertainty in the threshold (i.e. width of the box). The sum of these two probabilities places an upper bound on the probability ecosystem persists. With further deterioration (E), the lower bound on the probability of ecosystem persistence is zero (no dark shading) and the upper bound is the pale shaded area.

Spatial distribution. Finally, a description of ecosystem properties requires their extent to be specified and bounded at a given observational resolution [62]. The spatial element of ecosystem definition is best described through maps or inventories of locations (Table 1). Mapping is available for many ecosystem types in terrestrial, freshwater aquatic and marine benthic environments, either derived from remote sensing, biophysical distribution models or a combination of both (see examples in Appendix S2). The spatial features of some types of ecosystem, such as pelagic fisheries, are inherently uncertain and dynamic over relatively short time scales, and hence spatial data are scarce and distributions can only be described at very coarse levels of resolution. Given the diversity of methods and maps available, an important aspect of this element of description is to justify why a particular map base is an adequate representation of the ecosystem distribution.

Ecosystem Collapse and Risk Assessment

The protocol for Red Listing must synthesise the diverse evidence, causes, mechanisms and pathways of ecosystem decline within a generic risk assessment framework [63]. To estimate 'risk' – the probability of an adverse outcome over a specified time frame [64] – this framework must first define an endpoint to ecosystem decline (the adverse outcome). For species and populations, this endpoint is extinction, when the last individual dies [25]. Conceptually, species extinction appears to be a relatively discrete endpoint, although its measurement may be

uncertain (Fig. 1a–b). Extinction may be uncertain because, for example, individuals may escape detection [65]. For ecosystems, an analogous endpoint may be identified in terms of distribution size – when the last occurrence of an ecosystem disappears. However, closer examination reveals that the concept of a discrete endpoint (both for species and ecosystems) is problematic for several reasons that we discuss in the next section.

Uncertainties in the 'Endpoints' for Ecological Risk Assessment

The theory of risk assessment assumes a discrete endpoint or event (Fig. 1a–b) affecting the asset under evaluation [64]. Practical implementations of the theory, however, confront uncertainties in the definition of the asset itself, as well as endpoint threshold. For example, the boundaries of related species or ecosystem types are inherently vague [66]. Uncertainties include imperfect knowledge of character variation among individuals of species or occurrences of ecosystems, continuous rather than discrete patterns of natural variability between taxonomic units, and inconsistent taxon concepts that vary through time. These sources of uncertainty are likely greater for ecosystems than species, but they exist in both cases. Thus, the hazards addressed in a risk assessment are more accurately portrayed as bounded ranges than discrete endpoints (Fig. 1c–e).

The uncertainties become more conspicuous when considering endpoints in functional decline, than declines in distribution (Fig. 1) [12,13]. For ecosystems, many characteristic features of an ecosystem may be gone long before the last characteristic species

Table 2. Biotic and abiotic variables for assessing functional decline in the Aral Sea ecosystem, their reference values when the ecosystem was in a functional state (between 1911 and 1960) and bounded thresholds that define the collapsed state, assuming collapse occurred between 1976 and 1989.

	Functional reference state (1911–1960)	Bounded threshold of collapse (reference data 1976, 1989)
Fish species richness and commercial catch (t)	20, 44,000	4–10, 0
Sea volume (km³)	1,089	364–763
Sea surface area (km²)	67,499	39,734–55,700
Average salinity (g.l⁻¹)	10	14–30

Data from [78]. Further details in Appendix 2.5).

disappears from the last ecosystem occurrence ('assemblage extinction' of [53]). Some detrimental ecosystem changes may result from loss of individuals from the system, not loss of particular species [53]. In addition, ecosystems may not disappear, but rather transform into novel ecosystems with different characteristic biota and mechanisms of self-organisation [67]. Transition points from original to novel ecosystems, unlike theoretically discrete events, are inherently uncertain [66], though may still be estimated within plausible bounds (Fig. 1). An obvious analogue for this process in species is transformation by hybridisation [68], but more widespread vagueness in extinction becomes apparent when species concepts are viewed in the context of an artificial and continually developing taxonomy superimposed on dynamic constellations of genes of genotypes. Moreover, different ecosystems will have different points of transition to novel systems because they differ in resilience and natural variability [69,70,71], are threatened by different processes, and exhibit different symptoms of decline.

The definition of the endpoint to ecosystem decline needs to be sufficiently discrete to permit assessment of risk, but sufficiently general to encompass the broad range of contexts in which risk assessments are needed. To deal with this trade-off, we first propose a generic operational definition for an endpoint to ecosystem decline. Second, we provide guidance on how the operational definition of collapse may be translated for specific ecosystem types into an explicit threshold that recognises inherent uncertainties. Third, we propose a conceptual model of ecosystem risk as a basis for design of a protocol for assessing the risk of collapse.

Ecosystem Collapse: an Operational Definition

To acknowledge the contrasts with species extinctions, we propose the concept of "ecosystem collapse" as transition beyond a bounded threshold in one or more variables that define the identity of the ecosystem. Collapse is thus a transformation of identity, loss of defining features, and replacement by a novel ecosystem. It occurs when all occurrences lose defining biotic or abiotic features, and characteristic native biota are no longer sustained. For example, collapse may occur when most of the diagnostic components of the characteristic biota are lost from the system, or when functional components (biota that perform key roles in ecosystem organisation) are greatly reduced in abundance and lose the ability to recruit. Chronic changes in nutrient cycling, disturbance regimes, connectivity or other ecological processes (biotic or abiotic) that sustain the characteristic biota may also signal ecosystem collapse. Novel ecosystems may retain some or many biotic and abiotic features of the pre-collapse systems from which they were derived, but their relative abundances will differ,

they may be organised and interact in different ways and the composition, structure and/or function of the new system has moved outside the natural range of spatial and temporal variability of the old one. A collapsed ecosystem may have the capacity to recover given a long time scale, or with restoration, but in many systems recovery will not be possible.

In the next section, we illustrate how the operational definition of ecosystem collapse can be translated into practical applications. This is most easily done for ecosystems that have already collapsed and where time series data exist for relevant variables (Appendix S2.5). However, as shown in other case studies (Appendix S2), it will often be possible to infer characteristics of collapse from localised occurrences within the ecosystem distribution, even if the majority of the ecosystem remains extant and functional.

Transitions to collapse may be gradual, sudden, linear, non-linear, deterministic or highly stochastic [54,72,73,74,75]. These include regime shifts [72], but also other types of transitions that may not involve feedbacks. The dominant dynamic in an ecosystem will depend on abiotic or external influences (e.g. weather patterns or human disturbance), internal biotic processes (e.g. competition, predation, epidemics), historical legacies, and spatial context [76,77]. An ecosystem may thus be driven to collapse by any of several different threatening processes and through multiple alternative pathways [54]. Symptoms that an ecosystem is at risk of collapse may differ, depending on the characteristics that define the ecosystem identity, the nature of threatening processes and the pathways of decline that these generate.

A modern example of ecosystem collapse. The Aral Sea (see Appendix 2.5), the world's fourth largest continental water body, is fed by two major rivers, the Syr Dar'ya and Amu Dar'ya, in central Asia. Its characteristic native biota includes freshwater fish (20 species), a unique invertebrate fauna (>150 species) and shoreline reedbeds, which provide habitat for waterbirds including migratory species. Hydrologically, the sea was approximately stable during 1911–1960, with inflows balancing net evaporation [78]. Intensification of water extraction to support expansion of irrigated agriculture lead to shrinkage and salinisation of the sea. By 2005, only 28 aquatic species (including fish) were recorded, reed beds had dried and disappeared, the sea had contracted to a fraction of its former volume and surface area, and salinity had increased ten-fold. Consistent with our operational definition of ecosystem collapse, these changes suggest the Aral Sea had undergone a transformation of identity, lost many of its defining features (aquatic biota, reedbeds, waterbirds, hydrological balance and brackish hydrochemistry) and had been replaced by novel ecosystems (saline lakes and desert plains). Under this interpretation, collapse occurred before the volume and surface area of standing water declined to zero. Although the exact point of ecosystem collapse is uncertain, time series data for several variables are suitable for defining a functional reference state (prior to onset of change from 1960) and a bounded threshold of collapse (cf. Fig. 1c–e), assuming this occurred sometime during 1976–1989 when most of the biota disappeared (Table 2).

The choice of available variables for assessing the status of the ecosystem will depend on how closely they represent the ecosystem's defining features, the quantity and quality of the data, and the sensitivity of alternative variables to ecological change. Of those listed above, fish species richness and abundance may be the most proximal biotic variable to the features that define the identity of the Aral Sea ecosystem. Sea volume may be a reasonable abiotic proxy, because volume is functionally linked with salinity, which in turn mediates persistence of the characteristic freshwater/brackish aquatic fauna. Sea surface area is less

directly related to these features and processes, but can be readily estimated by remote sensing and may be useful for assessment when data are unavailable for other variables.

Collapse of the Aral Sea ecosystem may or may not be reversible. While it may be possible to restore the hydrological regime over a small part of the former sea [78], some components of the characteristic biota are apparently extinct (e.g. the Aral salmon, *Salmo trutta aralensis*), preventing reconstruction of the pre-collapse ecosystem.

Risk Assessment Model

Our risk assessment model (Fig. 2) groups symptoms of ecosystem collapse into four major types, and identifies the corresponding mechanisms that link the symptoms to the risk that an ecosystem will lose its defining features (characteristic native

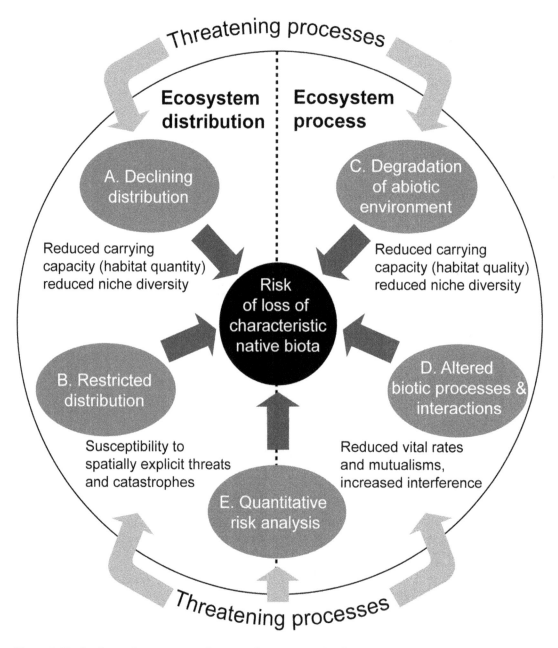

Figure 2. Mechanisms of ecosystem collapse, and symptoms of collapse risk.

Table 3. IUCN Red List criteria for ecosystems, version 2.0.

			Critically Endangered	Endangered	Vulnerable
A		Reduction in geographic distribution over ANY of following periods:			
	1	Present (over the past 50 years)	≥80%	≥50%	≥30%
	2a	Future (over the next 50 years)	≥80%	≥50%	≥30%
	2b	Future (over any 50 year period including the present and future)	≥80%	≥50%	≥30%
	3	Historic (since 1750)	≥90%	≥70%	≥50%
B		Restricted geographic distribution indicated by EITHER:			
	1	Extent of a minimum convex polygon enclosing all occurrences (Extent of Occurrence), OR	≤2,000 km^2	≤20,000 km^2	≤50,000 km^2
	2	The number of 10×10 km grid cells occupied (Area of Occupancy)	≤2	≤20	≤50
		AND at least one of the following (a-c):			
		(a) An observed or inferred continuing decline in EITHER:			
		i. a measure of spatial extent appropriate to the ecosystem; OR			
		ii. a measure of environmental quality appropriate to characteristic biota of the ecosystem; OR			
		iii. a measure of disruption to biotic interactions appropriate to the characteristic biota of the ecosystem			
		(b) Observed or inferred threatening processes that are likely to cause continuing declines in either geographic distribution, environmental quality or biotic interactions within the next 20 years			
		(c) Ecosystem exists at …	1 location	≤5 locations	≤10 locations
	3	A very small number of locations (generally fewer than 5) AND			
		prone to the effects of human activities or stochastic events within a very short time period in an uncertain future, and thus capable of collapse or becoming Critically Endangered within a very short time period			
C	1	Environmental degradation over the past 50 years based on change in an abiotic variable* affecting…	≥80% extent with ≥80% relative severity**	≥50% extent with ≥80% relative severity	≥50% extent with ≥50% relative severity
				≥80% extent with ≥50% relative severity	≥80% extent with ≥30% relative severity
					≥30% extent with ≥80% relative severity
	2	Environmental degradation over the next 50 years, or any 50-year period including the present and future, based on change in an abiotic variable affecting…	≥80% extent with ≥80% relative severity	≥50% extent with ≥80% relative severity	≥50% extent with ≥50% relative severity
				≥80% extent with ≥50% relative severity	≥80% extent with ≥30% relative severity
					≥30% extent with ≥80% relative severity
	3	Environmental degradation since 1750 based on change in an abiotic variable affecting…	≥90% extent with ≥90% relative severity	≥70% extent with ≥90% relative severity	≥70% extent with ≥70% relative severity
				≥90% extent with ≥70% relative severity	≥90% extent with ≥50% relative severity
					≥50% extent with ≥90% relative severity
D	1	Disruption of biotic processes or interactions over the past 50 years based on change in a biotic variable* affecting…	≥80% extent with ≥80% relative severity**	≥50% extent with ≥80% relative severity	≥50% extent with ≥50% relative severity
				≥80% extent with ≥50% relative severity	≥80% extent with ≥30% relative severity

Table 3. Cont.

		Critically Endangered	Endangered	Vulnerable
				≥30% extent with ≥80% relative severity
2	Disruption of biotic processes or interactions over the next 50 years, or any 50-year period including the present and future, based on change in a biotic variable affecting...	≥80% extent with ≥80% relative severity	≥50% extent with ≥80% relative severity	≥50% extent with ≥50% relative severity
			≥80% extent with ≥50% relative severity	≥80% extent with ≥30% relative severity
				≥30% extent with ≥80% relative severity
3	Disruption of biotic processes or interactions since 1750 based on change in a biotic variable affecting...	≥90% extent with ≥90% relative severity	≥70% extent with ≥90% relative severity	≥70% extent with ≥70% relative severity
			≥90% extent with ≥70% relative severity	≥90% extent with ≥50% relative severity
				≥50% extent with ≥90% relative severity
E	Quantitative analysis that estimates the probability of ecosystem collapse to be...	≥50% within 50 years	≥20% within 50 years	≥10% within 100 years

These supercede an earlier set of four criteria [12]. Refer to Appendix S1 for definitions of terms.
*see text for guidance on selection of variable appropriate to the characteristic native biota of the ecosystem.
**see text and Fig. 6 for explanation of relative severity of decline.

biota and/or ecological processes). Two of the four mechanisms produce distributional symptoms (Fig. 2): A) ongoing declines in distribution, which reduce carrying capacity for dependent biota; and B) restricted distribution, which predisposes the system to spatially explicit threats. Two other mechanisms produce functional symptoms (Fig. 2): C) degradation of the abiotic environment, reducing habitat quality or abiotic niche diversity for component biota; and D) disruption of biotic processes and interactions, resulting in the loss of mutualisms, biotic niche diversity, or exclusion of some component biota by others. Interactions between two or more of these four contrasting mechanisms may produce additional symptoms of transition towards ecosystem collapse. Multiple mechanisms and their interactions may be integrated into a simulation model of ecosystem dynamics to produce quantitative estimates of the risk of collapse (E). These five groups of symptoms form the basis of ecosystem Red List criteria (Table 3).

Protocol structure. The risk assessment protocol comprises five rule-based criteria based on thresholds for distributional and functional symptoms represented in the risk model (Fig. 2, Table 3). Symptoms may be measured by one or more proxy variables. These may be generic or specific to particular ecosystems (see text on respective criteria for guidance on variable selection). The criteria and thresholds assign each ecosystem to one of three ordinal categories of risk (Table 3, Fig. 3), or else one of several qualitative categories.

An ecosystem under assessment should be evaluated using all criteria for which data are available. Overall threat status is the highest level of risk returned by any of the criteria (Fig. 3), since risk is determined by the most limiting factor [25]. The quantitative categories of risk [12] mirror those of the IUCN Red List of Threatened Species (IUCN 2001): Critically Endan-

gered (CR); Endangered (EN); and Vulnerable (VU). These are complemented by several qualitative categories that accommodate 1) ecosystems that just fail to meet the quantitative criteria for the three threatened categories (NT, Near Threatened); 2) ecosystems that unambiguously meet none of the quantitative criteria (LC, Least Concern); 3) ecosystems for which too few data exist to apply any criterion (DD, Data Deficient); and 4) ecosystems that have not yet been assessed (NE, Not Evaluated). An additional category (CO, Collapsed) is assigned to ecosystems that have collapsed throughout their distribution, the analogue of the extinct (EX) category for species [6].

Time scales. The criteria assess declines over three time frames: current, future, and historic (Fig. 4). Current declines are assessed over the past 50 years: recent enough to capture current trends, but long enough to reliably diagnose directional change, distinguish it from natural fluctuations in most instances and to plan management responses. Causes of decline are often uncertain but, taking a precautionary approach, the protocol assumes that current declines indicate future risks irrespective of cause.

Assessment of future declines requires predictions about changes over the next 50 years or any 50-year period including the present and future (Fig. 4). Past declines may provide a basis for such predictions, but future declines may be predicted even when the ecosystem is currently stable. Such predictions require a defensible assumption about the pattern of future change (i.e. accelerating, constant, decelerating). Plausible alternative models of change should be explored [79], but a constant proportional rate of decline is often a reasonable default assumption for a range of ecosystems (e.g. [80]).

Assessments of historical declines are essential for ecosystems containing biota with long generation lengths and slow population turnover [25]. Even where future rates of decline abate, historical

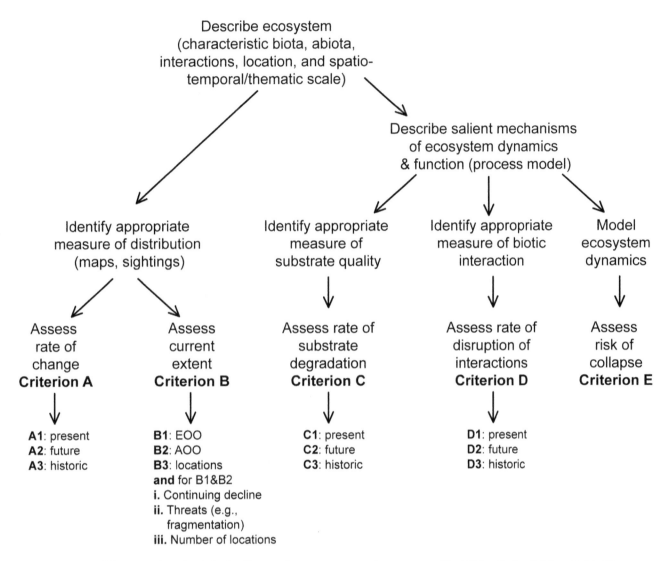

Figure 3. Protocol for assessing the risk of collapse of an ecosystem using proposed Red List criteria v2.0 (see Table 3).

reductions in distribution or function may predispose an ecosystem to additional threats [81,82], and reduce its ability to absorb adverse changes [68]. Historic declines are assessed relative to ecosystem status at a notional reference date of 1750 (Fig. 4), corresponding approximately with the earliest onset of industrial-scale exploitation of ecosystems, although the actual onset varies worldwide. Some anthropogenic changes occurred prior to 1750 [83], but knowledge of earlier distributions, trends and their causes is limited. Distribution models with environmental predictors may be used to estimate historic declines based on the difference between the current state of an ecosystem and its expected state in the absence of anthropogenic effects.

Decline thresholds. The ordinal categories of risk are delimited by different thresholds of decline. Our rationale for setting these thresholds is partly grounded in theory and partly pragmatic, recognizing that: i) theory provides a qualitative basis for ordered thresholds for decline, but offers limited guidance for setting their absolute values; and ii) our aim is to rank ecosystems into informative ordinal categories of risk, rather than estimate precise probabilities of collapse.

Species-area relationships [84] provide theoretical guidance for estimating loss of biota with declining area of available habitat.

However, generic use of species-area relationships across many ecosystems and large scales is problematic for several reasons. Firstly, species loss cannot simply be calculated by reversing species accumulation curves [85]: the area in which the last individual of a species disappears (extinction) is always larger than the sample area needed to detect the first individual of a species. Secondly, the slope (z), of the species-area relationship varies empirically from 0.1 to 0.25, depending on the taxonomic groups assessed [84], habitat quality [86], habitat heterogeneity [87], mainland-island context [84] and time lags in reaching equilibrium [82,88]. A third problem is that application of species-area relationships to landscapes and seascapes does not account for the patchiness of species occurrence within ecosystem types [89]. Moreover, some relationships exhibit context-dependent threshold behaviour that differs between taxonomic groups and landscape types [90,91]. Fourthly, species-area relationships predict only species richness, not their abundance, which may affect ecosystem functions [53]. Species-area models are therefore unlikely to support universal threshold values of decline for assessing ecosystem status.

It is noteworthy that the relationship between biodiversity and ecosystem function, when averaged over many cases, has a similar

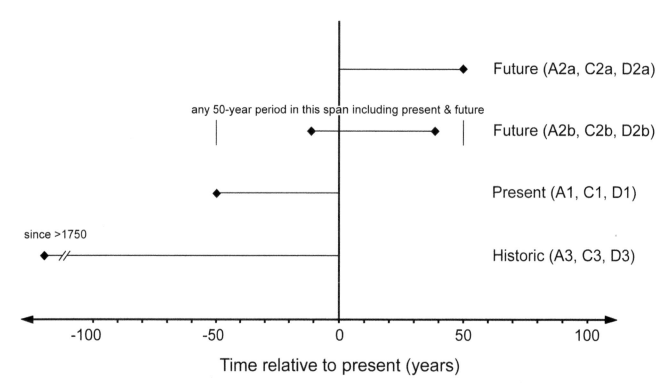

Figure 4. Time scales for assessment of change under criteria A, C and D.

monotonic form to species-area relationships and also varies in slope [31]. Thus, in the absence of a clear theoretical foundation for setting particular thresholds for criteria involving declines in area or function (A, C, and D), we set threshold values at relatively even intervals for current and future declines (Vulnerable 30%, Endangered 50%, Critically Endangered 80%). The spread of thresholds between zero and 100% seeks to achieve an informative, rather than highly skewed ranking of ecosystems among the categories, while the lowest threshold of 30% recognises that an evidence of an appreciable decline in ecosystem distribution or function is necessary to support listing in a threatened category. These base thresholds are consistent with thresholds for population reduction in species Red List criteria (IUCN 2001). We set higher thresholds for historic declines (50%, 70%, 90%) because times frames are longer. Declines within 5–10% of VU thresholds may warrant listing as NT (Fig. 5), although we propose no quantitative thresholds for this category. Below, we explore the sensitivity of risk assessment outcomes to variation in these thresholds.

Collapse thresholds. Each of the five criteria implies a threshold of collapse (Fig. 1). For criteria based on spatial extent (A and B), ecosystems may be generally assumed to have collapsed if their distribution declines to zero (Fig. 1a–b) - when the ecosystem has undergone transformation throughout its entire range. However, use of the zero threshold will depend on the variables and maps used to represent the ecosystem distribution, and some ecosystems may collapse before their mapped distribution declines to zero (e.g. Table 2).

For criteria based on functional variables (C and D), a range of values will typically define collapse for a given variable (Fig. 1c–e). This range should be bounded between the minimum possible value, where there is no doubt that the ecosystem has collapsed, and a plausible maximum value based on observations of localised cases where the ecosystem appears to have moved beyond its natural range of variation (defined in the description of its

characteristic native biota and processes), and as a result has lost characteristic native biota (see Appendix S2 for examples). A similar approach can be applied when simulation models are used to estimate the risk of collapse under criterion E. The collapsed state(s) should be identified among those represented in the model and bounded thresholds of relative abundance and/or persistence should be specified to identify the bounds of natural variation in the system.

The Risk Assessment Criteria

The five risk assessment criteria are summarised in Table 3 and Appendix S1 contains a glossary of terms applied in the criteria and supporting concepts. Below we discuss the theoretical rationale that underpins each one and offer guidance for choosing and estimating the variables required to assess them.

Criterion A. Decline in Distribution

Theory. Declining distribution is an almost universal element of existing ecosystem risk assessment protocols [13] and is analogous to Caughley's declining population paradigm [92], as both represent diminishing abundance of biota. The diversity of species persisting within an ecosystem is positively related to the area or volume of substrate available [93]. Conversely, as ecosystem area declines, so do carrying capacities for component species, niche diversity and opportunities for spatial partitioning of resources and avoidance of competitors, predators and pathogens [87,94,95]. These area-related changes will increase extinction risks for component species and reduce an ecosystem's ability to sustain its characteristic biota (Fig. 2). As ecosystem area declines, the resulting loss of biota depends on its spatial pattern in relation to threats and conservation measures [96,97]. Although sampling effects preclude reversal of the quantitative species-area model [85], the qualitative relationship holds even for species that only

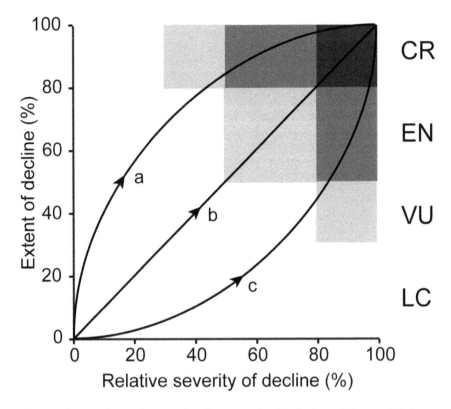

Figure 5. Contrasting pathways of environmental or biotic degradation and their corresponding risk classifications under criteria C and D. (a) initially widespread and benign degradation, later increasing in severity. (b) severity and extent of degradation increase at similar rates. (c) localised but severe degradation, later becoming more widespread. Ecosystems that just fail to meet the thresholds for Vulnerable status (e.g. extremely severe (>80%) decline in environmental quality over 20–30% of distribution, or severe (>30%) decline over 70–80% of distribution) may be assigned Near Threatened (NT) status.

lose unoccupied habitat, because such losses diminish opportunities for colonisation and rescue to compensate stochastic extirpations and declines [98].

Estimation. Rates of decline in ecosystem distribution will typically be estimated from time series of maps (e.g. [80]), field observations [65] or range maps constructed from point locations (e.g. [99]). Potential spatial proxies for ecosystem distributions include field observations of organism assemblages, climate, substrate, topography, bathymetry, ocean currents, flood regimes, aquifers or some synthesis of these that can be justified as valid representations of the distribution of ecosystem biota or its niche space. Vegetation mapping [100] and remote sensing [23] provide useful proxies for terrestrial, freshwater and benthic marine ecosystems [101]. The case studies (Appendix S2) provide a diversity of examples of such maps. For marine ecosystems, maps of physical factors such as sea floor characteristics, ocean currents, water temperatures and water chemistry may also be appropriate [49,102,103]. In some subterranean, freshwater and marine ecosystems, trends in the depth dimension may be appropriate proxies of declines in distribution (e.g. Table 2), so long as they reflect trends in carrying capacity and niche diversity for characteristic biota.

Current reductions in distribution may be calculated directly if data are available for 50 years ago and the present, or through an annual rate as a basis for cautious extrapolation. Spatial models [104] may be used for projecting expected distributions into the recent past (criterion A1, Table 3), future (criterion A2) or to estimate historic anthropogenic change (criterion A3) [105].

Criterion B. Restricted Distribution

Theory. Many processes that threaten ecosystems are spatially autocorrelated (clustered). Examples include catastrophes or disturbance events [106,107], localised invasions of alien species [108] and regional climate changes [74,109,110]. Risks posed by such processes are spread across multiple independent patches in widely distributed ecosystems, but not in ecosystems with geographically restricted distributions [13]. The primary role of criterion B is to identify ecosystems whose distribution is so restricted that they are at risk of collapse from the concurrence of threatening events or processes [13,79]. It also serves as an assessment of occupied habitat for component biota which, through carrying capacity, is positively related to population viability irrespective of exposure to catastrophic events [64]. These concepts are analogous to Caughley's (1994) small population paradigm [25,92], and are incorporated into most existing risk assessment protocols [13].

Estimation. Two metrics, Extent of Occurrence (EOO) and Area of Occupancy (AOO), represent conceptually different aspects of species range size [111] and are also relevant to ecosystems (Table 3). EOO (criterion B1) measures the ability to spread risks over a contiguous area that encloses all occurrences using a minimum convex polygon, whereas AOO (criterion B2) measures the ability to spread risks among occupied patches with a count of occupied grid cells [53,79,112]. The same measurement protocols are appropriate to entities with depth dimensions or linear patterns of distribution [25]. In some cases, spatial data may be insufficient to estimate EOO or AOO, but there is evidence that a small number of plausible threatening events may cause an ecosystem to become Critically Endangered within the near future. Such ecosystems may be listed as Vulnerable under criterion B3 if they occupy few 'locations' relative to the extent of threatening events (Appendix S1).

Estimates of AOO are highly sensitive to both spatial and thematic grain [13,79,113]. Ecosystems may be classified so broadly or mapped so coarsely that they never meet thresholds for threatened categories or, conversely, so narrowly or finely that they always qualify for threatened status [13]. To reduce bias, all estimates of AOO for Red List assessment must be standardized to the same spatial grain. We recommend 10×10 km grid cells for estimating ecosystem AOOs (in contrast to the 2×2 km grids recommended for species assessments; [79]), first because ecosystem boundaries are inherently vague (sensu [66]), so it is easier to determine that an ecosystem occurrence falls within a larger grid cell than a smaller one. Second, larger cells may be required to diagnose the presence of ecosystems characterized by processes that operate over large spatial scales, or diagnostic features that are sparse, cryptic, clustered or mobile (e.g. pelagic or artesian systems). Last, larger cells allow AOO estimation even when high resolution data are limited. These considerations therefore suggest that a larger cell size is appropriate for ecosystems than recommended for species [79]. A potential limitation of AOO estimates based on large grain sizes is that they may be inflated for ecosystems with many small, dispersed patches (e.g. forest fragments, small wetland patches), yet such occurrences may not substantially offset risks. To reduce this effect, we recommend that cells are counted as occupied only if the ecosystem covers more than 1 km^2 (1%) of cell area.

Thresholds and subcriteria. Critically Endangered, Endangered and Vulnerable ecosystems are delineated by AOO thresholds of two, 20 and 50 grid cells, respectively (Table 3). EOO thresholds were an order of magnitude larger (Table 3) because, like species, ecosystems generally extend across larger areas than they actually occupy [6]. We recognise that such thresholds are somewhat arbitrary and below, we explore the sensitivity of risk assessment outcomes to variation in the thresholds. However, the proposed thresholds are based on our collective experience on the extent of wildland fires, extreme weather events, chemical spills, disease epidemics, land conversion and other spatially explicit threats. Studies on the risks posed by spatial processes of varying extent are needed across a variety of ecosystems to inform the adequacy of these values.

To be eligible for listing in a threat category under criterion B, an ecosystem must also meet at least one of three subcriteria that address various forms of decline. These subcriteria distinguish restricted ecosystems at appreciable risk of collapse from those that persist over long time scales within small stable ranges [114,115]. Only qualitative evidence of decline is required to invoke the subcriteria, but declines must i) reduce the ability of an ecosystem to sustain its characteristic native biota; ii) be non-trivial in magnitude; and iii) be likely to continue into the future (Appendix S1). These declines may be in ecosystem distribution or processes (abiotic or biotic). Evidence of past declines is not essential, but future declines may be inferred from serious and imminent threats or occurrence at few locations, indicating limited capacity to spread risks [79].

Criterion C: Environmental Degradation

Theory. Environmental (abiotic) degradation may diminish the ability of an ecosystem to sustain its characteristic native biota by changing the variety and quality of environmental niche space available to individual species. This interpretation relies on measurement of abiotic variables and excludes biotic mechanisms

of degradation. Most existing protocols conflate the assessment of biotic and abiotic declines in ecosystem function [13]. In contrast, our risk assessment model defines separate assessment pathways (criteria C and D, Fig. 2) because the threats, their causes, effects and mechanisms of functional decline differ fundamentally between biotic and abiotic degradation, and hence so do the variables needed to assess them.

A reformulation of the species-area relationship [86] provides a theoretical basis for degradation criteria by incorporating the influence of habitat quality on the number of species able to persist in a given area. This model predicts bird species richness by including a habitat complexity score relative to an optimal value. We generalise this to an index of 'relative severity' of degradation, representing the ratio of observed change in environmental suitability (for ecosystem biota) over a given time to the amount of change that would cause an ecosystem to collapse (Fig. 6). Theoretically, suitability is aggregated across all characteristic biota, but in practice may be estimated from key environmental variables that regulate ecosystem behaviour (e.g. river flows for riparian wetlands, see examples in Appendix S2).

Criterion C (Table 3) is structured to account for ecosystems undergoing environmental degradation with contrasting scenarios of severity and extent (Fig. 5). Thus, ecosystems are only eligible for listing as Critically Endangered if environmental change that threatens the persistence of their characteristic biota is both extremely severe ($\geq 80\%$ relative severity) and extremely extensive ($\geq 80\%$ of the distribution). In contrast, those undergoing extremely severe but localised degradation or less severe degradation over very extensive areas may be eligible for listing in lower threat categories (Fig. 5).

Estimation. We suggest four requirements to assess risks posed to ecosystems by environmental degradation. First, there must be plausible evidence of a causal relationship between a process of environmental change and loss of characteristic native biota (Fig. 2). For example, an assessment of wetland degradation based on change in water quality would require evidence that decline in water quality was associated with loss of wetland biota, at least in comparable ecosystem types. Development of simple diagrammatic process models can help to make explicit the diagnosis of salient processes that influence transitions between functional and degraded ecosystem states, as well as the characteristics that differentiate the states [54,56]. Hence, these models serve the minimum requirements for inferring appropriate measures of environmental degradation for risk assessment (see examples in Appendix S2).

Second, assessing abiotic degradation requires suitable spatial and scalar variables for estimating the extent and severity of degradation. The characteristics of the ecosystem, environmental dependencies of biota and agents of degradation will determine which variables are relevant. The most suitable will be those with the most proximal cause-effect relationships and the greatest sensitivity to loss of biota. Approaches that apply generic indices across functionally contrasting ecosystems are unlikely to assess degradation accurately because salient processes may differ between ecosystems. Furthermore, aggregation of multiple variables could confound different mechanisms and directions of environmental change, making the index less sensitive to degradation than individual variables. Table 4 lists examples of potentially suitable abiotic variables for different ecosystems, while Appendix S2 provides more detailed justifications of variable selection for specific ecosystem types. For some ecosystems, it is noteworthy that measures of environmental heterogeneity may be more appropriate than absolute measures, because declines in the number of limiting resources (niche dimension) reduce species

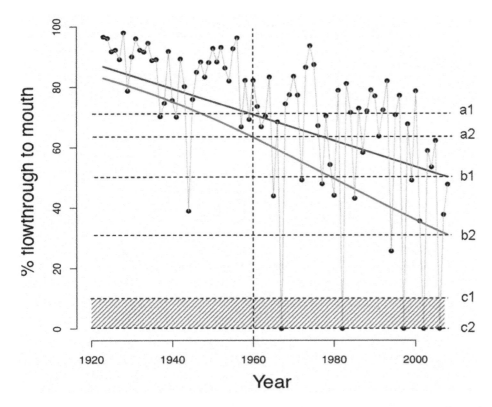

Figure 6. Estimation of relative severity of environmental degradation (criterion C) or disruption of biotic interactions (criterion D). Example using stream flowthrough data as percent of mean unregulated flows (aqua line joining filled circles) for the Murray River adapted from [57], see Appendix S2.8. There is uncertainty in both the rate of decline in flowthrough (two alternative regression lines) and the level of flowthrough at which the water-dependent ecosystem would collapse (shaded area). The threshold of collapse is the level of stream flowthrough that would result in widespread tree death and replacement of forest vegetation (most likely by shrubland). This was estimated to occur when mean flowthrough (as estimated by long-term regression) falls to 0–10% of unregulated flow levels (shown as a bounded estimate c1-c2, dashed lines), as widespread tree dieback began to occur when flowthrough was zero in several year of the past decade (see Appendix S2.8 for process model and justification). Based on a best-fit Gaussian regression model of the flowthrough data (dark blue line), the mean flowthrough fell from 71% in 1960 (dotted line a1) to 50% in 2009 (dotted line b1). A beta regression model (red line) gave an improved fit to the data and indicates a decline in mean flowthrough from 63% in 1960 (a2) to 31% in 2009 (b2). A standardised estimate of the relative severity of hydrological degradation over the past 50 years = $100 \times (b-a)/(c-a)$. The minimum plausible estimate = $100 \times (b1-a1)/(c1-a2) = 100 \times (71-50)/(71-0) = 30\%$ and the maximum plausible estimate = $100 \times (b2-a2)/(c2-a1) = 100 \times (63-31)/(63-10) = 60\%$. Based on uncertainty in the flowthrough regression models and collapse threshold, a bounded estimate of hydrological degradation in this ecosystem is therefore 30–60% over the past 50 years.

diversity in a range of terrestrial, freshwater and marine ecosystems [95].

Third, assessing environmental degradation requires calculation methods to compare observed or projected changes against the criteria. Assessors may either estimate the extent of degradation (as % of ecosystem distribution) that exceeds a threshold level of severity (Fig. 5) or estimate the average severity of degradation across the entire ecosystem distribution (100% of extent). 'Relative severity' measures the proportional progress of an ecosystem on a trajectory to collapse over the time frame of assessment, and is essential for comparing risks across ecosystems undergoing different types of degradation. It can be calculated by range-standardising the raw values of the degradation variable between its initial value and its collapse threshold (Fig. 6). This requires an assumption about the level of degradation that corresponds with collapse (Table 2), and a functional form for interpolation (e.g. linear). Comparisons with reference sites may justify these assumptions [116].

Finally, estimating, inferring or projecting the severity and extent of degradation over specific time frames may require extrapolation of trends from available time series. This requires assumptions about whether degradation is constant, accelerating,

or decelerating (see criterion A), based on an understanding of the mechanism of decline and its historical and spatial context. Assessors also need to evaluate whether the available data are sufficiently representative of prevailing conditions to permit extrapolation, preferably with statistical inference (but subjective reasoning may play a greater role when sample sizes are too small). Where time series data are unavailable, it may be possible to infer changes in degradation using space-for-time substitution sampling with appropriate reference sites [117,118].

Criterion D: Disruption of Biotic Processes and Interactions

Theory. The persistence of biota within ecosystems depends on biotic processes and interactions (Fig. 2), including competitive, predatory, facilitatory, mutualistic, trophic and pathogenic processes, as well as interactions between organisms and their physical environment, habitat fragmentation, mobile links (e.g. seasonal migration), species invasions and direct exploitation by humans. There is a growing body of theory and empirical evidence that biodiversity loss reduces the capacity of ecosystems to capture resources, produce biomass, decompose organic matter and recycle carbon, water and nutrients, and also that biodiversity

Table 4. Examples of variables potentially suitable for assessing the severity of environmental degradation under criterion C.

Degradation process	Example variables	Sources
Desertification of rangelands	Proportional cover of bare ground, soil density, soil compaction indices, remote sensing landcover indices	[159,160]
Eutrophication of soils, freshwater streams or lakes	Levels of dissolved or soil nitrogen, phosphorus, cations, oxygen, turbidity, bioassay	[15]
De-humidification of cloud forests	Cloud cover, cloud altitude	[161]
Deforestation by acid rain	Rain water chemistry	[62]
Homogenisation of microhabitats	Diversity of micro-terrain features, spatial variance in inundation depth and duration	[162]
Changed water regime or hydroperiod	Field-based monitoring of stream flow volume, or piezometric water table depth; remote sensing of spatial extent of surface water, frequency and depth of inundation	[57]
Salinisation of soils or wetlands	Field monitoring of salinity of soils or groundwater, remote sensing of ground surface albido	[163]
Sedimentation of streams, coral reefs	Sediment accumulation rates, sediment load of streams, discharge, turbidity of water column, frequency and intensity of sediment plume spectral signatures	[164]
Structural simplification of benthic marine ecosystems (e.g. by bottom trawling)	Microrelief, abundance of benthic debris, trawling frequency and spatial pattern	[165]
Sea level rise	Acoustic monitoring of sea level, extent of tidal inundation	[166]
Retreat of ice masses	Remote sensing of sea ice extent	[167]

loss reduces the stability of these functions through time [30]. Both the identity and diversity of organisms within a system control its functioning, firstly because key taxa make disproportionate contributions to particular functions, and secondly because niche partitioning and positive species interactions promote complementary contributions to function from individual species [30].

Feedback interactions underpin self-organisation and are crucial to ecosystem resilience, the ability to absorb environmental change while maintaining structure, characteristic biota and processes [119]. Conversely, significant disruptions to biotic processes and interactions can cause collapse, regime shift and re-organisation into a new entity that is unable to sustain the biota of the original system [35,74,120,121]. Diamond [122] identified trophic cascades caused by disruption to interactions as one of five major threats to biodiversity. Subsequent work has sought to identify factors that promote this mechanism of ecosystem collapse [123,124], although non-trophic interactions also play important roles [125,126].

Certain types of ecosystems may be especially sensitive to disruption of biotic processes and interactions. These include systems with strong top-down trophic regulation [58,124,127,128], systems with many mutualistic or facilitation interactions [126,129], systems that are strongly dependent on mobile links [130] and systems where disturbance regimes impose top-down regulation and positive feedbacks operate between the biota and the disturbance [131,132].

Estimation. Assessment of criterion D must address the same four requirements as criterion C: i) plausible evidence of the causes or mechanisms of functional decline; ii) selection of appropriate biotic variables for assessing declines; iii) range standardisation to estimate relative severity; and iv) calculations and justifiable assumptions to estimate declines over relevant time frames. Process models again provide a useful framework for interpretation and explicit justification of analytical choices. A broad set of variables are potentially useful for assessing biotic processes and associated functional declines (Table 5). We briefly review some strengths and weaknesses of alternatives below and present detailed examples of assessment in Appendix S2.

Species loss reduces ecosystem function and resilience to ecosystem collapse and reduces the possible range of alternative

ecological organizations [31,120]. Species richness is the simplest and most generic measure of this process (Table 5), but its sensitivity may be limited if declines in some species are lagged or offset by increases in others that do not perform similar functions [16]. Also, the functional consequences of species loss may not be apparent. Ecosystem collapse often involves changes in species composition and dominance [74]. These variables avoid some pitfalls of species richness, although it may be difficult to discriminate functional decline from natural variability in composition and dominance.

Problems with generic measures may be mitigated by variables that are more proximal to biotic mechanisms that maintain ecosystem resilience and characteristic biota [133]. Partitioning component species into functional types or guilds [134] allows more direct analysis of declines in function and resilience through trends in functional diversity, redundancy and complementarity [33,64,135,136,137,138]. The abundance, biomass or dominance of key native or alien species may be useful measures of functional decline (Table 5), so long as there is plausible evidence of their functional roles and their influence on the persistence of characteristic native biota. Declines in large herbivores and large predators, for example, may drastically affect the dynamics and functioning of ecosystems with top-down regulation [124,128,139]. Invasion of alien species may transform ecosystems through interactions as competitors, predators, pathogens or ecosystem engineers [108,140].

Measures of interaction diversity, such as the structure and size of interaction networks, provide another perspective on functional decline (Table 5). Decoupling of interactions may reduce diversity by preventing some species from completing their life cycles [126,129]. Trophic diversity (Table 5), a special case of interaction diversity where interactions are directional and hierarchical [141], can mediate co-existence, resilience and function in contrasting ecosystems [15,58,139,142].

Spatial dynamics of biotic interactions influence ecosystem resilience and function through exchanges across heterogeneous landscapes and seascapes [130]. Movements of organisms involve transfer of nutrients and genes, and may initiate local reorganization through episodic predation and ecosystem engineering. These exchanges provide spatial insurance for sustaining ecosys-

Table 5. Examples of biotic variables potentially suitable for assessing the severity of disruption to biotic interactions under criterion D.

Variable	Role in ecosystem resilience and function	Example
Species richness (number of species within a taxonomic group per unit area)	Ecological processes decline at an accelerating rate with loss of species [168]. Species richness is related indirectly to ecosystem function and resilience through its correlations with functional diversity, redundancy and complementarity (see below)	Response of graminoid diversity and relative abundance to varying levels of grazing in grassland [135].
Species composition and dominance	Shifts in dominance and community structure are symptoms of change in ecosystem behaviour and identity	Shift in diet of top predators (killer whales) due to overfishing effects on seals, caused decline of sea otters reduced predation of kelp-feeding urchins, causing their populations to explode with consequent collapse of giant kelp, structural dominants of the benthos [58]. See Appendix S2.
Abundance of key species (ecosystem engineers, keystone predators and herbivores, dominant competitors, structural dominants, transformer invasive species)	Invasions of certain alien species may alter ecosystem behaviour and identity, and make habitat unsuitable for persistence of some native biota. Transformer alien species are distinguished from benign invasions that do not greatly influence ecosystem function and dynamics	Invasion of crazy ants simplifies forest structure, reduces faunal diversity and native ecosystem engineers [108]. Invasion of arid Australian shrublands and grasslands by Buffel Grass makes them more fire prone and less favourable for persistence of native plant species [169,170].
Functional diversity (number and evenness of types)	High diversity of species functional types (e.g. resource use types, disturbance response types) promotes co-existence through resource partitioning, niche diversification and mutualisms [71]. Mechanisms similar to functional complementarity (see below).	High diversity of plant-derived resources sustains composition, diversity and function of soil biota [171], Fire regimes promote coexistence of multiple plant functional types [134]. Appendix S2.
Functional redundancy (number of taxa per type; within- and cross-scale redundancy; see (Allen et al. 2005)	Functionally equivalent minor species may substitute for loss or decline of dominants if many species perform similar functional roles (functional redundancy). Low species richness may be associated with low resilience and high risks to ecosystem function under environmental change [71,135].	Response of bird communities to varying levels of land use intensity [138].
Functional complementarity (dissimilarity between types or species)	Functional complementarity between species (e.g. in resource use, body size, stature, trophic status, phenology) enhances coexistence through niche partitioning and maintenance of ecosystem processes [172]	High functional complementarity within both plant and pollinator assemblages promotes recruitment of more diverse plant communities [125].
Interaction diversity (interaction frequencies and dominance, properties of network matrices)	Interactions shape the organisation of ecosystems, mediate evolution and persistence of participating species and influence ecosystem-level functions, e.g. productivity [173]	Overgrazing reduced diversity of pollination interactions [129].
Trophic diversity (number of trophic levels, interactions within levels, food web structure)	Compensatory effects of predation and resource competition maintain coexistence of inferior competitors and prey. Loss or reduction of some interactions (e.g. by overexploitation of top predators) may precipitate trophic cascades via competitive elimination or overabundance of generalist predators	Diverse carnivore assemblages (i.e. varied behaviour traits and densities) promote coexistence of plant species [142], decline of primary prey precipitates diet shifts and phase shifts [174].
Spatial flux of organisms (rate, timing, frequency and duration of species movements between ecosystems)	Spatial exchanges among local systems in heterogeneous landscapes provide spatial insurance for ecosystem function [143]. Exchanges may involve resources, genes or involvement in processes [130]	Herbivorous fish and invertebrates migrate into reefs from seagrass beds and mangroves, reducing algal abundance on reefs and maintaining suitable substrates for larval establishment of corals after disturbance [175].
Structural complexity (e.g.complexity indices, number and cover of vertical strata in forests, reefs, remote sensing indices)	Simplified architecture reduces niche diversity, providing suitable habitats for fewer species, greater exposure to predators or greater competition for resources (due to reduced partitioning)	Structurally complex coral reefs support greater fish diversity [176], structurally complex woodlands support greater bird diversity [86].

tem biota, both through spatial averaging and functional compensation [143,144]. Measures of disruption to these processes include changes in identity and frequency of species movements, and measures of fragmentation (Table 5).

Finally, niche diversity in some ecosystems depends on structural complexity generated by components of the biota itself (Table 5). For example, vegetation structure is often used as a measure of habitat suitability for forest and woodland fauna [86], while reef rugosity is similarly used to evaluate habitat suitability for fish and some marine invertebrates [145]. As well as being salient representations of diversity in a range of ecosystems, data on structural complexity can relative inexpensive to obtain in the field, and some indices lend themselves to remote sensing.

Criterion E. Quantitative Estimates of Risk of Ecosystem Collapse

Theory and estimation. A diverse range of simulation models of ecosystem dynamics allow the probability of ecosystem collapse to be estimated directly over the same 50-year future period as other criteria [59,60,136,146,147,148,149]. These models permit exploration of interactions and potential synergies between multiple mechanisms of collapse. This distinguishes direct risk estimation from the other criteria, each of which assess separate mechanisms through particular symptoms of risk (Fig. 2). Even where available data preclude construction of quantitative simulation models, criterion E provides a useful anchor for risk assessment and an overarching framework for other criteria, as its analogue does in Red List criteria for species [25]. Although development of simulation models was beyond the scope of this paper, we demonstrate criterion E with an existing model in a case study on the Coorong Lagoon in Appendix S2.

Case Studies

Sample Ecosystems

Twenty ecosystems were selected for assessment based on the authors' areas of expertise, spanning five continents and three ocean basins (full details of assessments in Appendix S2). Although non-random, the selection encompassed terrestrial, subterranean, continental aquatic and marine aquatic environments in Europe, Africa, Asia, Australasia and the Americas and represented a wide range of thematic scales, threatening processes, data availability and levels of risk. Each ecosystem was assessed using the protocol in Fig. 3. The ecosystems assessed are summarised in Table 6.

Data Availability

Data were available to assess all five criteria in one ecosystem, four criteria in five ecosystems, three criteria in seven ecosystems and two criteria for the remainder (Table 6). Data were most commonly available to assess criterion B, followed by A, C and D, with only one ecosystem, the Coorong Lagoon, assessed for E (Fig. 7). The number of assessable subcriteria varied between ecosystems from two to 12, with at least seven of the 13 subcriteria assessed in half of the case studies (Table 6). All but four of the ecosystems (80%) had sufficient data to assess at least one distributional criterion (A or B) and one functional criterion (C or D).

The majority of terrestrial and freshwater case studies assessing criteria A and B used vegetation maps as spatial proxies to estimate ecosystem distributions, while some of the marine case studies used specialised map products derived from remote sensing. Estimates of current change in distribution were derived from time series of maps or imagery, almost all of which required reasoned assumptions to justify interpolation or extrapolation to the required 50-year time frame. Historical changes in distribution were most commonly inferred by comparing a contemporary map with a model of environmentally suitable areas which were assumed to be occupied by the ecosystem prior to human transformation of the landscape. This approach was less suitable for marine ecosystems, which were generally Data Deficient in criterion A3. In three ecosystems (Coastal upland swamps, River Red Gum forests, Cape fynbos), models of environmental suitability were used to project future changes in distribution, with outputs of alternative plausible models used to estimate uncertainty in the projections.

Eleven of the case studies used explicit process models to guide selection of functional variables for assessment of criteria C and D. Only one of these models was quantitative, permitting simulations

to estimate risks of collapse under criterion E, although data appear sufficient to support construction of such models in at least two other case studies (1 and 8). A variety of abiotic proxy variables were used to assess environmental degradation, primarily in freshwater and marine ecosystems, including water flows and extraction rates, groundwater flows (subterranean/freshwater) nitrogen levels (both freshwater and marine ecosystems), climatic moisture, water volume, salinity, sea surface temperatures and ocean acidity. Proxy variables used to assess criterion D included the abundance of structurally important groups of species (resprouting shrubs, corals, kelp, seagrass), mobile links (birds), meso-predators (sea otters, fish), sensitive species (plankton), invasive species and threatened species. In a few cases, the available data were insufficient to make an assessment, but the identification of the proxy highlighted future needs.

Assessment Outcomes

The outcomes of assessment varied from Least Concern to Collapsed (Table 6), with the overall status supported by multiple subcriteria for all but four of the ecosystems. In the four ecosystems for which overall status was supported by a single subcriterion, another subcriterion was assessed at the next lowest category of risk. Three ecosystems that were assessed as Least Concern or Collapsed were supported by 8–11 subcriteria. All of the criteria except E determined the overall status in multiple ecosystems, with criterion B yielding the highest threat in a lower proportion of ecosystems than A, C and D (Fig. 7). Nine of the ecosystem types selected for case studies had been assessed by government agencies or non-government organisations using local listing criteria. For eight of these nine case studies, the IUCN protocol produced the same threat status as those produced by local authorities. The status of the remaining ecosystem differed by only one category.

Sensitivity Analysis of Thresholds

A sensitivity analysis was carried out on the thresholds in all criteria using data from the 20 case studies. Thresholds were adjusted by ±5%, ±10%, ±15% and ±20% of current values i) for each individual subcriterion; ii) for all subcriteria in combination within each criterion; and iii) across all criteria in combination. This represents a plausible range of alternative thresholds, since larger adjustments would result in overlap between categories. Variation to thresholds by a given proportion across all criteria in combination resulted in a change in status for a slightly larger proportion of ecosystems (Figure 8a). For half of the ecosystems that changed status, however, the changes were within the bounds of uncertainty for the original assessment. The proportion of ecosystems that changed status outside the bounds of uncertainty were approximately commensurate with the proportional adjustment to thresholds. For example a 5% change in thresholds produced a change in status in approximately 5% of ecosystems, while a 20% change in thresholds produced a change in status for approximately 20–25% of ecosystems, depending on whether thresholds were increased or decreased. Although the sample size is limited, the results suggest moderate sensitivity of overall risk assessment outcomes to the thresholds, particularly as the case studies used for this analysis cover a wide variety of ecosystem types and data availability.

Individually, criteria A, C and D displayed similar levels of sensitivity to variation in their threshold values (allowing for different levels of data availability), and this was similar to the sensitivity of the overall risk status when all five criteria were combined (Figs. 8b, 8d, 8e cf. 8a). Criterion B was relatively insensitive, with only 5–10% of ecosystems changing status outside the bounds of uncertainty when thresholds were adjusted by

Table 6. Summary of trial assessments for 17 ecosystems from freshwater (F), terrestrial (T), marine (M) and subterranean (S) environments.

	Local threat status	IUCN status	# criteria assessed	# subcriteria assessed	# subcriteria supporting overall status	Spatial criteria assessed	Functional criteria assessed	Criteria determining overall status
1 Coastal sandstone upland swamps, Australia (F)	EN	EN-CR	4	9	2	+	+	A2,C2
2 Raised bogs, Germany	CR	CR	3	6	2	+	+	A3,C3
3 German tamarisk pioneer vegetation, Europe (F)	EN	EN	2	5	3	+		A1,A3, B2a,b
4 Swamps, marshes and lakes in the Murray-Darling Basin, Australia (F)	NE	EN-CR	4	10	2	+	+	D1,D3
5 Aral Sea, Uzbekistan and Kazakhstan (F)		CO	4	12	9	+	+	A1-3, C1-3, D1-3
6 Reedbeds, Europe (F)	LC	VU	4	8	3	+	+	A1,A3,D1
7 Gonakier forests of Senegal River floodplain (F)		CR	3	6	2	+	+	A1,A3
8 Floodplain Ecosystem of river red gum and black box, south-eastern Australia (F)	NE	VU	4	12	3	+	+	A2,C1,C2
9 Coolibah - Black Box woodland, Australia (F/T)	EN	EN	3	7	1	+	+	C1
10 Semi-evergreen vine thicket, Australia (T)	EN	EN	2	2	2	+	+	A3,B2a
11 Tepui shrubland, Venezuela (T)	LC	LC	3	8	8	+	+	A1-3,B1-3, D1-3
12 Granite gravel fields & sandplains, New Zealand (T)	LC	LC	4	11	11	+	+	A1-3,B1-3, C1-3,D1-3
13 Cape Sand Flats Fynbos, South Africa (T)	CR	CR	2	6	1	+	+	B1a,b
14 Tapia Forest, Madagascar (T)	NE	EN	2	4	1	+	+	A3
15 Great Lakes Alvar (T)		VU-EN	3	5	1	+	+	A3
16 Giant kelp forests, Alaska (M)	NE	EN-CR	2	4	2	+	+	D1,D3
17 Caribbean coral reefs (M)	NE	EN-CR	2	5	2	+	+	D1,D3
18 Seagrass meadows, South Australia (M)	NE	EN-CR	3	6	2	+	+	A1,C1
19 Coorong lagoons, Australia (F/M)	NE	CR	5	9	4	+	+	B1a,b,C2, D1,E
20 Karst rising springs, South Australia (C/F)	NE	CR	3	7	3	+	+	B1b,C1,C2

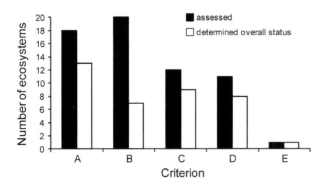

Figure 7. Number of ecosystems assessed for each criterion and number for which each criterion determined overall status.

±20% (Fig. 8c). The only ecosystem assessable under criterion E (case study 19, Appendix S2) did not change status when criterion E thresholds were varied by up to 20% (Fig. 8f). The sensitivity of individual subcriteria (not shown) was similar to the criteria to which they belong.

Performance of the Protocol

Several aspects of the case studies show that the IUCN Red List criteria for ecosystems are workable, robust and sufficiently general for application to wide range of ecosystems types and threatening processes. Firstly, the overall status was supported by assessments of multiple subcriteria in 90% of the case studies. This high level of concordance among criteria suggests that assessments are robust because outcomes are unlikely to be very sensitive to missing data.

Secondly, no one criterion had a consistently dominant or subordinate effect on overall status across the full set of case studies. This suggests strong complementarity among criteria. Collectively, they are able to detect symptoms that may signal the susceptibility of an ecosystem to any of several contrasting threatening processes.

Thirdly, close correspondence between Red List status and prior assessments carried out by local experts suggest that the IUCN criteria should not produce markedly different outcomes to most listing processes that currently operate in national and regional jurisdictions.

Fourthly, although poorly studied ecosystems were under-sampled in our analysis, the case studies show that suitable data can be obtained from a range of sources and that defensible inferences may be drawn from appropriate use of proxies, various methods of estimation and scaling up.

Several aspects of protocol performance may be attributed to their rule-based structure. This structure promotes the ensemble properties of criteria, minimises the impact of missing data and avoids assumptions that different symptoms are additive or interchangeable in their effect on overall risk of ecosystem collapse [112]. A potential disadvantage of a rule-based structure is that it may underestimate risk if data on the most limiting criteria are lacking or if there are synergistic interactions between different mechanisms of threat [150]. Such interactions can be built into simulation models and used to assess risks of collapse under criterion E.

Discussion

Generality and Consistency

Our assessments of widely contrasting ecosystems from terrestrial, subterranean, freshwater and marine environments demonstrate the generality of the Red List criteria. A key feature of our risk assessment model (Figs. 1 and 2) is its generic framework for selecting and assessing ecosystem-specific biotic and abiotic variables to estimate the relative severity of declines in ecosystem function. Range standardisation of severity allows functional changes to be assessed in a wide range of ecosystems against a common set of thresholds. It also forces assessors to be explicit about their choice of functional variable and its threshold values that signal ecosystem collapse.

The common set of thresholds of decline and distribution size that delimit different categories of risk promotes consistency of risk assessments across contrasting terrestrial, subterranean, freshwater and marine ecosystems. Current theory provides limited guidance for setting the precise values of these thresholds. Our choice of thresholds was aimed at promoting informative risk categories based on relatively even intervals of decline, alignment with thresholds of decline in the species Red List protocol, consistency with the monotonic relationships for species - area and biodiversity - ecosystem function, and a broad understanding of the spatial extent of threatening processes. Although these pragmatic principles could also be met by slightly different threshold values, risk assessment outcomes were shown to be only moderately sensitive to variations in decline thresholds and relatively insensitive to variations in thresholds of distribution size. In the most extreme cases, the proportional change in risk classifications was only slightly greater than the proportional adjustment of the thresholds.

Although the flexibility to select appropriate variables for assessment underpins the generality of the protocol, this may have trade-offs if selections are poorly justified. These trade-offs may affect the consistency of assessments if, for example, different assessors select different proxy variables to assess the same or closely related ecosystem(s). An alternative risk assessment method could limit such inconsistency by prescribing one or a few mandatory generic variables to assess functional change (e.g. species richness, productivity, aggregated indices of condition, health or landscape geometry), but only by sacrificing alternative variables that are more proximal to causes and/or more sensitive to functional change. Moreover, a failure to apply ecosystem-specific mechanistic interpretations to trends in generic variables runs a risk of perverse assessment outcomes.

Some inconsistencies between assessments are an inevitable consequence of a risk model that seeks broad generality by incorporating flexibility to select ecosystem-specific measures of function. However, these inconsistencies can be partially offset, firstly by governance processes and standards that promote collaboration and critical evaluation of assessment outcomes (see below), and secondly by using methods to deal with uncertainties described above. Thirdly, the use of cause/effect process models to interpret salient processes and their proxies should mitigate inconsistency, especially if they are critically reviewed, either through peer-reviewed literature or though a structured elicitation process [19,151]. These models provide a useful basis and context for distinguishing natural variability from functional decline, and help to translate general ecosystem concepts into usable tools [42].

The use of standardised measures of distribution in criterion B also contributes to generality of the protocol and mitigates some of its sensitivity to spatial scale [13]. The agreement between our assessments and those of local authorities for both broadly and

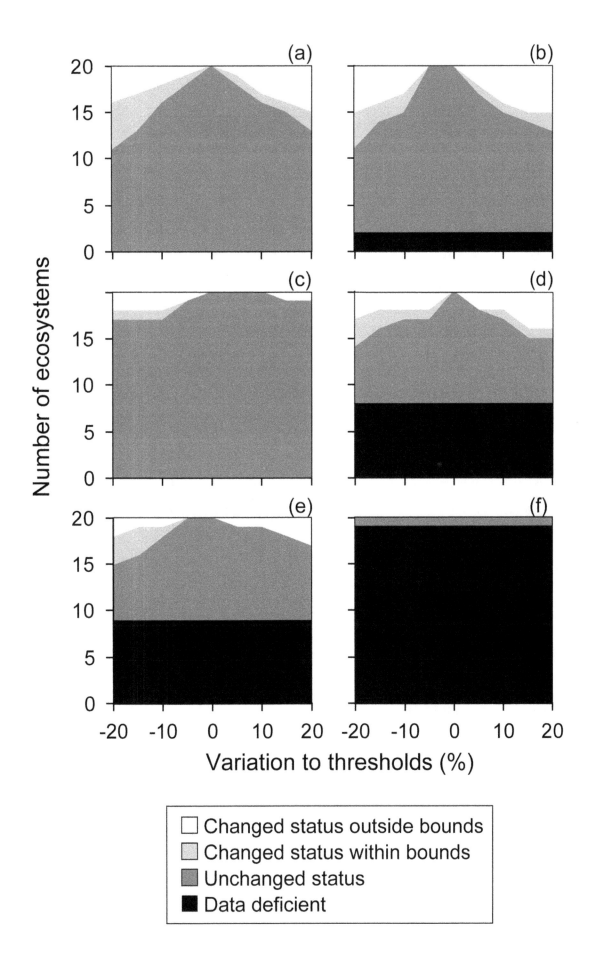

Figure 8. Sensitivity of risk assessment outcomes (relative to uncertainty bounds of the original assessment) to variation in threshold values for (a) all five criteria in combination; (b) criterion A only; (c) criterion B only; (d) criterion C only; (e) criterion D only; and (f) criterion E only.

narrowly defined ecosystems suggests some robustness to variation in thematic resolution. Nevertheless, risk assessments may be exposed to methodological artefacts if units are defined broadly or too finely. Data will often be more uncertain, fragmentary and more limited as the thematic resolution of assessment units increases and the available data are consequently subdivided among more units. Similarly, if the spatial domain of assessment is too small to consider relevant spatial processes, the outcomes of assessments may simply reflect patch dynamics. Further work is needed to define the limits of scale at which the criteria may be validly applied, and to develop methods to reduce scale-sensitive bias in the assessments as those limits are reached. This will support applications at fine thematic scales, which are sometimes needed for land use planning under national regulatory and legal frameworks (e.g. [43]).

Uncertainty

Assessments of ecosystem risk will always carry some uncertainty due to incomplete knowledge. This includes measurement uncertainty related to data availability, boundary vagueness and system variability, as well as model uncertainty (including selection of functional variables, see below) due to imperfect understanding of processes. Risk assessments of ecosystems will generally be less certain than species assessments (Fig. 1), largely because of conceptual generalities required to accommodate assessments of a broad range of ecosystems (see below). Some components of measurement uncertainty, such as detectability, however, may be greater in magnitude for many species than ecosystems.

Uncertainties can be incorporated into risk assessment using bounded estimates (Fig. 6; Appendix S2), fuzzy arithmetic, structured elicitation or Bayesian approaches [19]. Model uncertainty may be accommodated by carrying out multiple assessments based on plausible alternative process models [66]. Very high levels of uncertainty may preclude meaningful assessments of any of the criteria, in which case an assessment will produce a 'Data Deficient' outcome. However, close collaboration between spatial scientists and process ecologists should ensure that both distributional and functional symptoms of risk are addressed as comprehensively as possible.

Assessment Units

Unlike species, a widely accepted global classification of ecosystems is currently lacking. Development of a global taxonomy and classification of ecosystems would strengthen the consistency and comparability of assessments between regions and terrestrial/marine realms. It would also help resolve the limits of thematic scaling discussed above. The principal difficulties in delineating units of assessment stem from conceptual uncertainties in the nature of ecosystem properties, with conflicting discrete and continuum models both having strengths and limitations [43]. Abiotic elements of ecosystems are characteristically continuous, creating uncertain boundaries, although zones of transition may be identified where spatial turnover is high relative to adjacent areas, creating the appearance of discrete units at particular scales [152]. Further uncertainties stem from boundary dynamism or divergence between compositional, physical and functional boundaries [45,62].

In comparison, the global taxonomy for species appears well established and plays an important role in defining units for risk

assessment. In recent decades, however, development of cladistic methods and advent of molecular phylogenies are driving a major reconstruction of classifications at multiple levels to resolve polyphyletic taxa. Ongoing alpha taxonomic activity continually increases the number of described taxa, often resulting in new circumscriptions of existing taxa affected by splitting or lumping. Furthermore, the current operational taxonomic units are based on different morphological, biological or evolutionary species concepts, depending on the major taxonomic groups to which they belong, partly for pragmatic reasons and partly due to historical legacies. Successive Red Data Books and Red Lists have thus developed under substantial taxonomic dynamism and inconsistency. This suggests that Red Lists can be functional and reliable conservation tools despite uncertainties in the underlying classification, even though some changes in listings occur solely as a consequence of taxonomic changes [153].

We suggest that development of a global taxonomy for ecosystems can proceed contemporaneously with risk assessment. Indeed, the shortcomings of existing regional taxonomies underscore the need to describe characteristic biota, abiotic features, distribution and an ecological process model as integral components of ecosystem risk assessment. Ideally, the taxonomic framework should be hierarchical, elucidating relationships between assessment units defined at different scales and integrating elements of existing work at global, regional and national levels across terrestrial, subterranean, freshwater and marine environments biomes [48,49,51,100,154,155,156]. Such a framework would permit assessment at multiple thematic scales to suit different needs, including subglobal applications that provide essential support for local conservation planning [157].

Governance

Developing a Red List of ecosystems will involve ongoing questions about ecosystem description, variable selection, data analysis and model development. This requires a governance structure that promotes technical support and rigorous peer review. Preparation of interpretive guidelines (cf. [79]) and regional training initiatives will build individual and institutional capacity to support a global network of assessors and scientific reviewers, similar to the species specialist groups and the Standards and Petitions Committee within IUCN's Species Survival Commission (see http: //www. iucn.org/about/work/programmes/species/about_ssc/specialist_groups/directory_specialist_groups/).

Conclusion

The Red List criteria for ecosystems will establish a consistent, robust, practical and theoretically grounded international standard for risk assessment of biodiversity, complementing the Red List criteria for species. A global Red List can raise awareness of conservation needs in governments, industries and communities worldwide. However, guidelines are also needed to support assessment at regional and national scales, where much conservation action is planned and implemented. A Red List of ecosystems will firstly strengthen global capacity to report on and monitor the status of biodiversity under internationally agreed Aichi targets [39]. Secondly, it will inform priorities and decisions in planning for land and water use, establishment and management of protected areas, economic development and investments under different governance regimes. The latter includes local community projects and international finance of major development projects that are evaluated against environmental risk standards (http: //www.equator-principles.com/). The separate task of setting priorities for these actions also requires inputs on irreplaceability of biodiversity features, cultural valuations, plasticity of demand for ecosystem services and the potential for investments to reduce risks of decline [40,158]. Finally, an understanding of key services contributed by each ecosystem and the relationship between the symptoms of risk and delivery of services will help the Red List inform sustainable use of ecosystem services. Forging these links will help to avoid scenarios such as the collapse of the Aral Sea ecosystem, which has lead to collapse of a viable fishing industry and declines in human health associated with dust and chemical aerosols liberated from the dry sea bed [78].

Many of the mechanisms and symptoms of species vulnerability are relevant to ecosystems, because species are integral parts of ecosystems. Yet ecosystems embody processes and higher-order components of biodiversity that are difficult or impossible to account for in species-by-species assessment. Whereas species risk assessment rests on population theory, ecosystem risk assessment must draw from a wider array of inter-related theories that deal with continua, niches, fractal geometry, succession, resilience, ecological integrity, biodiversity-ecosystem function and insurance, as well as population theory. The success of ecosystem risk assessment therefore rests on a robust synthesis of conservation planning and process ecology to translate theoretical foundations into a practical assessment protocol that can be applied to a wide variety of ecosystems by specialist assessors with differing backgrounds and limited data.

Acknowledgments

Participants of six international workshops hosted by the Zoological Society of London (UK), NatureServe (Washington DC, USA), the Smithsonian Institution (Washington DC, USA), Tour du Valat (Arles, France), the Centre of Excellence for Environmental Decisions (Melbourne, Australia) and Centre de Suive Ecologique (Dakar, Senegal). The process to develop Red List criteria for ecosystems was launched with a resolution at the fourth IUCN (International Union for Conservation of Nature) World Conservation Congress in 2008, and consolidated with another resolution adopted by the fifth World Conservation Congress in 2012. French and Spanish translations of this article, as well as other key documents and publications, may be found at the website of the IUCN Red List of Ecosystems (www.iucnredlistofecosystems.org). We invite the members of the global conservation community to translate this publication into other languages and make it available at this website as well.

Author Contributions

Conceived and designed the experiments: DAK JPR KMR EN KA AA MA SB AB EGB JSB MJB RB TMB MAB PC FAC FE DFL PGF RJH MJ RTK REL RM MAM PP BP TJR UR MDS. Performed the experiments: DAK JPR SB AB JSB MJB RB FE PGF RJH RTK REL RM JM MAOM PP BP UR MDS. Analyzed the data: DAK JPR KMR EN MA SB AB JSB MJB RB FE PGF RJH RTK REL RM JM MAOM PP BP UR MDS SZM. Wrote the paper: DAK JPR KMR EN KA MA SB AB EGB JSB MJB RB TMB MAB PC FAC FE DFL PGF RJH MJ RTK REL RM MAM PP BP TJR UR MDS.

References

1. Vitousek PM, Mooney HA, Lubchenco J, Melillo JM (1997) Human domination of Earth's ecosystems. Science 277: 494–499.
2. Dirzo R, Raven PH (2003) Global state of biodiversity and loss. Annual Review of Environment and Resources 28: 137–167.
3. Hoekstra JM, Boucher TM, Ricketts TM, Roberts C (2005) Confronting a biome crisis: global disparities of habitat loss and protection. Ecology Letters 8: 23–29.
4. Butchart SHM, Walpole M, Collen B, van Strien A, Scharlemann JPW, et al. (2010) Global biodiversity: indicators of recent declines. Science 328: 1164–1168.
5. Ayensu E, Claasen D van R, Collins M, Dearing A, Fresco L, et al. (1999) International ecosystem assessment. Science 286: 685–686.
6. IUCN (2001) Red List categories and criteria. IUCN: Gland.
7. Baillie JEM, Hilton-Taylor C, Stuart SN (2004) IUCN Red List of Threatened Species. A Global Species Assessment. IUCN: Gland and Cambridge.
8. Butchart SHM, Stattersfield AJ, Bennun LA, Shutes SM, Akçakaya HR, et al. (2004) Measuring global trends in the status of biodiversity: Red List Indices for birds. PLoS Biology 2: 2294–2304.
9. McCarthy MA, Thompson CJ, Garnett ST (2008) Optimal investment in conservation of species. Journal of Applied Ecology 45: 1428–1435.
10. Rodrigues ASL, Pilgrim JD, Lamoreux JF, Hoffmann M, Brooks TM (2006) The value of the IUCN Red List for conservation. Trends in Ecology and Evolution 21: 71–76.
11. Noss RF (1996) Ecosystems as conservation targets. Trends in Ecology and Evolution. 11: 351.
12. Rodriguez JP, Rodriguez-Clark K, Baillie JE, Ash N, Benson J, et al. (2011) Establishing IUCN Red List Criteria for Threatened Ecosystems. Conservation Biology 25: 21–29.
13. Nicholson E, Keith DA, Wilcove DS (2009) Assessing the conservation status of ecological communities. Conservation Biology 23: 259–274.
14. Rodríguez JP, Rodríguez-Clark KM, Keith DA, Barrow EG, Benson J, et al. (2012) IUCN Red List of Ecosystems. Sapiens 5: 6–70.
15. Carpenter SR (2003) Regime shifts in lake systems: patterns and variation. Excellence in ecology series. Ecology Institute: Olderdorf/Luhe.
16. Srivastava DS, Vellend M (2005) Biodiversity – ecosystem function research: Is it relevant to conservation? Annual Review of Ecology, Evolution, and Systematics 36: 267–294.
17. Loreau M (2010) Linking biodiversity and ecosystems: towards a unifying ecological theory. Philosophical Transactions of the Royal Society B 365: 49–60.
18. Akçakaya HR, Ferson S, Burgman MA, Keith DA, Mace GM, et al. (2000) Making consistent IUCN classifications under uncertainty. Conservation Biology 14: 1001–1013.
19. Burgman MA (2005) Risks and decisions for conservation and environmental management. Cambridge University Press: Cambridge.
20. Stoddard JL, Herlihy AT, Peck DV, Hughes RM, Whittier TR, et al. (2008) A process for creating multimetric indices for large-scale aquatic surveys. Journal of the North American Benthological Society 27: 878–891.
21. Tierney GL, Faber-Langendoen D, Mitchell BR, Shriver G, Gibbs J (2009) Monitoring and evaluating the ecological integrity of forest ecosystems. Frontiers in Ecology and the Environment 7: 308–316.
22. Patrick WS, Spencer P, Link J, Cope J, Field J, et al. (2010) Using productivity and susceptibility indices to assess the vulnerability of United States fish stocks to overfishing. Fisheries Bulletin 108: 305–322.
23. Curran,L, Trigg SN (2006) Sustainability science from space: quantifying forest disturbance and land use dynamics in the Amazon. Proceedings of the National Academy of Science 103: 12663–12664.
24. Lindenmayer DB, Likens GE, Andersen A, Bowman D, Bull M, et al. (2012) The importance of long-term studies in ecology. Austral Ecology 37: 745–757.
25. Mace GM, Collar N, Gaston KJ, Hilton-Taylor C, Akçakaya HR, et al. (2008) Quantification of extinction risk: IUCN's system for classifying threatened species. Conservation Biology 22: 1424–1442.
26. Blab J, Riecken U, Ssymank A (1995) Proposal on a criteria system for a National Red Data Book of Biotopes. Landscape Ecology 10: 41–50.
27. Benson JS (2006) New South Wales vegetation classification and assessment: Introduction – the classification, database, assessment of protected areas and threat status of plant communities. Cunninghamia 9: 331–381.
28. Rodriguez JP, Balch JK, Rodriguez-Clark KM (2007) Assessing extinction risk in the absence of species-level data: quantitative criteria for terrestrial ecosystems. Biodiversity and Conservation 16: 183–209.
29. Riecken U (2002) Novellierung des Bundesnaturschutzgesetzes: Gesetzlich geschützte Biotope nach Paragraph 30. Natur und Landschaft 77: 397–406.
30. Cardinale BJ, Duffy JE, Gonzalez A, Hooper DU, Perrings C, et al. (2012) Biodiversity loss and its impact on humanity. Nature 486: 59–67.

31. Hooper DU, Adair EC, Cardinale BJ, Byrnes JEK, Hungate BA, et al. (2012) A global synthesis reveals biodiversity loss as a major driver of ecosystem change. Nature 486: 105–U129.

32. Tilman D, Reich PB, Knops J, Wedin D, Mielke T, et al. (2001) Diversity and productivity in a longterm grassland experiment. Science 294: 843–845.

33. Heemsbergen DA, Berg MP, Loreau M, van Haj JR, Faber JH, et al. (2004) Biodiversity effects on soil processes explained by interspecific functional dissimilarity. Science 306: 1019–1020.

34. Hooper DU, Chapin FS, Ewel JJ, Hector A, Inchausti P, et al. (2005) Effects of biodiversity on ecosystem functioning: A consensus of current knowledge. Ecological Monographs 75: 3–35.

35. Thebault E, Loreau M (2005) Trophic interactions and the relationship between species diversity and ecosystem stability. The American Naturalist 166: E95–E114.

36. Danovaro R, Gambi C, Dell'Anno A, Corinaldesi C, Fraschetti S, et al. (2008) Exponential decline of deep-sea ecosystem functioning linked to benthic biodiversity loss. Current Biology 18: 1–8.

37. Isbell F, Calcagno V, Hector A, Connolly J, Harpole WS, et al. (2011) High plant diversity is needed to maintain ecosystem services. Nature 477: 199–202.

38. Hector A, Bagchi R (2007) Biodiversity and ecosystem multifunctionality. Nature 448: 188–191.

39. CBD (1992) The Convention on Biological Diversity. CBD: Montréal.

40. Cowling RM, Egoh B, Knight AT, O'Farrell PJ, Reyers B, et al. (2008) An operational model for mainstreaming ecosystem services for implementation. Proceedings of the National Academy of Sciences of the United States of America 105: 9483–9488.

41. Tansley AG (1935) The use and abuse of vegetational concepts and terms. Ecology 16: 284–307.

42. Pickett STA, Cadenasso ML (2002) The ecosystem as a multidimensional concept: meaning, model, and metaphor. Ecosystems 5: 1–10.

43. Keith DA (2009) The interpretation, assessment and conservation of ecological communities and ecosystems. Ecological Management and Restoration 10: S3–S15.

44. Wiens JA (1989) Spatial scaling in ecology. Functional Ecology 3: 385–397.

45. Jax K, Jones CG, Pickett STA (1998) The self-identity of ecological units. Oikos 82: 253–264.

46. Whittaker RH (1972) Evolution and measurement of species diversity. Taxon 21: 213–251.

47. Willis AJ (1997) The ecosystem: An evolving concept viewed historically. Functional Ecology 11: 268–271.

48. Olson DM, Dinerstein E, Wikramanayake ED, Burgess ND, Powell GVN, et al. (2001) Terrestrial Ecoregions of the World: A New Map of Life on Earth. Bioscience 51: 933–938.

49. Spalding MD, Fox HE, Allen GR, Davidson N, Ferdaña ZA, et al. (2007) Marine ecoregions of the world: a bioregionalization of coastal and shelf areas. Bioscience 57: 573–583.

50. Faber-Langendoen D, Tart DL, Crawford RH (2009) Contours of the revised U.S. National Vegetation Classification standard. Bulletin of the Ecological Society of America 90: 87–93.

51. Jennings MD, Faber-Langendoen D, Loucks OL, Peet RK, Roberts D (2009) Standards for associations and alliances of the U.S. National Vegetation Classification. Ecological Monographs 79: 173–199.

52. Chytrý M, Tichý L, Holt J, Botta-Dukát Z (2002) Determination of diagnostic species with statistical fidelity measures. Journal of Vegetation Science 13: 79–90.

53. Gaston KJ, Fuller RA (2007) Commonness, population depletion and conservation biology. Trends in Ecology and Evolution 23: 14–19.

54. Hobbs RJ, Suding KN (2009) New models for ecosystem dynamics and restoration. Island Press: Washington DC.

55. Westoby M, Walker B, Noy-Meir I (1989) Opportunistic management for rangelands not at equilibrium. Journal of Range Management 42: 266–274.

56. Briske DD, Fuhlendorf SD, Smeins FE (2005) State-and-transition models, thresholds, and rangeland health: a synthesis of ecological concepts and perspectives. Rangeland Ecology and Management 58: 1–10.

57. Mac Nally R, Cunningham SC, Baker PJ, Horner GJ, Thomson JR (2011) Dynamics of Murray-Darling floodplain forests under multiple stressors: The past, present, and future of an Australian icon. Water Resources Research 47: W00G05.

58. Estes JA, Doak DF, Springer AM, Williams TM (2009) Causes and consequences of marine mammal population declines in southwest Alaska: a food-web perspective. Philosophical Transactions of the Royal Society of London - Series B 364: 1647–1658.

59. Rumpff L, Duncan DH, Vesk PA, Keith DA, Wintle BA (2011) State-and-transition modelling for Adaptive Management of native woodlands. Biological Conservation 144: 1224–1236.

60. Lester RE, Fairweather PJ (2011) Creating a data-derived, ecosystem-scale ecological response model that is explicit in space and time. Ecological modelling 222: 2690–2703.

61. Fulton EA, Link JS, Kaplan IC, Savina-Rolland M, Johnson P, et al. (2011) Lessons in modelling and management of marine ecosystems: the Atlantis experience. Fish Fisheries 12: 171–188.

62. Likens GE (1992) The ecosystem approach: its use and abuse. Ecology Institute, Oldendorf: Luhe.

63. Hallenbeck WH (1986) Quantitative risk assessment for environmental and occupational health. Lewis Publishers: Chelsea.

64. Burgman MA, Ferson S, Akcakaya HR (1993) Risk assessment in conservation biology. Chapman and Hall: London.

65. Solow AR (2005) Inferring extinction from a sighting record. Mathematical Biosciences 195: 47–55.

66. Regan HM, Colyvan M, Burgman MA (2002) A taxonomy of and treatment of uncertainty for ecology and conservation biology. Ecological Applications 12: 618–628.

67. Hobbs RJ, Arico S, Aronson J, Baron JS, Bridgewater P, et al. (2006) Novel ecosystems: theoretical and management aspects of the new ecological world order. Global Ecology and Biogeography 15: 1–7.

68. Rhymer JM, Simberloff D (1996) Extinction by hybridization and introgression. Annual Review of Ecology and Systematics 27: 83–109.

69. Carpenter S, Walker B, Anderies M, Abel N.(2001) From metaphor to measurement: resilience of what to what? Ecosystems 4: 765–781.

70. Folke C, Carpenter SR, Walker B, Scheffer M, Elmqvist T, et al. (2004) Regime shifts, resilience and biodiversity in ecosystem management. Annual Review of Ecology and Systematics 35: 557–581.

71. Allen CR, Gunderson L, Johnson AR (2005) The use of discontinuities and functional groups to assess relative resilience in complex systems. Ecosystems 8: 958–966.

72. Connell JH, Slatyer RO (1977) Mechanisms of succession in natural communities and their role in community stability and organisation. American Naturalist 111: 1119–1144.

73. Underwood AJ, Fairweather PJ (1989) Supply-side ecology and benthic marine assemblages. Trends in Ecology and Evolution 4: 16–20.

74. Scheffer M, Carpenter SR, Foley JA, Folke C, Walker B (2001). Catastrophic shifts in ecosystems. Nature 413: 591–596.

75. del Morel R (2007) Limits to convergence of vegetation during early primary succession. Journal of Vegetation Science 18: 479–488.

76. Holling CS (2001) Understanding the complexity of economic, ecological, and social systems. Ecosystems 4: 390–405.

77. Pickett STA, Cadenaao ML, Bartha S (2001) Implications from the Buell-Small Succession Study for vegetation restoration. Applied Vegetation Science 4: 41–52.

78. Micklin P (2006) The Aral Sea crisis and its future: An assessment in 2006. Eurasian Geography and Economics 47: 546–567.

79. IUCN (2011) Guidelines for the application of IUCN Red List categories and criteria. Version 9.0. Red List Standards and Petitions Subcommittee of the Species Survival Commission, IUCN: Gland.

80. Keith DA, Orscheg C, Simpson CC, Clarke PJ, Hughes L, et al. (2009) A new approach and case study for estimating extent and rates of habitat loss for ecological communities. Biological Conservation 142: 1469–1479.

81. Tilman D, May RM, Lehman CL, Nowak MA (1994) Habitat destruction and the extinction debt. Nature 371: 65–66.

82. Hahs AK, McDonald MJ, McCarthy MA, Vesk PA, Corlett RT, et al. (2009) A global synthesis of plant extinction rates in urban areas. Ecology Letters 12: 1165–1173.

83. Rackham O (1986) The History of the Countryside. Dent: London.

84. Rosensweig ML (1995) Species diversity in space and time. Cambridge, New York.

85. He F, Hubell SP (2011) Species–area relationships always overestimate extinction rates from habitat loss. Nature 473: 368–371.

86. Huth N, Possingham HP (2011) Basic ecological theory can inform habitat restoration for woodland birds. Journal of Applied Ecology 48: 293–300.

87. Shi JM, Ma KM, Wang JF, Zhao JZ, He K (2010) Vascular plant species richness on wetland remnants is determined by area and habitat heterogeneity. Biodiversity and Conservation 19: 1279–1295.

88. Sang A, Teder T, Helm A, Partel M (2010) Indirect evidence for an extinction debt of grassland butterflies half century after habitat loss. Biological Conservation 143: 1405–1413.

89. Barbone E, Basset A (2010) Hydrological constraints to macrobenthic fauna biodiversity in transitional waters ecosystems. Rendiconti lincei.- Sscienze fisiche e naturali 21: 301–314.

90. Andren H (1996) Population responses to habitat fragmentation: statistical power and the random sample hypothesis. Oikos 76: 235–242.

91. Radford JQ, Bennett AF, Cheers GJ (2005) Landscape-level thresholds of habitat cover for woodland-dependent birds. Biological Conservation 124: 317–337.

92. Caughley G (1994) Directions in conservation biology. Journal of Animal Ecology 63: 215–244.

93. MacArthur RH, Wilson EO (1967) The theory of island biogeography. Princeton University Press: Princeton.

94. Lapointe NW, Corkum LD, Mandrak NE (2010) Macrohabitat of fishes in shallow waters of the Detroit River. Journal of Fish Biology 76: 446–466.

95. Harpole WS, Tilman D (2007) Grassland species species loss resulting from reduced niche dimension. Nature 446: 791–793.

96. McKnight MW, White PS, McDonald RI, Lamoreaux JF, Sechrest W, et al. (2007) Putting beta-diversity on the map: broad-scale congruence and coincidence in the extremes. PLOS Biology 5: e272.

97. Visconti P, Pressey RL, Segan DB, Wintle BA (2010) Conservation planning with dynamic threats: The role of spatial design and priority setting for species' persistence. Biological Conservation 143: 756–767.

98. Hanski I (1998) Metapopulation dynamics. Nature 396: 41–49.

99. Burgman MA, Fox JD (2003) Bias in species range estimates from minimum convex polygons: implications for conservation and options for improved planning. Animal Conservation 6: 19–28.

100. Mucina L, Rutherford M. 2006. The vegetation of South Africa, Lesotho and Swaziland. South African National Biodiversity Institute: Pretoria.

101. Yamano H, Tamura M (2004) Detection limits of coral reef bleaching by satellite remote sensing: Simulation and data analysis. Remote Sensing of Environment 90: 86–103.

102. Todd BJ, Kostylev VP, Shaw J (2006) Benthic habitat and sun-illuminated seafloor topography, Browns Bank, Scotian Shelf, offshore Nova Scotia. Geological Survey of Canada.

103. Costello MJ, Cheung A, de Hauwere NJ (2010) Surface area and the seabed area, volume, depth, slope, and topographic variation for the world's seas, oceans, and countries. Environmental Science and Technology 44: 8821–8828.

104. Elith J, Graham CH, Anderson RP, Dudík M, Ferrier S, et al. (2006) Novel methods improve prediction of species' distributions from occurrence data. Ecography 29: 129–151.

105. Essl F, Dullinger S, Moser D, Rabitsch W, Kleinbauer I (2012) Vulnerability of mires under climate change: imlpications for nature conservation and climate change adaptation. Biodiversity and Conservation 21: 655–669.

106. Peterson CH, Rice D, Short JW, Esler D, Bodkin J, et al. (2003) Long-Term Ecosystem Response to the Exxon Valdez Oil Spill. Science 302: 2082–2086.

107. Blodgett N, Stow DA, Frankin J, Hope AS (2010) Effect of fire weather, fuel age and topography on patterns of remnant vegetation following a large fire event in southern California, USA. International Journal of Wildland Fire 19: 415–426.

108. Green PT, O'Dowd DJ, Abbott KL, Jeffery M, Retallick K, et al. (2011) Invasional meltdown: invader-invader mutualism facilitates a secondary invasion. Ecology Letters 92: 1758–1768.

109. Mearns AJ, Reisch DJ, Oshida PS, Ginn T (2010) Effects of pollution on marine organisms. Water environment research 82: 2001–2046.

110. Klemas V (2010) Tracking oil slicks and predicting their trajectories using remote sensors and models: case studies of the Sea Princess and Deepwater Horizon oil spills. Journal of Coastal Research 26: 789–797.

111. Gaston KJ (1993) Rarity. Chapman and Hall: London.

112. Keith DA (1998) An evaluation and modification of World Conservation Union Red List criteria for classification of extinction risk in vascular plants. Conservation Biology 12: 1076–1090.

113. Hartley S, Kunin WJ (2003) Scale dependency of rarity, extinction risk, and conservation priority. Conservation Biology 17: 1559–1570.

114. Byrne M, Hopper SD (2008) Granite outcrops as ancient islands in old landscapes: evidence from the phylogeography and population genetics of Eucalyptus caesia (Myrtaceae) in Western Australia. Biological Journal of the Linnean Society 93: 177–188.

115. Vrijenhoek RC (2010) Genetic diversity and connectivity of deep-sea hydrothermal vent metapopulations. Molecular Ecology 19: 4391–4411.

116. Comin FA (2010) Ecological restoration: A global challenge. Cambridge Universtiy Press: Cambridge.

117. Pickett STA (1989) Space-for-time substitution as an alternative to long-term studies. In: Likens GE, editor. Long-term studies in ecology: approaches and alternatives. Springer-Verlag: New York 110–135.

118. Fukami T, Wardle DA (2005) Long-term ecological dynamics: reciprocal insights from natural and anthropogenic gradients. Proceedings of the Royal Society B 272: 2105–2115.

119. Holling CS (1973) Resilience and stability of ecological systems. Annual Review of Ecology and Systematics 4: 1–23.

120. Peterson G, Allen CR, Holling CS (1998) Ecological resilience, biodiversity, and scale. Ecosystems 1: 6–18.

121. Larsen TH, Williams NM, Kremen C (2005) Extinction order and altered community structure rapidly disrupt ecosystem functioning. Ecology Letters 8: 538–547.

122. Diamond JM (1989) The Present, Past and Future of Human-Caused Extinction Philosophical Transactions of the Royal Society of London - Series B 325: 469–477.

123. Thebault E, Huber V, Loreau M (2007) Cascading extinctions and ecosystem functioning: contrasting effects of diversity depending on food web structure. Oikos 116: 163–173.

124. Fritz H, Loreau M, Chamaillé-Jammes S, Valeix M, Clobert J (2011) A food web perspective on large herbivore community limitation. Ecography 34: 196–202.

125. Fontaine C, Dajoz I, Meriguet J, Loreau M (2006) Functional diversity of plant-pollinator interaction webs enhances the persistence of plant communities. PLoS Biology 4: 129–135.

126. Goudard A, Loreau M (2008) Nontrophic interactions, biodiversity, and ecosystem functioning: An interaction web model. American Naturalist 171: 91–106.

127. Schmitz OJ, Hambäck PA, Beckerman AP (2000) Trophic cascades in terrestrial ecosystems: a review of the effects of carnivore removals on plants. American Naturalist 155: 141–153.

128. Ripple WJ, Beschta RL (2004) Wolves and the ecology of fear: can predation risk structure ecosystems? Bioscience 54: 755–766.

129. Vázquez DP, Simberloff D (2003) Changes in interaction biodiversity induced by an introduced ungulate. Ecology Letters 6: 1077–1083.

130. Lundberg J, Moberg F (2003) Mobile link organisms and ecosystem functioning: implications for ecosystem resilience and management. Ecosystems 6: 87–98.

131. Staver AC, Bond WD, Stock WD, van Rensberg SJ, Waldram MS (2009) Browsing and fire interact to suppress tree density in an African savanna. Ecological Applications 19: 1909–1919.

132. Araoz E, Grau HR (2010) Fire-mediated forest encroachment in response to climatic and land-use change in subtropical Andean treelines. Ecosystems 13: 992–1005.

133. Lindenmayer DB, Likens GE (2001) Direct measurement versus surrogate indicator species for evaluating environmental change and biodiversity loss. Ecosystems 14: 47–59.

134. Keith DA, Holman L, Rodoreda S, Lemmon J, Bedward M (2007) Plant Functional Types can predict decade-scale changes in fire-prone vegetation. Journal of Ecology 95: 1324–1337.

135. Walker B, Kinzig A, Langridge J (1999) Plant Attribute Diversity, Resilience, and Ecosystem Function: The Nature and Significance of Dominant and Minor Species. Ecosystems 2: 95–113.

136. Yachi S, Loreau M (1999) Biodiversity and ecosystem productivity in a fluctuating environment: The insurance hypothesis. Proceedings of the National Academy of Sciences of the United States of America 96: 1463–1468.

137. Loreau M (2000) Biodiversity and ecosystem functioning: recent theoretical advances. Oikos 91: 3–17.

138. Fischer J, Lindenmayer DB, Blomberg SP, Montague-Drake R, Felton A, et al. (2007) Functional richness and relative resilience of bird communities in regions with different land use intensities. Ecosystems 10: 964–974.

139. Walters C, Kitchell JF (2001) Cultivation/depensation effects on juvenile survival and recruitment. Implications for the theory of fishing. Canadian Journal of Fisheries and Aquatic Sciences 58: 39–50.

140. Molnar JL, Gamboa RL, Revenga C, Spalding MD (2008) Assessing the global threat of invasive species to marine biodiversity. Frontiers in Ecology and the Environment 6: 485–492.

141. Duffy JE, Cardinale BJ, France KE, McIntyre PB, Thébault E, et al. (2007) The functional role of biodiversity in ecosystems: incorporating trophic complexity. Ecology Letters 10: 522–538.

142. Calcagno V, Sun C, Schmitz OJ, Loreau M (2011) Keystone Predation and Plant Species Coexistence: The Role of Carnivore Hunting Mode. The American Naturalist 177: E1–E13.

143. Loreau M, Mouquet N, Gonzalez A (2003) Biodiversity as spatial insurance in heterogeneous landscapes. Proceedings of the National Academy of Science 100: 12765–12770.

144. Lindenmayer DB, Fischer J (2006) Habitat fragmentation and landscape change. Island Press: Washington DC.

145. Alvarez-Filip L, Dulvy NK, Gill JA, Côté IM, Watkinson AR (2009) Flattening of Caribbean coral reefs: region-wide declines in architectural complexity. Proceedings of the Royal Society B 276: 3019–3025.

146. Lee JK, Park RA, Mausel PW (1992) Application of geoprocessing and simulation modeling to estimate impacts of sea level rise on the northeast coast of Florida. Photogrammetric Engineering and Remote Sensing 58: 1579–1586.

147. King KJ, Cary GJ, Bradstock RA, Chapman J, Pyrke A, et al.(2006) Simulation of prescribed burning strategies in south-west Tasmania, Australia: effects on unplanned fires, fire regimes, and ecological management values. International Journal of Wildland Fire 15: 527–540.

148. Scholze M, Knorr W, Nigel W, Arnell NW, Prentice IC (2006) A climate-change risk analysis for world ecosystems. Proceedings of the National Academy of Sciences of the United States of America 103: 13116–13120.

149. Midgley GF, Davies ID, Albert CH, Altwegg R, Hannah L, et al. (2010) BioMove – an integrated platform simulating the dynamic response of species to environmental change. Ecography 33: 612–616.

150. Burgman MA, Keith DA, Walshe TV (1999) Uncertainty in comparative risk analysis for threatened Australian plant species. Risk Assessment 19: 585–598.

151. Czembor CA, Vesk PA (2009) Incorporating between-expert uncertainty into state-and-transition simulation models for forest restoration. Forest ecology and management 259: 165–175.

152. Jacquez GM, Kaufmann A, Goovaerts P (2008) Boundaries, links and clusters: a new paradigm in spatial analysis?. Environmental and Ecological Statistics 15: 403–419.

153. Keith DA, Burgman MA (2004) The Lazarus effect: can the dynamics of extinct species lists tell us anything about the status of biodiversity? Biological Conservation 117: 41–48.

154. Costello MJ (2009) Distinguishing marine habitat classification concepts for ecological data management. Marine Ecology Progress Series 397: 253–268.

155. Kontula T, Raunio A (2009) New method and criteria for national assessments of threatened habitat types. Biodiversity and Conservation 18: 3861–3876.

156. Ramirez-Llodra E, Brandt A, Danovaro R, De Mol B, Escobar E, et al. (2010) Deep, diverse and definitely different: unique attributes of the world's largest ecosystem. Biogeosciences 7: 2851–2899.

157. Miller RM, Rodriguez JP, Aniskowicz-Fowler T, Bambaradeniya C, Boles R, et al. (2007) National threatened species listing based on IUCN Criteria and Regional Guidelines: current status and future perspectives. Conservation Biology 21: 684–696.

158. Possingham HP, Andelman S, Burgman MA, Medellin RA, Master LL, et al. (2002) Limits to the use of threatened species lists. Trends in Ecology and Evolution 17: 503–507.

159. Zhao HL, Zhao XL, Zhou RI, Zhang TH, Drake S (2005) Desertification processes due to heavy grazing in sandy rangeland, Inner Mongolia. Journal of Arid Environments 63: 309–319.

160. Ludwig JA, Bastin GE, Chewings VH, Eager RW, Liedloff AC (2007) Leakiness: a new index for monitoring the health of arid and semiarid landscapes using remotely sensed vegetation cover and elevation data. Ecological Indicators 7: 442–454.

161. Pounds JA, Fogden MPL, Campbell JH (1999) Biological response to climate change on a tropical mountain. Nature 398: 611–615.

162. Cabezas A, Comin FA, Begueria S, Trabucchi M (2009) Hydrologic and landscape changes in the Middle Ebro River (NE Spain): implications for restoration and management. Hydrology and Earth System Sciences 13: 1–12.

163. Mitternicht GI, Zinck JA (2003) Remote sensing of soil salinity: potential and constrainst. Remote Sensing of the Environment 85: 1–20.

164. Rogers CS (1990) Responses of coral reefs and reef organisms to sedimentation. Marine Ecology Progress Series 62: 185–202.

165. Watling L, Norse EA (1998) Disturbance of the seabed by mobile fishing gear: A comparison to forest clearcutting. Conservation Biology 12: 1180–1197.

166. Hannah J, Bell RG (2012) Regional sea level trends in New Zealand. Journal of Geophysical Research 117: C01004.

167. Hong S, Shin I (2010) Global Trends of Sea Ice: Small-Scale Roughness and Refractive Index. Journal of Climate 23: 4669–4676.

168. Cardinale BJ, Matulich KL, Hooper DU, Byrnes JE, Duffy E, et al. (2011) The functionalo role of producer diversity in ecosystems. American journal of botany 98: 572–592.

169. Clarke PJ, Latz PK, Albrecht DE (2005) Long-term changes in semi-arid vegetation: Invasion of an exotic perennial grass has larger effects than rainfall variability. Journal of Vegetation Science 16: 237–248.

170. Miller G, Friedel M, Adam P, Chewings V (2010) Ecological impacts of buffel grass (Cenchrus ciliaris L.) invasion in central Australia – does field evidence support a fire-invasion feedback? The Rangeland Journal 32: 353–365.

171. Eisenhauer N, Milcu A, Sabais ACW, Bessler H, Brenner J, et al. (2011) Plant Diversity Surpasses Plant Functional Groups and Plant Productivity as Driver of Soil Biota in the Long Term. PLoS ONE 6: e16055.

172. Cardinale BJ, Wright JP, Cadotte MW, Carroll IT, Hector A, et al. (2007) Impacts of plant diversity on biomass production increase through time because of species complementarity. Proceedings of the National Academy of Science of the United States of America 104: 18123–18128.

173. Thompson JN (1997) Conserving interaction biodiversity. in O. R. Pickett STA, In: Shachak M, Likens GE, editors. The Ecological basis of conservation: heterogeneity, ecosystems, and biodiversity. Chapman and Hall: New York.

174. Springer AM, Estes JA, van Vliet GB, Williams TM, Doak DF, et al. (2003) Sequential megafauna collapse in the North Pacific Ocean: an ongoing legacy of industrial whaling? Proceedings of the National Academy of Science 100: 12223–12228.

175. Moberg F, Folke C (1999) Ecological goods and services of coral reef ecosystems. Ecological Economics 29: 215–233.

176. Arias-González JE, Acosta-González G, Membrillo N, Garza-Pérez JR, Castro-Pérez JM (2012) Predicting spatially explicit coral reef fish abundance, richness and Shannon–Weaver index from habitat characteristics. Biodiversity and Conservation 21: 115–130.

17

Polycyclic Aromatic Hydrocarbons in Coastal Sediment of Klang Strait, Malaysia: Distribution Pattern, Risk Assessment and Sources

Seyedeh Belin Tavakoly Sany[1,2]*, **Rosli Hashim**[1], **Aishah Salleh**[1], **Majid Rezayi**[2,3], **Ali Mehdinia**[4], **Omid Safari**[5]

1 Institute of Biological Sciences University of Malaya, Kuala Lumpur, Malaysia, **2** Food Science and Technology Research Institute, ACECR Mashhad Branch, Mashhad, Iran, **3** Chemistry Department, Faculty of Science, University Malaya, Kuala Lumpur, Malaysia, **4** Department of Marine Science, Marine Living Group, Iranian National Institute for Oceanography, Tehran, Iran, **5** Faculty of Natural Resources and Environment, Ferdowsi University of Mashhad, Mashhad, Iran

Abstract

Concentration, source, and ecological risk of polycyclic aromatic hydrocarbons (PAHs) were investigated in 22 stations from surface sediments in the areas of anthropogenic pollution in the Klang Strait (Malaysia). The total PAH level in the Klang Strait sediment was 994.02±918.1 µg/kg dw. The highest concentration was observed in stations near the coastline and mouth of the Klang River. These locations were dominated by high molecular weight PAHs. The results showed both pyrogenic and petrogenic sources are main sources of PAHs. Further analyses indicated that PAHs primarily originated from pyrogenic sources (coal combustion and vehicular emissions), with significant contribution from petroleum inputs. Regarding ecological risk estimation, only station 13 was moderately polluted, the rest of the stations suffered rare or slight adverse biological effects with PAH exposure in surface sediment, suggesting that PAHs are not considered as contaminants of concern in the Klang Strait.

Editor: Karl Rockne, University of Illinois at Chicago, United States of America

Funding: This work and survey components were supported by the High Impact Research Grant (UM.C/625/1/HIR/162) from the Ministry of Higher Education (Malaysia) and University Malaya Research Grant (RP004A-SUS). Study designs and data collections were done by the academic staff at the Institute of Biological Science, University Malaya, Kuala Lumpur, Malaysia. Analyses undertaken and the decision to prepare and publish this manuscript were made by Institute staff in conjunction with the Chemistry Department of University Malaya. The funders had no role in study design, data collection and analysis, decision to publish, or preparation of the manuscript.

Competing Interests: The authors have declared that no competing interests exist.

* E-mail: belintavakoli332@gmail.com

Introduction

Polycyclic aromatic hydrocarbon (PAH) contamination is a major hazard that is a concern for aquatic life in marine sediments, particularly in areas close to anthropogenic sources [1,2]. Many PAHs are at the same time persistent, bioaccumulative, and toxic for humans and aquatic organism [3,4,5,6,7].

In environmental research, the aromatic fraction of C_{11}–C_{22} was selected as being representative of aromatic hydrocarbon compounds for the purpose of assessment of ecological and human risks. This fraction is associated with the release of petroleum products to the environment, and is potentially toxic due to its mobility and stability in sediments [8,9]. Physicochemical properties of these fractions are provided in the Supplementary data (Table S1).

Similar to other pollutants, the sources of PAHs are divided into major groups; anthropogenic and lithogenic. Anthropogenic sources of PAHs originated from pyrogenic and petrogenic sources. Pyrogenic PAHs are usually made up of high molecular weight PAHs with 4–6 rings that includes, fluoranthene (Fla), pyrene (Py), benzo(a)anthracene (BaA), chrysene (Chy), benzo(b)fluoranthene (BbF), benzo(k)fluoranthene (BkF), benzo(a)pyrene (BaP), dibenzo(a,h)anthracene (DibA), benzo(g,h,i)perylen (BghiP) and Indeno[1,2,3,(c,d)]pyrene; (InP) [2,10]. The pyrogenic are

mainly detected in incomplete combustion of organic compounds, such as fossil fuels (heating oil, cooking, coal burning, vehicle emissions, waste tire), and biomass burning (fireplace, controlled burning) [2,11,12]. Pyrogenic sources are thought to be more thermodynamically stable and toxic than petrogenic sources due to their high concentrations of non-alkylated PAHs [11,13,14].

Petrogenic PAHs involve naphthalene, acenaphthylene, acenaphthene, and fluorine, and belong to the alkyl-substituted PAHs or low molecular weight PAHs, possessing 2–3 rings. These hydrocarbons are thought to have originated from oil spills from fresh or used crankcase oil, crude and fuel oil, chronic or accidental leakages of marine and land pipelines, and domestic and industrial wastes [2,10,12,14].

The Klang Strait is surrounded by the west coast of Malaysia and the Straits of Malacca in Southeast Asia. The strategic location of this strait has made it into one of the busiest shipping routes in the world, corresponding to huge economic demands from the Middle East and the Far East [15]. This strait also experienced the rapid development of industrialization, urbanization, and motorization over the past few decades. Thus, this area is under constant threat from multiple sources of energy, such as petroleum [16,17]. In this research, a hypothesis is defined based on the serious threat posed by PAHs contamination to the Klang Strait.

Hence, this study tries to estimate the concentration and distribution of PAHs in coastal sediments from the Klang Strait to identify the possible sources of PAHs in the Port, and to conduct an ecological risk assessment to recognize the possible adverse ecological effects on the biological community due to the exposure to PAHs concentrated in Klang Strait sediments.

Ecological assessment of PAHs compounds in this study has been highlighted via two main problems: the first problem was related to the scarcity of background and updated databases on the PAHs concentrations in Klang Strait sediments. This problem is due to the limitations in collecting samples such as strong currents, high traffic density of shipping activity. The second problem was related to the lack of SQGs (Sediment Quality Guidelines) for coastal waters of Malaysia, since no SQGs were available to assess biological effects. Therefore, the results from this study will be applied in the form of the managerial tools in order to control the pollution occurrence and protect living organisms to assure the safe future of the marine environment of Peninsular Malaysia. Moreover it can be practical as background data for future studies.

Materials and Methods

Ethics Statement

The Ethical Review Committee of University Malaya approved this study. All necessary permits were obtained for the described field study from University Malaya and Port Klang Authorities (Permit Number UM.G/KB4/6/1). No specific permission was required for water and sediment surveys as they were conducted at these stations. Likewise, the field studies did not involve endangered or protected species, and no lethal sampling was conducted.

Study area and sample collection

The Klang Strait is divided into three main Ports, most of which are well-sheltered by several mangrove islands and mudflats, forming natural enclosures. Experimental samples were collected from three sites (North Port, South Port and West Port) in the Klang Strait, which is located on the western coastal region (03° 00'N to 101° 24'E) of Peninsula Malaysia at the north end of the Malacca Strait, and over 573 km^2 [18].

The sites were chosen based on in their unique activities. The North Port is a terminal of container to transship the coal, oil, and other chemical products, while the South Port is a terminal jetty for fishing boats, ferries, and yachts, and is linked to the Klang river, which is known as the most polluted river in Malaysia [15,19].

The West Port is a main container terminal for massive cargo transport by ocean ships. Similarly, an industrial complex, including cement industries and a food, and palm oil factories have been constructed at the West Port [15].

In this research, 22 stations were selected from three transects parallel to the coastline, with three different distances (Figure 1), and a station was selected as a control point 21 km from the Klang Strait at a remote location. The sampling stations were adjacent coastal industrial outlets and busy shipping lanes. The sampling stations were identified a progressive number: six up-gradient stations in North Port (Station 1–6), six stations in South Port close to Klang River (station 7–15), and nine down-gradient stations in West Port (station 16–21) (Table 1).

The study area lies within the humid tropical part with rainy (North monsoon, November to March) and dry seasons (south monsoon, April to October). In the Klang Strait, the average salinity is 30.25‰ (±1.36), the average temperature is 30.04°C (±0.62), and the average surface dissolved oxygen (DO) is

5.38 mg/l (±0.17), while the monthly average surface and bottom pH values were between 7.85 to 8.25. The sediment samples were taken with a Peterson grab sampler (0.07 m^2) at two dates per season during November 2011 until October 2012. The top 10 cm of the sediment sample was removed with a stainless steel spoon for subsequent analysis. All of the samples were packed into aluminum boxes and transported on ice to the laboratory. The samples were air-dried, crushed, and sieved (<2 mm) to remove residual roots and stones, and immediately stored at −20°C in pre-cleaned amber glass bottles until further treatments. For PAHs analysis, three replicates of grab samples were used to investigate the spatial and temporal distributions.

Experimental method

Chemical and reagents. A standard solution of 16 USEPA priority PAHs (Table 2) and a five-surrogate standard (acenaphthene-d10, chrysene-d12, perylened12, naphthalene-d8 and phenanthrene-d10) were purchased from Ultra Scientific, Inc. (North Kingstown, RI, USA). A standard reference material (SRM) was obtained from the National Institute of Standards and Technology (NIST, Gaithersburg, MD, USA). Neat (99%) hexamethylbenzene was purchased from Aldrich Chemical Company (Milwaukee, WI, USA). All solvents (acetone, dichloromethane, methanol and hexane) applied in the analyses were of analytical grade and redistilled twice.

The alumina (120–200 mesh) and Silica gel (80–100 mesh) were extracted for a period 72 hours in a Soxhlet apparatus, activated in the oven at 150°C and 180°C for a period 12 hours, respectively, and then deactivated by distilled water at a ratio of 3% (m/m). The deionised water was obtained from a Milli-Q system.

Extraction and fractionation. For the extraction procedure, about 5 g of dried sediment samples were Soxhlet-extracted by dichloromethane (150 ml) for a period of 72 hours. A mixture of surrogate standard was added to all of the samples as a recovery surrogate standards. Elemental sulfur was removed from the extracts using activated copper granules. After extraction, the extract was concentrated up using a rotary vacuum evaporator. The volume of the extract was adjusted to 2–3 ml, and solvent-exchanged into 10 ml n-hexane, which caused it to further decrease to approximately 1–2 ml.

Clean up and fraction procedures were performed with a 1:2 alumina/silica gel column. PAHs were obtained by eluting it with 60 ml of hexane/dichloromethane (1:1). PAHs fraction were re-concentrated using a rotary evaporator, while the volume was concentrated to up to 1 ml. The fraction was further reduced to 0.2 ml under a stream of filtered purified nitrogen gas. An aliquot of 0.2 ml of each extract was applied to gas chromatography–mass spectrometer (GC–MS) analysis. Likewise, hexamethylbenzene was added as an internal standard prior to the GC–MS analysis.

Instrumental analysi. The PAHs were estimated using a Hewlett–Packard 5890 series gas chromatograph interfaced with 5972 mass-selective detector (MSD) in the selective ion-monitoring (SIM) mode. A fused silica capillary column (50 m, 0.32 mm, 0.17 µm) coated with HP-5MS (film thickness 0.25 lm) was used for separation, with helium as a carrier gas at a flow rate of 2 ml/min, a head pressure of 12.5 psi, and a linear velocity of 39.2 cm/s at 290°C.

The temperature was programmed to vary from 80 to 290°C at 4°C/min, and was held at the final temperature for 30 minutes. The injector temperature and transfer line temperature were 250°C and 180°C, respectively. The mass spectrometer operated at electron energy of 70 eV, with an ion source temperature of 250°C.

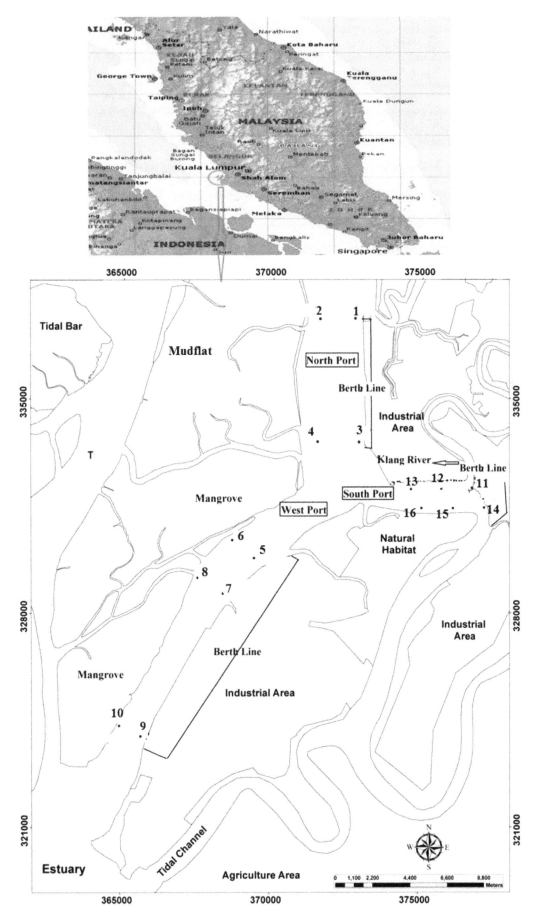

Figure 1. Location of the sampling stations in Klang Strait.

The interface temperature and injection were programmed at 290°C. The oven temperature was initially isothermal at 80°C for 5 minutes, programmed from 80 to 290°C at 3°C/min, and held at the final temperature (290°C) for 30 minutes. A 1 ml sample was manually injected in the split-less injector with a 1-minute solvent delay. The mass spectrometer was operated at electron impact (EI) of 70 eV, with an ion source temperature of 250°C. The ranges of mass scanning were between m/z 50 and m/z 500.

Quality assurance. For every set of samples, a procedural method blanks (solvent), sample duplicates, spiked matrices (standards spiked into solvent), and standard reference material (SRM1941) were used to assess quality control and assurances. Individual PAHs were quantified according to the retention time and m/z ratio of PAHs mixed standard (Sigma), while a standard calibration curve was used to calibrate the concentrations of each PAH.

The method detection limits (MDLs) for the investigated organic pollutants were determined according to USEPA [20]. The MDL of PAHs were estimated based on six analyses of the minimal solutions (Σ16PAH) of the standard PAH mixture, which was subjected to the full analytical procedure in the range of 0.0641 ng/g to 1.0183 ng/g. These values were within the acceptable range of the EPA method [20]. The percentage of relative standard deviation (RSD) for all of the investigated PAHs were lower than 20.0% in the fortification experiment and in the

replicates used to determine the MDL. The reported results were corrected with the recoveries of the surrogate standards. The average recoveries of surrogate standards varied from 63.28–96.75. The recoveries of all the PAHs and RSD falls within the range of 78.4–95.2% and 2.6–13.5%, respectively, which met the acceptance criteria of the EPA method [20]. A detailed recovery and RSD values were provided in the Supplementary data (Table S1).

Measurement of TOC and sediment grain size. The sediment grain size was determined using a multi-wavelength particle size analyzer (model LS 13 320), and the results were divided into sand (>64 μm), silt (2 μm<size<64 μm), and clay (<2 μm) fractions for the determination of PAHs in the contaminated sediments [21].

Freeze-dried sediment samples were grounded, and was treated with 10% (v/v) HCl to remove carbonates from the sediment samples. After that, the samples were dried at 60°C in an oven. Total organic carbon (TOC) was determined in the surface sediments using a carbon analyzer (Horbia Model 8210).

PAHs source identification

Ratio values such as an Ant/Ant+Phn and a Fla/Fla+Pyr had been used to distinguish between pyrogenic and petrogenic sources [22,23]. Sediments with Ant/Ant+Phn ratio <0.100 were mainly contaminated by petrogenic inputs, while a ratio >0.100 indicates

Table 1. Physicochemical description of sampling stations.

Sites	Station number	Latitude	Longitude	Silt and clay (<64 μm)%	TOC (%)	Depth (m)	Salinity (‰)	Description of stations
North Port	1	3° 3'1.49"N	101°21'18.70"E	58.20	41.79	14.3	30.15	Liquid and dry berth line
	2	3° 3'1.33"N	101°20'56.04"E	49.63	50.36	20.5	30.18	Remote
	3	3° 3'1.47"N	101°20'33.11"E	73.77	26.22	10.3	31.24	Mangrove
	4	3° 0'53.11"N	101°21'20.25"E	59.78	40.21	13.5	30.81	Container berths
	5	3° 0'52.64"N	101°20'58.04"E	50.89	49.10	21.6	31.02	Remote
	6	3° 0'52.49"N	101°20'34.54"E	65.19	34.80	11.2	31.36	Mangrove
West Port	7	2°58'44.00"N	101°19'21.02"E	53.57	46.42	12.5	30.86	Dry berth and Cement factory outlets
	8	2°58'54.12"N	101°19'9.06"E	45.96	54.03	19.5	30.98	Remote
	9	2°59'3.12"N	101°18'58.38"E	63.42	36.57	7.8	30.86	Mangrove
	10	2°58'6.34"N	101°18'48.14"E	56.33	43.66	13.3	30.44	Liquid berth and palm oil factory oulets
	11	2°58'14.90"N	101°18'34.56"E	41.10	58.89	20.3	30.58	Remote
	12	2°58'23.07"N	101°18'20.99"E	70.81	29.18	8.8	30.79	Mangrove
	13	2°55'34.43"N	101°17'18.76"E	52.31	47.68	15.5	30.51	Container berths
	14	2°55'39.38"N	101°17'7.57"E	50.69	49.30	21.1	30.63	Remote
	15	2°55'45.02"N	101°16'55.55"E	70.36	29.63	6.8	30.77	Mangrove
South Port	16	2°59'59.08"N	101°23'18.88"E	95.39	4.60	7.5	26.10	Dry berths, Klang river
	17	2°59'58.17"N	101°22'45.35"E	93.16	6.83	10.5	26.12	Klang river
	18	2°59'57.66"N	101°22'12.45"E	64.69	35.30	12.4	30.11	Semi-urban
	19	2°59'38.25"N	101°23'32.32"E	69.50	30.49	10.3	29.45	Liquid berth
	20	2°59'37.70"N	101°22'57.93"E	69.72	30.27	11.3	29.54	Mangrove
	21	2°59'37.23"N	101°22'23.78"E	57.73	42.26	10.4	30.50	Mangrove
Control Point	22	3° 6'55.95"N	101°12'44.70"E	51.60	48.39	17.5	31.54	Remote

Table 2. Concentration of PAHs (µg/kg dw) in surface sediments of Klang Strait.

Station	Nap	Acy	Ace	Flr	Phn	Ant	Fla	Pyr	BaA	Chy	BbF	BkF	BaP	DibA	Bghip	Inp	∑PAHs
1	ND	ND	2.1	34.2	47.5	100.4	ND	482.8	613.0	ND	ND	ND	ND	ND	ND	ND	1280.4
2	1.3	ND	ND	9.7	13.3	12.7	ND	131.3	31.5	ND	ND	ND	ND	ND	ND	ND	199
3	1.7	ND	8.1	118.5	89.0	58.8	42.7	62.9	ND	ND	ND	ND	ND	ND	ND	ND	381.7
4	0.4	16.32	ND	1.1	10	59.8	12.5	12	66.7	50.0	852.8	ND	1495.8	190.2	ND	99.8	2852.2
5	0.8	ND	0.3	16.8	12.6	12	10	195.6	ND	ND	11.0	ND	ND	ND	ND	ND	259.1
6	4.0	ND	0.4	57.3	36.2	37.9	26.0	31.8	179.4	ND	ND	ND	ND	ND	ND	42.3	415.2
7	1.3	ND	ND	27.6	127.6	267.8	10.3	12.7	204.0	ND	ND	47.6	ND	ND	ND	ND	698.9
8	5.7	ND	1.0	39.0	6.8	71.4	13.5	15.8	50.5	ND	ND	ND	122.1	ND	64.2	77.4	467.4
9	1.2	ND	ND	7.7	5.1	8.2	ND	28.1	191.3	198.0	ND	ND	ND	67.8	ND	ND	507.4
10	0.8	ND	ND	23.7	5.0	10.1	ND	236.5	278.8	ND	ND	ND	ND	ND	ND	68.7	623.7
11	0.7	ND	0.7	48.8	15.6	37.8	17.6	25.1	29.5	ND	ND	367.6	ND	ND	ND	ND	543.4
12	2.0	ND	1.0	39.8	24.8	233.9	21.5	341.5	774.7	ND	ND	1.3	1.0	7.3	ND	ND	1448.8
13	16.1	645.3	14.8	520.8	225.6	308.0	166.5	243.8	390.9	ND	11.6	28.6	825.2	ND	49.4	ND	3446.9
14	3.0	ND	0.5	33.8	75.8	181.3	15.8	14.0	146.3	ND	11.3	ND	59.9	ND	ND	31.8	573.5
15	2.2	ND	ND	8.7	9.6	3.8	ND	ND	78.3	ND	12.8	ND	ND	130.2	ND	348.4	594.0
16	109.0	ND	1.2	120.6	66.3	100.5	92.2	99.7	974.8	116.0	ND	ND	716.8	ND	ND	ND	2397.1
17	9.6	ND	4.0	225.6	94.3	501.2	ND	226.2	1292.0	549.2	ND	ND	ND	ND	ND	ND	2902.1
18	1.3	ND	ND	20.8	8.5	11.6	ND	119.3	191.8	ND	378.6	ND	ND	ND	ND	ND	731.9
19	2.2	ND	ND	35.6	39.7	33.3	16.5	17.1	267.3	ND	10.2	ND	103.9	ND	ND	ND	525.8
20	2.1	ND	ND	37.1	38.6	30.2	16.3	17.0	226.8	ND	9.3	ND	89.9	ND	ND	ND	467.5
21	2.4	ND	0.6	37.2	16.7	16.5	16.0	11.2	194.6	ND	22.0	ND	ND	129.6	ND	ND	483.5
22	1.8	ND	ND	12.5	1.7	4.1	ND	ND	56.8	ND	23.4	ND	ND	0.0	ND	ND	100.3
Mean	7.7	30.04	1.6	67.13	43.6	95.5	21.2	105.7	282.7	41.5	61.0	20.2	155.8	23.9	5.2	32.0	994.02
SD	22.6	137.39	3.453	131.0	53.01	142.3	38.0	144.0	363.8	121.7	191.9	77.5	377.7	53.7	16.7	76.1	918.1
Min	ND	ND	ND	1.1	1.7	3.8	10	11.2	ND	ND	ND	ND	ND	ND	ND	ND	100.3
Max	109.0	645.3	14.8	520.8	225.6	501.2	169	599	1292.0	549.2	852.8	367.6	1495.8	190.2	64.2	348.4	3446.9
*TEL	34.6	5.87	6.71	21.2	86.7	46.9	113	153	74.8	108	320	280	88.8	6.22	430	*	1684
*PEL	391	128	88.9	144	544	245	1494	1398	693	846	1880	1620	763	135	1600	*	16770

TELs (threshold effects level) and PELs (probable effects level), ** <MDL, below the method detection limit; ND, not detect.

a dominance of combustion [10,22,23]. A Fla/Fla+Pyr ratio of 0.500 is usually defined as the petroleum/combustion transition point. This boundary appears to be less definitive than 0.100 for Ant/Ant+Phn.

The Fla/Fla+Pyr ratio is below 0.4 for most petroleum samples, between 0.400–0.500 for liquid fossil fuel (vehicle and crude oil) combustion, and exceeds 0.500 in kerosene, grass, most coals, and wood combustion [22,23].

Ecological risk assessment

A Screening Level Ecological Risk Assessment (SLERA) was performed to evaluate the possibility of adverse ecological effects according to the framework developed by the United States Environmental Protection Agency (USEPA) [11,24,25]. The average concentration of individual PAHs in Klang Strait was used as the exposure concentration in a conservative assumption.

The sediment quality guideline, which was previously developed, was applied in the risk characterization step. In the present study, a consensus approach based on MacDonald et al. was used according to the following specific values of TEL (threshold effects level) and PEL (probable effects level) [26,27]. In the risk characterization step, the hazard quotient (HQ) was calculated

based on the maximum concentration of each individual PAHs (MCPAHs) and the estimated consensus based sediment quality guideline (CCPEL) according to the following equation [11,27]:

$$HQ = \frac{MCPAHs}{CCPEL} \qquad (1)$$

The PEL quotient (PELq) is a unique factor to describe the contamination effect of PAHs on biological organisms according to the analyses of chemical data and matching toxicity from 1,068 sediment samples from coastal and estuaries water in the USA [28]. The PELq factor is the average of the ratios between the PAHs concentration in the sediment sample and the related PEL value [11,29]. PELq factors were divided into four categories, which can be used to describe the sediment as non-adverse effect (PELq<0.1), slightly adverse effect (0.1<PELq>0.5), moderately effect (0.5<PELq>1.5), and heavily effect (PELq>1.5) [11,30]. This factor is practical for comparing different historical episodes or other study areas, and to facilitate the decision maker's work in sediment quality assessment [31].

The Nemerow composite index (Nemerow Index) for all of the sediments samples was estimated based on [32]. This method used the individual contamination index, which directly reflect the sediments' contamination degree. The equation is as follows:

$$P = \sqrt{\frac{(P_{iave})^2 + (P_{iave})^2}{2}} \qquad (2)$$

Where $P_{i\ max}$ is the maximum concentration of the individual contamination indices, while $P_{i\ ave}$ is the average concentration of the individual contamination indices. When $P_i = C_i/S_i$, P_i is the individual contamination index of PAHs contaminates i, C_i is the determined concentration of PAHs from at least 5 sampling points, and S_i is the required standard based on the evaluation criteria. In the present study, the adopted evaluation criteria were specific values of the sediment quality guideline, as Malaysian coastal waters have no stipulated guideline on PAHs in sediments. The Nemerow pollution index (P) is classified as the quality of sediment into 5 pollution grades safety domain (P<0.7), precaution domain (0.7<P≤1.0), slightly polluted domain (1.0<P≤2.0), moderately polluted domain (2.0<P≤3.0), and seriously polluted domain (P>3.0) [32,33].

Statistical analysis

Statistical analyses were performed using Microsoft Excel and SPSS 17 software (SPSS, Chicago, IL) for statistical data evaluation. Multivariate techniques are sensitive to the distribution of non-parametric geochemical data. Thus, evaluating normal distribution and data transformation are essential to appropriately pretreat data. The data were arranged based on the samples and estimated values, and converted into a single matrix formed by the data points (cases) and the concentration values of the 16 PAHs analyzed (variables), forming a [22×16] data matrix. The methods were selected based on the results of the Shapiro-Wilk Normality test, the levene test for homogeneity of variances, and the Bartlett's test of equal variances. Normally, the distributed data were evaluated via one-way ANOVA, with an alpha level of 0.05. Data that did not pass the tests of normality and homogeneity was evaluated with the Kruskal-Wallis one-way nonparametric AN-OVA. The differences were determined to be significant, where p<0.05. The data transformation was also done using the Box–Cox method with an optimal transformation for data that were not normally distributed. This method was performed based on the Kannel et al. 2007 to reduce the mean squared error [71]. Statistical tests, multivariate analyses, diagnostic ratios, and risk calculations include non-detect (or left-censored) data with substitution by zero. The nonparametric correlation method (Kendall's tau-b) was used to obtain the correlation coefficient and the significance of the correlation among physicochemical parameters in the sediments. Multivariate techniques, such as cluster analysis (CA) and principal components analysis (PCA) were used to estimate structure and relationships in a multivariate data. PCA is also known as a dimensional reduction because this method is able to decrease the dimensionality of the primary set of data (measured PAH contents in sediment samples) and compress data into a lower dimensional matrix (principal components) [34]. Specifically, the KMO and Bartlett's test of sphericity were applied to extract all factors with eigenvalues that exceed 1, and were rotated by the Varimax method. CA analysis was used to classify a set of data into different groups based on similarity. The geo-statistical analysis was performed using Surfer 8 software (GPS value of stations) based on the geospatial methods to better understand the contaminant pathways and to provide a comprehensive contour map of the spatial distribution of contaminants over a large range.

Results and Discussion

Spatial distribution of PAHs in sediments

In the Klang Strait, fine-grained sediments are predominated in almost all stations, with the highest amount of fine sediment in the vicinity of stations near the mangrove edge and the mouth of the Klang River (Table 1). TOCs' distribution was synchronous with that of fine-grain-sized sediment in most parts of the study areas (with the notable exception of stations 9 and 12) (Table 1). The details related to the distribution and concentration of TOC in the Klang strait was recorded by Tavakoly Sany et al. in 2013 [35]. Generally, the PAH concentration showed an insignificant correlation with the percentage of TOC and fine-grained sediments (p<0.05, r = 0.15 and r = 0.11) in the surface sediments of the Klang Strait. Thus, it might be probable that fine-grained sediments and TOC are not effective parameters in controlling the distribution of PAHs in the Klang Strait.

In the Klang Strait, the concentration of total PAHs ranged from 100.3 to 3446.9 μg/kg dw, with an average concentration of 994.02±918.1 μg/kg dw (Table 2). The highest concentration of PAHs was estimated at station 13, while the lowest concentration was observed at the control point. The PAH contamination was estimated for surface sediments all over the strait, with higher concentrations observed at the stations close to the berth line (except station 12), especially in front of the container terminal in the West Port (station 13: 3446.9 μg/kg dw), in front of the dry and liquid terminal in the North Port (station 1: 1280.4 μg/kg dw and station 4: 2851.2 μg/kg dw), and in stations located adjacent to the mouth of the Klang River in the South Port (16: 2397.1 μg/kg dw and 17: 2902.1 μg/kg dw). PAH concentrations were generally lower in more remote stations and mangrove side, except for station 12, which had an elevated concentration of PAHs (1448.8 μg/kg dw) (Figure 2A). A Kruskal–Wallis test showed that there are significant differences (p<0.05, df = 21, sig = 0.003) among the concentrations of PAHs at all stations.

The concentrations of high-molecular-weight PAHs (HMWPAHs) are usually estimated in order to assess the combustion values [2,10,11]. In our study, combustion values ranged from 80.3 μg/kg dw to 2767.8 μg/kg dw, with an average concentration of 767.2 μg/kg dw (Table 3). High-molecular-weight PAHs represented between 27.66% and 97.08% of the total concentration of PAHs, with a mean value of 77.18%. Moreover, the combustible PAHs constituted a significant portion of the total PAHs (more than 50%) at all stations, with the exception of stations 13 (in front of the container terminal) and 7 (close to cement outlets) in the West Port, and stations 3 in the North Port (Figure 2b).

Some high-molecular-weight PAHs, such as BaA, Chy, BbF, BkF, BaP, InP and DibA are regarded as known as toxic PAHs, due to their mutagenic and carcinogenic effects on humans and other organisms [11,36]. Their concentration made up an average of 61.25% of the total concentration of PAHs, ranging from 0–97.3%. These concentrations were highest in stations 1, 4, 13, 15, 16, and 17, whereas at certain remote stations (2, 3, 5 and 22), the toxics' PAH concentrations were significantly lower.

Distribution maps were used to better visualize of the distribution of contamination at the spatial scales (Figure 2C). At the South Port, the distributions of PAH compound (PAHs combust and PAH toxic) sediments exhibited a homogeneous pattern of decreasing concentration in the north-to-south direction, and

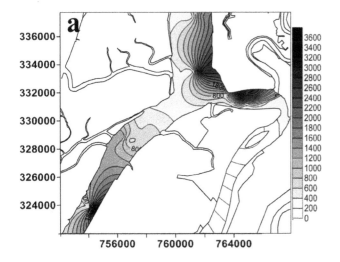

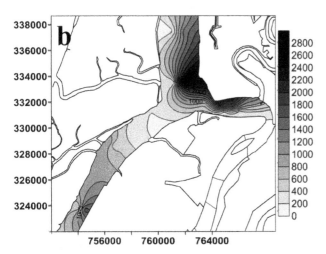

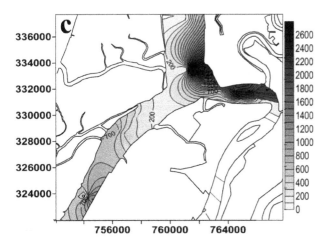

Figure 2. Spatial variations of PAHs compounds in surface sediment of the Klang Strait (a: Total PAHs, b: PAHs combust, c: PAHs toxic).

concentrations were high at the stations 16 and 17, which are parallel to the mouth of the Klang River (Figure 2 and Table 3). This pattern supports the view that the Klang River may be the primary source of contamination in the Klang Strait, and

influences the concentration and distribution of PAHs as water and suspended solids were easily exchanged between the South Port and the polluted Klang River, which contains industrial effluents and untreated municipal waste. Water currents in the vicinity of the South Port are weak; therefore, there is enough time for the deposit of organic components by suspended solids to enter surface sediments. Organic compounds are not easily deposited onto bottom sediments where there are strong water currents [15].

At both the West Port and North Port, the spatial distribution of PAHs exhibited a decreasing east–west gradient. These stations were located close to the near shore area, and are thus strongly influenced by port activities, especially at stations 4, 13 and 14, which were close to the terminal containers in the North and West Port. Additionally, land-based run-offs directly release organic compounds in the vicinity of these stations. Further research revealed significant differences in the sources and concentration of PAHs in sediment samples of near-shore vs. offshore areas. PAH concentrations in sediment samples collected from near-shore sites (city hinterland) were greater than the PAH concentrations in offshore areas [11,30,37,38,39]. This implies that PAH concentrations in near-shore areas are influenced by lateral transport, such as run-off and the transportation of water due to daily rainfall [12,40]. In 2005, Ikaneka indicated that heavy rainfall and floods significantly contributed to the pollution of marine sediments. Moreover, the results of their work indicated that near-shore areas received both burnt material (pyrogenic) and oil products (petrogenic), whereas offshore stations were primarily influenced by the burnt materials [9,24,40].

Overall, the distinct pattern in the distributions of measured parameters revealed that multiple sources contributed significantly to the PAH loads in the Klang Strait. These sources include the large-scale inflow from industries, such as the palm oil, cement, and food manufacturers that are located along the coastline of the North and West Ports, vessel-based discharges, Klang River outflow, land-based run-off, sedimentation, and siltation.

Source identification

Several methods were used to identify the source of PAHs according to the correlation analysis, diagnostic PAHs ratios, PCA analysis, and significant differences in the concentration of the individual PAHs.

A correlation analysis (Table S2) revealed a significant correlation between some individual PAHs related to pyrogenic and petrogenic compounds. This implied that the sediment samples are polluted by anthropogenic sources of PAHs because unpolluted sediment samples do not exhibit an oil fingerprint, and only some PAH compounds may be found in this sediment [12,40].

Mahyar et al. in 2010 studied the characteristics and possible origins of PAHs in the developed and developing areas around the coastal waters of Peninsular Malaysia [12,40]. The collected sediment core samples were used to assess the historical profile of PAHs from 1875 to 2007. Their research agrees well with our results, as because they revealed that older sediment samples (1875–1899) did not exhibit an oil fingerprint, and that only some PAH compounds were found. Sediment samples that are polluted by oil products generally exhibit high concentrations of phenanthrene and its methylated derivative. Thus, there is a high correlation between crude oil and petrogenic and pyrogenic compounds, whereas this correlation in old sediment samples with natural inputs (with no oil fingerprints) ranges from low to negative values [12,40].

Ratios of PAHs, such as Ant/Ant+Phn and Fla/(Fla+Pyr) had been applied to provide an accurate estimation of PAH sources

Table 3. Molecular indices of PAHs, content in surface sediments of Klang Strait.

Station	$\sum$ PAH $_{combust}$	$\sum$ PAH $_{toxic}$	Ant/Ant+Phn	Fla/Fla+Pyr	L/H-PAH
1	1195.8	613.0	0.68	0.00	0.15
2	162.8	31.5	0.49	0.00	0.23
3	105.6	0.0	0.40	0.40	2.61
4	2767.8	2755.3	0.85	0.51	0.03
5	206.6	11.0	0.49	0.05	0.20
6	279.4	221.7	0.51	0.45	0.49
7	274.6	251.6	0.68	0.45	1.55
8	343.5	250.0	0.91	0.46	0.36
9	485.2	457.1	0.61	0.00	0.05
10	584.1	347.6	0.67	0.00	0.07
11	439.8	397.1	0.71	0.41	0.24
12	1046.8	784.3	0.90	0.08	0.29
13	1715.9	1256.3	0.58	0.41	1.07
14	279.2	249.3	0.71	0.53	1.05
15	569.7	569.7	0.28	0.00	0.04
16	2099.5	1907.6	0.60	0.48	0.19
17	2267.4	2041.3	0.86	0.00	0.41
18	689.7	570.4	0.58	0.00	0.06
19	415.0	381.5	0.46	0.49	0.27
20	359.4	326.1	0.44	0.49	0.30
21	410.1	382.8	0.50	0.59	0.18
22	80.3	80.3	0.71	0.00	0.25
Mean	767.2	631.1	0.70	0.16	0.33
Standard deviation	233.2	213.8	0.22	0.28	0.32
Minimum	80.3	11	0.28	0.4	0.04
Maximum	2767.8	2755.3	1	1	2.61

$\sum$ PAH $_{combust}$: Combustion value of PAHs calculated as sum of concentration of high-molecular-weight PAHs include Fla, Pyr, BaA, Chy, BbF, BkF, BaP, InP, DibA, BghiP; $\sum$ PAH $_{toxic}$: Toxic value of PAHs calculated as sum of concentration of BaA, Chy, BbF, BkF, BaP, InP and DibA; L/H: ratio of low molecular weight hydrocarbon (**n**-C10–**n**-C14) to high molecular weight hydrocarbon (**n**-C16–**n**-C22).

(Table 3) [41]. The ratio of Ant/Ant+Phn was greater than 0.1 for all stations, implying that the origin of the PAHs in the sediments of the Klang Strait is primarily combustion. However, the ratio of Fla/(Fla+Pyr) showed the mixed pattern of pyrogenic and petrogenic sources in most stations. In the present study, ratio analyses showed the contradictions in certain stations.

In several researches, PAH ratios have been considered for "ideal" samples dominated by a single source, such as smoke, vehicle emission, and wood. However, some authors indicated that the limits between the diagnostic PAHs ratios are imprecise, as most environmental samples contain PAHs from mixed sources, none of which dominates the PAH profile [2,40,41]. Likewise, the validity of many diagnostic PAHs ratios has been examined for source-distinction based on the source material in tropical Asian waters [12,14,42].

Although these ratios are usable for the differentiation of pyrogenic and petrogenic sources, they are not definitive, since several exceptions have been found. For example, Saha et al. in 2009 reported that, although all of the crude oil samples have Ant/(Ant+Phn)<0.1 and Phn/Ant>10, implying a petrogenic source, one crude oil sample (Miri) has Anh/(Anh+Phn)>0.1 and Phn/Ant<10, showing a pyrogenic source, and one wood sample has Anth/(Anh+Phn)<0.1; implying a petrogenic sources of PAHs

[12,14,41]. Hence, these PAHs ratios are not authoritative enough to be exact; they only provide a rough idea for sources-distinction. These contradictions are most likely due to inconsistencies associated with the conventional ratios, large range of thermodynamic stability (discrimination ability) of different parent PAHs, and different homolog distributions between pyrogenic and petrogenic sources [2,14,42]. Despite these exceptions, the diagnostic PAHs ratios are now routinely analyzed. Several researchers suggest using more indices and cross-plot of diagnostic PAHs ratios to provide a more robust interpretation [10,41,43].

In the present study, the PAH ratio of Fla/(Fla+Pyr) was plotted against Ant/Ant+Phn for all of the sampling stations in Klang Strait (Figure 3). Figure 3 showed a mixed pattern of contamination from pyrogenic and petrogenic sources (petroleum combust) at some of the stations (3, 6, 7, 8, 11, 13, 16, 19 and 20). Simultaneously, wood combustion fingerprints (more frequent PAHs representing wood, coal and grass combustion) were strongly present in stations 4, 14, and 21. Other stations (1, 2, 5, 9, 10, 12, 15, 17, 18 and 22) have primarily originated from combustion sources with a significant contribution of petroleum inputs. Furthermore, An Ant/(Ant+Phn) ratio >0.10 almost always corresponds to a Fla/(Fla+Pyr) ratio >0.40, supporting the conclusion that combustion products are a main component of the contamination [10,14,41].

The purpose of principal component analysis (PCA) is to represent the total variation of the PAHs data with a minimum number of the factor loadings. By critically quantifying the percentage of contribution of individual PAH sources and identifying the occurrence of the sources responsible, each factor can be realized [2,23,44]. As shown in Table 4, PCA analysis classified the dataset of PAHs into five principal components (PCs) that control 81.89% of the variability in the data.

The first factors (PC1) explained a total variance of 26.73% in the data. These factors were strongly weighted the context of Acy, Ace, Flr, Phn and Fla, and were believed to be from petrogenic sources of PAHs, and were also been identified as tracers for volatilization or spill of petroleum products [2,12,14,23,41,42]. Phn and Fla are the thermodynamically stable triaromatic isomer, and their high concentration are regarded as characteristic of petrogenic sources [14,23]. As indicated by a previous researcher, Phn and Fla were more abundant in crude oil, crankcase oil, and unburnt fuel, and originated from soot collected from diesel engines [12,41,45].

The second factor (PC2) responsible for 21.08% of the total variance was strongly related to BbF, BaP, DibA, Bghip, with moderate loading of InP and BkF. This factor are predominately composed of high molecular weight PAHs with 4–6 rings, and are basically known to be derived from the incomplete combustion and pyrolysis of fuel [44,46]. The source of this individual PAHs have been found in both vehicular emission sources (diesel and gas engine), road dust, and gasoline combusted ships [2,44]. InP and BkF have been found in gasoline vehicle soot, and both gas and diesel engine emissions [2,12,47,48]. Bghip and BaP have been identified as a tracer of auto emissions because there were determined to be rich in road dust from urban-areas and traffic tunnels. The abundance of these compounds was considered as an index for the determination of autoexhaust contribution of gasoline-engines. Therefore, this factor is selected to represent the combustion source of PAHs [2,12,14,41,48].

The third factor is responsible for 17.25% of the total variance, and is predominately composed of Ant, Pyr, Chy, and BaA (4-ring PAHs), with moderate loadings of Fla. More researchers have used Ant as the marker of wood combustion source [2] (Harrison et al., 1996). Also, several literature have identified Pyr, BaA, Chy, Fla (4-rings) and Bap as a tracer of auto emissions, as it was found to be enriched in road dust collected from Kuala Lumpur, Bangkok, Shanghai [2,12,14,47,48], and traffic tunnels [10,44,47,48]. The

Table 4. Rotated component matrix of 16 PAHs from the Klang Strait sediment.

PAHs	PC1	PC2	PC3	PC4	PC5
Nap	.095	0.000	0.138	**0.954**	−0.034
Acy	**.957**	0.037	0.009	0.006	0.012
Ace	**.918**	0.061	.185	−0.013	0.031
Flr	**.910**	−0.042	0.332	0.114	0.069
Phn	**.824**	−0.106	0.322	0.091	.124
Ant	.429	0.063	**.814**	−0.064	.429
Fla	**.866**	0.017	.466	0.465	0.07
Pyr	.201	−0.335	**.779**	−0.021	−0.236
BaA	.058	−0.142	**.861**	0.392	−0.112
Chy	−0.087	0.094	**.865**	0.029	0.043
BbF	−0.092	**0.871**	−0.016	−0.016	0.036
BkF	.012	0.521	−0.139	0.108	**0.852**
BaP	0.288	**0.777**	.026	.395	0.013
DibA	−0.152	**0.847**	−0.096	−0.137	−0.217
BghiP	0.185	0.584	−0.185	−0.058	0.173
InP	0.081	0.609	0.139	−0.221	−0.492
Variance (%)	26.73	21.08	17.25	9.66	7.17

*Rotation method: Varimax with Kaiser normalization.
*Bold loadings >0.70.

PAH profiles of road dust in these cities showed that the ratio of PAHs (4 rings) to PAHs (5–6 rings) exceeds 1 [12,47]. In the present paper, we showed that the ratio of PAHs (4 rings) to PAHs (5–6 rings) in the surface sediment, 0–126.44 with a mean of 7.805 is significantly greater than that in the road dust from Kuala Lumpur (1.58). This result indicates that there might be an additional PAH source (not traffic-related), leading to a higher ratio of these PAHs [2]. Several researchers have identified Pyr, BaA, Chy, and Fla as markers of coal combustion because these individual PAHs are dominant in coal-combustion profiles [2,12,14,47,48]. In Malaysia, coal is the main energy source, and is used widely for industrial and domestic and industrial purposes, such as cement, palm oil, tire, and power industry [2,12,14]. It is reasonable to assign these PAHs to coal combustion and traffic-related source of PAHs.

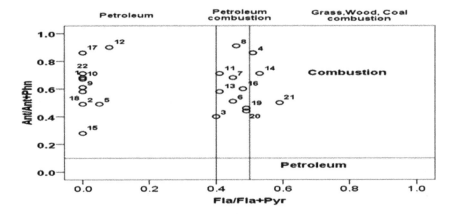

Figure 3. Plots of PAH isomer pair ratios for source identification (Ant/(Ant+Phy) vs Flu/(Flu+Pyr).

PC4 explained 9.66% of the total variance, with a high correlation with Nap. Nap is used as a tracer of oil spills, such as Acy and Ace. It can also originate from termite activities on vascular land plants and woody material in tropical areas and the Amazonian region [49,50]. The fifth factor (PC5) is responsible for 7.17% of the total variance and was strongly related to BkF, which are known to be derived from the incomplete combustion and pyrolysis of fuel [44,46].

The PCA analysis described 45.5% of the total variance (PC2, PC3 and PC5) in the PAH load, including individual PAHs that are known to originate from coal combustion and vehicular emissions or traffic-related sources.

Similarly, 36.39% of the total variance (PC1 and PC4) in PAHs load related to petrogenic sources and the variance in PAHs was unrelated to unknown and biogenic sources.

There are two origins for vehicular sources in the Klang Strait, one is the traffic-related PAHs in road dust entering the surface sediment by land-based runoff, others due to the exhaust emission from cargo vessels, ships, ferries, and fishing boats [2,12,14,47]. Furthermore, the cargo vessels play the main role as PAHs contributor in the Klang Strait compared to traffic related PAHs, as cargo vessels are responsible in transferring most coal to coal–fired power plants located along the West coastal water of Malaysia, such as Connaught bridge power, Kapar and Tanjung Bin, Manjung, and Jimah. Most fishing boats and ferries are still unequipped with catalytic converters [51,52].

Coal and wood combustions mainly originated from agricultural, industrial, and habitats activities. Recently, coal usage near the Klang Strait is concentrated in the Kapar Power Plant in the Klang Valley region with high consumption of coal (more than16 million tons). Additionally, It is also the only power plant in Malaysia with triple fuel firing capability (natural gas, oil, and coal). It started operating in in 1987 and was the first coal–fired power plant in Malaysia [51,52]. Moreover, Mahyar et al. in 2010 indicated that, in the near-shore and offshore areas near Klang, high levels of PAHs originated from pyrogenic sources, such as the combustion of fossil fuel and the burning of biomass and wood. The dominance of pyrogenic sources in this area is due to the products of combusted petroleum from industries and automobiles emissions [12,40].

Petrogenic sources include petroleum products and crude oil, which are identified as major fossil fuels being used in Malaysia (around 49% of total energy consumption) [14]. Zakaria et al. in 2002 reported that several sources lead to the input of petrogenic PAHs into Malaysian coastal water, namely spillage of unburnt fuel and waste crankcase oil, leakage of crankcase oils from vehicles, with subsequent land-based runoff, wastewater of industry outlets, and vessel maintenance [12,14,47].

In addition, other studies have reported that contamination from various seas contribute to increase petrogenic pollution in the coastal waters of western Malaysia [16,53]. For example, in 1999, Abdullah revealed that land-based run-off and contamination from an offshore oil field near Sumatra Island contributed to the petroleum contamination in the Malacca Strait, which is a narrow channel of marine water located between Sumatra Island, Indonesia, and Peninsular Malaysia. This strait is vulnerable to contamination caused by tanker operation and oil spills [16,53,54]. The Strait of Malacca also plays an important role in transporting pollution into the Klang Strait. Pauzi Zakaria in 2001 used a biomarker compound to identify the source of tar-balls off the coast of Malaysia.

They identified several petrogenic sources of contamination in the western coastal waters of Malaysia, such as accidental oil spills and tanker-derived sources in the Strait of Malacca and crude oils

originating from land-based run-off from human activities. Additionally, they clearly showed that the western coastal waters of Malaysia have received approximately 30% of their petroleum pollution from Middle East crude oil (MECO) and South-East Asian crude oil (SEACO), which was probably transported to Malaysia via marine currents, tankers and shipping discharges, including ballast water and tank-washing water [16].

Generally, the results of the PCA analyses are in concordance with the evidence from the pair isomer ratio of PAHs, which revealed a mixture of pyrogenic-and petrogenic-derived PAHs. PAHs contamination in the sediment in the Klang Strait are derived primarily from vehicular sources, coal and wood combustions, and crude oil and coal (petroleum combustion sources), which can originate from land-based run-off, oil spill, tanker operation, shipping activities, and industrial discharges such as effluents from cement, food and oil factories.

Ecological risk assessment

The results obtained from the ecological risk assessment of PAHs in the surface sediments of Klang Strait are summarized in Table 5. These results were arranged based on the concentration of PAHs measured at 22 stations. Generally, the risk assessment revealed that total PAHs are likely to cause slightly adverse effects on the biological communities at stations 4, 7, 12, 16, and 17 as these stations showed a value of $0.1 < PELq < 0.5$ and $1.0 < P \leq 2.0$. Additionally, moderately adverse biological effects were shown for station 13 ($0.5 < PELq < 1.5$, $2.0 < P \leq 3.0$), but the rest of the stations showed rare, adverse ecological effects due to the PAH exposure in surface sediments ($PELq < 0.1$, $P < 3$). The cluster analysis allows for a better understanding of the similarity between sampling stations and PAHs contamination (Figure 4). The dendrograms clearly classified the sampling stations into three main clusters based on adverse biological effects. Station 13 (cluster C) was completely separated from the other stations as it demonstrated the moderate adverse biological effect by PAHs compound, while stations 4, 16, and 17 are in the same group (cluster B), with slightly adverse effects. The other stations are grouped into cluster A, and exhibited non-adverse effects ($PELq < 0.1$, $P < 3$), with the exception of stations 7 and 12. However, the average toxicity effects of PAHs in stations 7 and 12 are in the slight range, but these stations were arranged in cluster A, because their values are closer to 0.1 compared to the stations in cluster B. Based on a comparison with the sediment quality guidelines, only certain individual PAHs, such as Acy, Flr, Ant, BaA, BaP and DibA were associated with adverse biological effects, and only at stations 4 (close to the container terminal in the North Port), 7 (cement outlet), 13 (container terminal in the West Port), 12 (around the mangrove forest in the West Port) 16, and 17 (vicinity of the mouth of the Klang River) (Table 5).

Moreover, the first three components of PCA analysis are strongly related to these individual PAHs that controls 63.36% of changes in PAHs contamination in surface sediment of the Klang Strait. According to the cluster and PCA classification, stations 4, 7, 13, 16 and 17 can be regarded as the main entrances for loading individual PAHs, especially Acy, Flr, Ant, BaA, BaP and DibA.

The present study shows that generally, the sediment is slightly polluted with PAHs ($PELq = 0.16$ and $P = 1.01$) in the Klang Strait. According to the results of ecological risk, it can be concluded that PAHs are not a primary pollution concern in the Klang Strait, and that only station 13 can be considered as a vulnerable station in the Klang area.

In this research, our hypothesis was defined based on the serious threat posed by petroleum contamination to the Klang Strait due

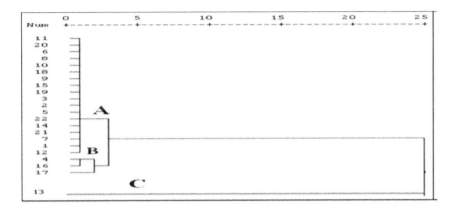

Figure 4. Cluster analyses to classify stations based on adverse biological effect of PAHs in surface sediment.

to several evidences, which was reported in previous literature [16,40,53,55].

The highest value for PAH contamination was estimated to be 1700 ng/g d.w. in near-shore Klang sediment samples from 1976–1999, with a predominant input of burnt materials and combusted petroleum discharges from industries, oil spill events and ship ballasting or bilge pumping. From 1976–1999, rapid development of the West Port and maritime-related industries that caused greater numbers of ocean going ships to enter the strait from the

South China Sea and to the Far East to transport goods. Similarly, in 1997, a collision between two tankers in the straits of Singapore caused 25000 tons of heavy fuels oil to be released into the marine environment, of which 700 tons of oil were released into the Malacca strait, negatively affecting the coastal area of the Klang Strait [12,40].

According to the present study, the hypothesis was rejected due to the fact that in most of the stations, the average PAHs concentration was significantly lower than the threshold effects

Table 5. Ecological risk calculated for individual PAHs in the surface sediment of different station.

Stations	Nap	Acy	Ace	Flr	Phn	Ant	Fla	Pyr	BeA	Chy	BbF	BkF	BaP	DibA	BghiP	PELq	P
1	0.00	0.00	0.02	0.23	0.08	0.36	0.00	0.41	0.8	0.00	0.00	0.00	0.00	0.00	0.00	0.09	0.92
2	0.00	0.00	0.00	0.07	0.02	0.05	0.00	0.09	0.05	0.00	0.00	0.00	0.00	0.00	0.00	0.02	0,77
3	0.00	0.00	0.06	0.82	0.16	0.24	0.03	0.04	0.00	0.00	0.00	0.00	0.00	0.00	0.00	0.09	0,83
4	0.00	0.00	0.00	0.01	0.02	0.24	0.01	0.01	0.10	0.06	0.45	0.00	**1.96**	**1.41**	0.00	**0.28**	**1,56**
5	0.00	0.00	0.00	0.12	0.02	0.05	0.01	0.14	0.00	0.00	0.01	0.00	0.00	0.00	0.00	0.02	0,79
6	0.01	0.00	0.00	0.40	0.07	0.15	0.02	0.02	0.26	0.00	0.00	0.00	0.00	0.00	0.00	0.06	0,83
7	0.00	0.00	0.00	0.19	0.23	**1.09**	0.01	0.01	0.29	0.00	0.00	0.03	0.00	0.00	0.00	**0.12**	**1.02**
8	0.01	0.00	0.01	0.27	0.01	0.29	0.01	0.01	0.07	0.00	0.00	0.00	0.16	0.00	0.04	0.06	0,85
9	0.00	0.00	0.00	0.05	0.01	0.03	0.00	0.02	0.28	0.23	0.00	0.00	0.00	0.50	0.00	0.08	0,86
10	0.00	0.00	0.00	0.16	0.01	0.04	0.00	0.17	0.40	0.00	0.00	0.00	0.00	0.00	0.00	0.05	0,90
11	0.00	0.00	0.01	0.34	0.03	0.15	0.01	0.02	0.04	0.00	0.00	0.23	0.00	0.00	0.00	0.06	0,87
12	0.01	0.00	0.01	0.28	0.05	0.95	0.01	0.24	**1.12**	0.00	0.00	0.00	0.00	0.05	0.00	**0.18**	**1,14**
13	0.04	**7.26**	0.12	**4.31**	0.41	**1.26**	0.11	0.17	0.56	0.00	0.01	0.02	**1.08**	0.00	0.03	**1.03**	**2.03**
14	0.01	0.00	0.00	0.23	0.14	0.74	0.01	0.01	0.21	0.00	0.01	0.00	0.08	0.00	0.00	0.10	0,88
15	0.01	0.00	0.00	0.06	0.02	0.02	0.00	0.00	0.11	0.00	0.01	0.00	0.00	0.96	0.00	0.08	0,89
16	0.28	0.00	0.01	0.84	0.12	0.41	0.06	0.07	**1.41**	0.14	0.00	0.00	**1.07**	0.00	0.00	**0.30**	**1,42**
17	0.02	0.00	0.03	**1.57**	0.17	**2.45**	0.00	0.16	**2.15**	0.65	0.00	0.00	0.00	0.00	0.00	**0.48**	**1,57**
18	0.00	0.00	0.00	0.14	0.02	0.05	0.00	0.09	0.28	0.00	0.20	0.00	0.00	0.00	0.00	0.05	0,93
19	0.01	0.00	0.00	0.25	0.07	0.14	0.01	0.01	0.39	0.00	0.01	0.00	0.14	0.00	0.00	0.07	0,87
20	0.01	0.00	0.00	0.26	0.07	0.12	0.01	0.01	0.33	0.00	0.00	0.00	0.12	0.00	0.00	0.06	0,85
21	0.01	0.00	0.00	0.25	0.03	0.07	0.01	0.01	0.26	0.00	0.01	0.00	0.00	0.96	0.00	0.09	0.86
22	0.00	0.00	0.00	0.09	0.00	0.02	0.00	0.00	0.08	0.00	0.01	0.00	0.00	0.00	0.00	0.01	0,74
Mean	0.02	0.33	0.01	0.50	0.08	0.41	0.01	0.08	0.42	0.05	0.03	0.01	0.21	0.18	0.00	0.16	1,01

*Nemerow pollution index (P).
*The PEL quotient (PELq).

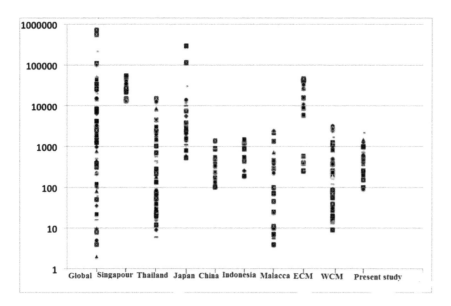

Figure 5. Total PAHs concentration (ng/g dw) in Klang Strait sediments in comparison to those of the reported concentrations for global, Southeast Asian countries, and Malaysian costal zones. Data for global sediments and others areas are derived from several literatures [12,14,40,44,45,52,55,56,58,59].

level (TELs and PELs), and rarely reached a level likely to cause adverse biological effects.

Several studies recorded a significant decline in PAH discharges around the west coastal water of Malaysia after 2000 due to the establishment of an integrated management programme, weathering, and meteorological conditions [16,40,53,55].

An integrated management programme was established in 1996, where several responsible organizations contributed to solving the environmental problems in Malaysia's marine environments. These organizations ratified specific regulations related to marine pollution to control petroleum and chemical contaminations in Malaysian coastal waters. These regulations focused on strategies that were based on international agreements to prevent and control pollution from ships, platform draining, and industrial inputs [16].

The weathering process can greatly deplete the concentration of PAH in a marine environment. The L/H-PAH ratio is accepted as a practical method for assessing weathering based on the differences between the low and high molecular weights of PAH compounds [16,40,53,55]. These results show that the L/H-PAH ratio was below 1 at most of the stations, and exceeds 1 only at stations 3, 7, 13, and 14 (Table 3). Generally, a low ratio of L/H-PAHs is attributed to the high resistances of the high-molecular-weight PAHs to microbial degradation, which is consistent with other studies. Such a low ratio could also be due to the high solubility and volatility of the low-molecular-weight PAHs, which would lead to their depletion [16,53,56].

In addition, there is also evidence of a greater depletion of low-molecular-weight PAHs by the weathering process, which is clearly indicated by Pauzi Zakaria (2001) [16]. He revealed that the tar-ball samples from western Malaysian coastal waters had undergone significant weathering, as their L/H-PAH ratios were much lower (0.23–1.48) than the ratios of crude oil samples (8–44 for SEACO crude oil and 12.13–20.3 for MECO crude oil) [16]. In the present study, this ratio ranged from 0.02–2.61, which was significantly lower than that of the crude oil in this region. Several possibilities have been described to explain why the PAH compositions had a lower ratio of L/H in the western coastal

waters of Malaysia, such as wandering due to long-haul transportation, ballast water, and tank washings [16,53,56].

Additional research has indicated that meteorological conditions play a major role in controlling PAH concentrations over spatial and temporal scales, and significantly negative linear relationships were found between wind speeds, temperatures, and PAH concentrations, because atmospheric turbulence causes a dilution of contaminant concentrations and hastens the weathering process, especially if the wind speed is greater than 5.8 km h^{-1} and the temperature is exceeds than $20°C$ [30,57]. Thus, the higher temperature (approximately $30°C$) of the Klang Strait as a tropical area could potentially increase the depletion of PAHs.

Figure 5 shows the concentration of PAH in comparison with global, Southeast Asian countries, and Malaysian costal zones. Most values of PAHs concentrations in the present study, west coastal water of Malaysia, Malacca, Indonesia, China and Thailand ranged from tens to thousands of ng/g. Thus, PAHs values in Klang Strait sediments are comparable to these areas. On the other hand, the current PAHs concentration are 1–2 orders of magnitude lower than those in Japan, Singapore, and the eastern coastal water of Malaysia, as the value of PAHs in these countries ranged over thousands of ng/g. Thus, PAHs concentrations in Klang Strait can be categorized as low to moderate on a global scale.

Conclusion

Exploratory source identification of PAHs in Klang Strait sediment was based on the distribution pattern of the individual PAHs, principal component analysis, and paired isomer ratios. The results of the distribution pattern showed that the pyrogenic sources are the major source of PAHs in most stations. Petrogenic input appears be a main source of PAHs in station 13 (in front of the container terminal) and stations 7 (close to cement outlets) in the West Port, and station 3 in the North Port because the LMWPAHs constituted a significant portion of the total PAHs in these stations.

The results of diagnostic ratios showed that PAHs are derived from a mixture of sources (both petrogenic and pyrogenic), with high frequencies of pyrogenic sources in most stations. The PCA analysis method showed that the contributions of spill of oil products, traffic-related pollution, and coal and wood combustion are dominant in Klang Strait sediments.

The pyrogenic sources (traffic-related pollution and coal combustion) are responsible for 45.5% of the total variance in the Klang Strait sediment sediments. As well as pyrolytic input as a main source, petrogenic input was also regarded as a source for PAHs in the Klang Strait due to direct petrogenic sources from oil leaks (oil spill, urban run-off, tanker operation and shipping activities) in agreement with their location.

Despite the inherent limitations of this research, the adopted approach in the present study highlighted that no adverse biological effects are associated with the exposure to PAHs levels in the Klang Strait, and only the areas around terminal container (station 13) are moderately polluted. Several factors caused the decline of these contaminants, such as an integrated management programme and regulation, meteorological conditions, and weathering. This paper represents a baseline study for future studies, which may include the whole Malaysian estuary and coastal waters for the purpose of promoting our knowledge of the distribution of PAHs in coastal and estuary ecosystems.

Acknowledgments

We would like to acknowledge Professor A. Sasekumar, Professor Chong Ving Ching and Dr Wong Chee Seng and others involved in field collection of samples, for their support and inspiration. Special thanks to Chemistry Department of Marine Science, Iranian National Institute for Oceanography, for valuable contribution to laboratory analysis.

Author Contributions

Conceived and designed the experiments: SBTS RH AS AM. Performed the experiments: SBTS MR AM. Analyzed the data: SBTS OS. Contributed reagents/materials/analysis tools: AS RH. Wrote the paper: SBTS MR OS.

References

1. Veltman K, Huijbregts MAJ, Rye H, Hertwich EG (2012) Including impacts of particulate emissions on marine ecosystems in life cycle assessment: The case of offshore oil and gas production. Integr Environ Assess Manag 7: 678–686.

2. Liu Y, Chen L, Huang Q-h, Li W-y, Tang Y-j, et al. (2009) Source apportionment of polycyclic aromatic hydrocarbons (PAHs) in surface sediments of the Huangpu River, Shanghai, China. Sci Total Environ 407: 2931–2938.

3. Barkhordarian A (2012) Investigating the Influence of Anthropogenic Forcing on Observed Mean and Extreme Sea Level Pressure Trends over the Mediterranean Region. The Scientific World Journal 2012.

4. Zhou Q, Jiang G, Liu J (2002) Organotin pollution in China. The Scientific World Journal 2: 655–659.

5. Tavakoly Sany SB, Salleh A, Sulaiman AH, Tehrani GM (2012) Ecological risk assessment of poly aromatic hydrocarbons in the North Port, Malaysia. World Academy of Science, Engineering and Technology 69: 43–46.

6. Ahmadzadeh S, Kassim A, Rezayi M, Rounaghi GH (2011) Thermodynamic Study of the Complexation of p-Isopropylcalix [6] arene with Cs+ Cation in Dimethylsulfoxide-Acetonitrile Binary Media. Molecules 16: 8130–8142.

7. Rezayi M, Kassim A, Ahmadzadeh S, Yusof NA, Naji A, et al. (2011) Conductometric Determination of Formation Constants of tris (2-pyridyl) methylamine and Titanium (III) in Water-Acetonitryl Mixture. Int J Electrochem Sci 6: 4378–4387.

8. MADEP (2007) Sediment Toxicity of Petroleum Hydrocarbon Fractions,Prepared for Massachusetts Department of Environmental Protection Office of Research and Standards. 397 Washington Street, Duxbury, MA 02332: Battelle.

9. Tavakoly Sany SB, Hashim R, Salleh A, Safari O, Mehdinia A, et al. (2013) Risk assessment of polycyclic aromatic hydrocarbons in the West Port semi-enclosed basin (Malaysia). Environmental Earth Sciences: 1–14.

10. Riccardi C, Di Filippo P, Pomata D, Di Basilio M, Spicaglia S, et al. (2013) Identification of hydrocarbon sources in contaminated soils of three industrial areas. Sci Total Environ 450: 13–21.

11. Khairy MA, Kolb M, Mostafa AR, EL-Fiky A, Bahadir M (2009) Risk assessment of polycyclic aromatic hydrocarbons in a Mediterranean semi-enclosed basin affected by human activities (Abu Qir Bay, Egypt). J hazard mater 170: 389–397.

12. Zakaria MP, Takada H, Tsutsumi S, Ohno K, Yamada J, et al. (2002) Distribution of polycyclic aromatic hydrocarbons (PAHs) in rivers and estuaries in Malaysia: a widespread input of petrogenic PAHs. Environ Sci Technol 36: 1907–1918.

13. Beyer J, Jonsson G, Porte C, Krahn MM, Ariese F (2010) Analytical methods for determining metabolites of polycyclic aromatic hydrocarbon (PAH) pollutants in fish bile: A review. Environmental Toxicology and Pharmacology 30: 224–244.

14. Saha M, Togo A, Mizukawa K, Murakami M, Takada H, et al. (2009) Sources of sedimentary PAHs in tropical Asian waters: differentiation between pyrogenic and petrogenic sources by alkyl homolog abundance. Mar Pollut Bull 58: 189–200.

15. Tavakoly Sany SB, Salleh A, Suliman AH, Sasekumar A, Rezayi M, et al. (2012) Heavy metal contamination in water and sediment of the Port Klang coastal area, Selangor, Malaysia. Environmental Earth Sciences: 1–13.

16. Pauzi Zakaria M, Okuda T, Takada H (2001) Polycyclic aromatic hydrocarbon (PAHs) and hopanes in stranded tar-balls on the coasts of Peninsular Malaysia: applications of biomarkers for identifying sources of oil pollution. Mar Pollut Bull 42: 1357–1366.

17. Tehrani GM, Hshim R, Sulaiman AH, Tavakoly Sany SB, Salleh A, et al. (2012) Distribution Of Total Petroleum Hydrocarbons And Polycyclic Aromatic Hydrocarbons In Musa Bay Sediments (Northwest Of The Persian Gulf). Environment Protection Engineering 15: 115–128.

18. Yap (2005) Ecology of Klang Strait. In (eds: Sasekumar, A. and V.C. . Chong) Pollution. Kuala Lumpur: University of Malaya. pp. 225–235.

19. Naji A, Ismail A (2012) Sediment quality assessment of Klang Estuary, Malaysia. Aquat Ecosyst Health Manage 15: 287–293.

20. US-EPA (2010) United States Environmental Protection Agency, Method 610, Polynuclear Aromatic Hydrocarbons.

21. Rauret G (1998) Extraction procedures for the determination of heavy metals in contaminated soil and sediment. Talanta 46: 449–455.

22. Budzinski H, Mazeas O, Tronczynski J, Desaunay Y, Bocquene G, et al. (2004) Link between exposure of fish (Solea solea) to PAHs and metabolites: Application to the "Erika" oil spill. Aquat Living Resour 17: 329–334.

23. Wang C, Wang W, He S, Du J, Sun Z (2011) Sources and distribution of aliphatic and polycyclic aromatic hydrocarbons in Yellow River Delta Nature Reserve, China. Appl Geochem.

24. Dsikowitzky L, Nordhaus I, Jennerjahn TC, Khrycheva P, Sivatharshan Y, et al. (2011) Anthropogenic organic contaminants in water, sediments and benthic organisms of the mangrove-fringed Segara Anakan Lagoon, Java, Indonesia. Mar Pollut Bull 26: 1330–1336.

25. EPA (1991) Sediment quality guidelines. Chicago: United States Environmental Protection Agency, Draft report.Region V.

26. Macdonald DD, Carr RS, Calder FD, Long ER, Ingersoll CG (1996) Development and evaluation of sediment quality guidelines for Florida coastal waters. Ecotoxicology 5: 253–278.

27. United States Environmental Protection Agency, National Oceanic and Atmospheric Administration (2005) Predicting toxicity to Amphipods from sediment chemistry. Washington, DC: National Center for Environmental Assessment, Office of Research and Development.

28. Long ER, MacDonald DD, Smith SL, Calder FD (1995) Incidence of adverse biological rtile sulfide. Environ. Toxicol. effects within ranges of chemical concentrations in marine and estuarine sediments. Environ Manag 19: 81–97.

29. Fdez-Ortiz de Vallejuelo S, Arana G, De Diego A, Madariaga JM (2010) Risk assessment of trace elements in sediments: The case of the estuary of the Nerbioi–Ibaizabal River (Basque Country). J hazard mater 181: 565–573.

30. Montuori P, Triassi M (2012) Polycyclic aromatic hydrocarbons loads into the Mediterranean Sea: Estimate of Sarno River inputs. Mar Pollut Bull 64: 512–520.

31. McCready S, Birch G.F, Long E.R (2006) Metallic and organic contaminants in sediments of Sydney Harbour, Australia and vicinity–a chemical dataset for evaluating sediment quality guidelines. Environ Inte 32: 455–465.

32. Liang J, Chen C, Song X, Han Y, Liang Z (2011) Assessment of heavy metal pollution in soil and plants from Dunhua sewage irrigation area. Int J Electrochem Sci 6: 5314–5324.

33. Ogunkunle CO, Fatoba PO (2013) Pollution Loads and the Ecological Risk Assessment of Soil Heavy Metals around a Mega Cement Factory in Southwest Nigeria. Polish Journal of Environmental Studies 22: 487–49.

34. Primpas I, Karydis M (2010) Scaling the trophic index (TRIX) in oligotrophic marine environments. Environ Monit Assess: 1–13.

35. Tavakoly Sany SB, Salleh A, Rezayi M, Saadati N, Narimany L, et al. (2013) Distribution and Contamination of Heavy Metal in the Coastal Sediments of Port Klang, Selangor, Malaysia. Water Air Soil Pollut 224: 1–18.

36. Hale SE, Lehmann J, Rutherford D, Zimmerman A, Bachmann RT, et al. (2012) Quantifying the total and bioavailable PAHs and dioxins in biochars. Environ Sci Technol 46: 2830–2838.

37. Pan K, Wang WX (2011) Trace metal contamination in estuarine and coastal environments in China. Sci total environ 421–422: 4–16.

38. Saadati N, Abdullah MP, Zakaria Z, Tavakoly Sany SB, Rezayi M, et al. (2013) Limit of detection and limit of quantification development procedures for organochlorine pesticides analysis in water and sediment matrices. Chem Central J 7: 1–10.

39. Tavakoly Sany SB, Hashim R, Rezayi M, Salleh A, Safari O (2013) A review of strategies to monitor water and sediment quality for a sustainability assessment of marine environment. Environ Sci Pollut Res 21: 813–833.

40. Safari M, Zakaria MP, Mohamed CAR, Lajis NH, Chandru K, et al. (2010) The history of petroleum pollution in Malaysia; urgent need for integrated prevention approach. Environ Asia 3: 131–142.

41. Yunker MB, Macdonald RW, Vingarzan R, Mitchell RH, Goyette D, et al. (2002) PAHs in the Fraser River basin: a critical appraisal of PAH ratios as indicators of PAH source and composition. Organic Geochem 33: 489–515.

42. Guo Z, Lin T, Zhang G, Yang Z, Fang M (2006) High-resolution depositional records of polycyclic aromatic hydrocarbons in the central continental shelf mud of the East China Sea. Environ SciTechnol 40: 5304–5311.

43. Wagener AdL, Meniconi MdFtG, Hamacher C, Farias CO, da Silva GC, et al. (2012) Hydrocarbons in sediments of a chronically contaminated bay: The challenge of source assignment. Mar Pollut Bull 64: 284–294.

44. Zhang Y, Guo C-S, Xu J, Tian Y-Z, Shi G-L, et al. (2012) Potential source contributions and risk assessment of PAHs in sediments from Taihu Lake, China: Comparison of three receptor models. Water Research 46: 3065–3073.

45. Boonyatumanond R, Wattayakorn G, Togo A, Takada H (2006) Distribution and origins of polycyclic aromatic hydrocarbons (PAHs) in riverine, estuarine, and marine sediments in Thailand. Mar Pollut Bull 52: 942–956.

46. Chen H-y, Teng Y-g, Wang J-s (2012) Source apportionment of polycyclic aromatic hydrocarbons (PAHs) in surface sediments of the Rizhao coastal area (China) using diagnostic ratios and factor analysis with nonnegative constraints. Sci Total Environ 414: 293–300.

47. Boonyatumanond R, Murakami M, Wattayakorn G, Togo A, H T (2007) Sources of polycyclic aromatic hydrocarbons (PAHs) in street dust in a tropical Asian mega-city, Bangkok, Thailand. Sci Total Environ 384: 420–432.

48. Larsen IR, Baker J (2003) Source apportionment of polycyclic aromatic hydrocarbons in the urban atmosphere: a comparison of three methods. Environ Sci Technol 37: 1873–1881.

49. Okere U, Semple K (2012) Biodegradation of PAHs in 'Pristine'Soils from Different Climatic Regions. J Bioremed Biodegrad S 1: 2.

50. Ekpo BO, Oyo-Ita OE, Oros DR, Simoneit BRT (2011) Distributions and sources of polycyclic aromatic hydrocarbons in surface sediments from the Cross River estuary, SE Niger Delta, Nigeria. Environmonit assess: 1–11.

51. Alam L, Mohamed CAR (2011) Natural radionuclide of Po210 in the edible seafood affected by coal-fired power plant industry in Kapar coastal area of Malaysia. Environ Health 10: 43.

52. Koike T, Koike H, Kurumisawa R, Ito M, Sakurai S, et al. (2012) Distribution, source identification, and historical trends of organic micropollutants in coastal sediment in Jakarta Bay, Indonesia. J Hazard mater 217: 208–216.

53. Elias M, Wood AK, Hashim Z, Hamzah MS, Rahman SA, et al. (2009) Identification sources of polycyclic aromatic hydrocarbons (PAHs) in sediments from the straits of Malacca. Fresenius Environmental Bulletin 18: 843–847.

54. Mirsadeghi SA, Zakaria MP, Yap CK, Shahbazi A (2011) Risk assessment for the daily intake of polycyclic aromatic hydrocarbons from the ingestion of cockle (Anadara granosa) and exposure to contaminated water and sediments along the west coast of Peninsular Malaysia. J Environ Sci 23: 336–345.

55. Nasher E, Heng LY, Zakaria Z, Surif S (2013) Assessing the Ecological Risk of Polycyclic Aromatic Hydrocarbons in Sediments at Langkawi Island, Malaysia. The Scientific World Journal 2013.

56. Chandru K, Zakaria MP, Anita S, Shahbazi A, Sakari M, et al. (2008) Characterization of alkanes, hopanes, and polycyclic aromatic hydrocarbons (PAHs) in tar-balls collected from the East Coast of Peninsular Malaysia. Mar Pollut Bull 56: 950–962.

57. Neff JM, Stout SA, Gunster DG (2005) Ecological risk assessment of PAHs in sediments. Identifying sources and toxicity.Integr. Environ Assess Manage 1: 22–33.

58. Dsikowitzky L, Nordhaus I, Jennerjahn TC, Khrycheva P, Sivatharshan Y, et al. (2011) Anthropogenic organic contaminants in water, sediments and benthic organisms of the mangrove-fringed Segara Anakan Lagoon, Java, Indonesia. Mar Pollut Bull 62: 851–862.

59. Isobe T, Takada H, Kanai M, Tsutsumi S, Isobe KO, et al. (2007) Distribution of polycyclic aromatic hydrocarbons (PAHs) and phenolic endocrine disrupting chemicals in South and Southeast Asian mussels. Environ Monit and Assess 135: 423–440.

Permissions

The contributors of this book come from diverse backgrounds, making this book a truly international effort. This book will bring forth new frontiers with its revolutionizing research information and detailed analysis of the nascent developments around the world.

We would like to thank all the contributing authors for lending their expertise to make the book truly unique. They have played a crucial role in the development of this book. Without their invaluable contributions this book wouldn't have been possible. They have made vital efforts to compile up to date information on the varied aspects of this subject to make this book a valuable addition to the collection of many professionals and students.

This book was conceptualized with the vision of imparting up-to-date information and advanced data in this field. To ensure the same, a matchless editorial board was set up. Every individual on the board went through rigorous rounds of assessment to prove their worth. After which they invested a large part of their time researching and compiling the most relevant data for our readers.

The editorial board has been involved in producing this book since its inception. They have spent rigorous hours researching and exploring the diverse topics which have resulted in the successful publishing of this book. They have passed on their knowledge of decades through this book. To expedite this challenging task, the publisher supported the team at every step. A small team of assistant editors was also appointed to further simplify the editing procedure and attain best results for the readers.

Apart from the editorial board, the designing team has also invested a significant amount of their time in understanding the subject and creating the most relevant covers. They scrutinized every image to scout for the most suitable representation of the subject and create an appropriate cover for the book.

The publishing team has been an ardent support to the editorial, designing and production team. Their endless efforts to recruit the best for this project, has resulted in the accomplishment of this book. They are a veteran in the field of academics and their pool of knowledge is as vast as their experience in printing. Their expertise and guidance has proved useful at every step. Their uncompromising quality standards have made this book an exceptional effort. Their encouragement from time to time has been an inspiration for everyone.

The publisher and the editorial board hope that this book will prove to be a valuable piece of knowledge for researchers, students, practitioners and scholars across the globe.

List of Contributors

Binbin Wu, Guoqiang Wang, Changming Liu and Jin Wu
College of Water Sciences, Beijing Normal University, Key Laboratory of Water and Sediment Sciences, Ministry of Education, Beijing, China

Qing Fu
Chinese Research Academy of Environmental Sciences, Beijing, China

David Maxwell Suckling
Biosecurity Group, The New Zealand Institute of Plant and Food Research Ltd, Christchurch, New Zealand, 2 Better Border Biosecurity, Christchurch, New Zealand

René François Henri Sforza
European Biological Control Laboratory, USDA-ARS, Campus International de Baillarguet, Montferrier-sur-Lez, France

Lei Huang, Yuting Han and Jun Bi
State Key Laboratory of Pollution Control & Resource Reuse, School of the Environment, Nanjing University, Nanjing, P.R. China

Ying Zhou
Department of Environmental Health, Rollins School of Public Health, Emory University, Atlanta, Georgia, United States of America

Heinz Gutscher
Department of Psychology, University of Zurich, Zurich, Switzerland

Christopher J. Brown and Megan I. Saunders
The Global Change Institute and the School of Biological Sciences, The University of Queensland, St Lucia, Queensland, Australia

Hugh P. Possingham
Australian Research Council, Centre for Excellence for Environmental Decisions, School of Biological Sciences, The University of Queensland, St Lucia, Queensland, Australia

Anthony J. Richardson
Climate Adaptation Flagship, Commonwealth Scientific and Industrial Research Organisation, Marine and Atmospheric Research, Dutton Park, Queensland, Australia

Centre for Applications in Natural Resource Mathematics, School of Mathematics and Physics, The University of Queensland, St Lucia, Queensland, Australia

Mo-Qian Zhang, Hai-Qiang Guo, Xiao Xie, Ting-Ting Zhang, Zu-Tao Ouyang and Bin Zhao
Coastal Ecosystems Research Station of the Yangtze River Estuary, Ministry of Education Key Laboratory for Biodiversity Science and Ecological Engineering, Institute of Biodiversity Science, Fudan University, Shanghai, P.R. China

Reid Tingley and Richard Shine
School of Biological Sciences A08, University of Sydney, Sydney, New South Wales, Australia

Martina G. Vijver
Institute of Environmental Sciences (CML), Leiden University, Leiden, The Netherlands

Paul J. van den Brink
Alterra, Wageningen University and Research centre, Wageningen, The Netherlands
Wageningen University, Wageningen University and Research centre, Wageningen, The Netherlands

Nicolai Bodemer and Mirta Galesic
Center for Adaptive Behavior and Cognition, Max Planck Institute for Human Development, Berlin, Germany
Harding Center for Risk Literacy, Max Planck Institute for Human Development, Berlin, Germany

Azzurra Ruggeri
Center for Adaptive Behavior and Cognition, Max Planck Institute for Human Development, Berlin, Germany

Binbin Yu and Qixing Zhou
Key Laboratory of Pollution Processes and Environmental Criteria (Ministry of Education), College of Environmental Science and Engineering, Nankai University, Tianjin, China

Yu Wan
Department of Agricultural and Biological Engineering, University of Florida, Gainesville, Florida, United States of America

Wu Yang, Wei Liu, Junyan Luo and Jianguo Liu
Center for Systems Integration and Sustainability, Department of Fisheries and Wildlife, Michigan State University, East Lansing, Michigan, United States of America

Thomas Dietz
Center for Systems Integration and Sustainability, Department of Fisheries and Wildlife, Michigan State University, East Lansing, Michigan, United States of America
Environmental Science and Policy Program, Department of Sociology and Animal Studies Program, Michigan State University, East Lansing, Michigan, United States of America

Randy Swaty
The Nature Conservancy, Marquette, Michigan, United States of America

Kori Blankenship
The Nature Conservancy, Bend, Oregon, United States of America

Sarah Hagen and Joseph Fargione
The Nature Conservancy, Minneapolis, Minnesota, United States of America

Jim Smith
The Nature Conservancy, Jacksonville, Florida, United States of America

Jeannie Patton
The Nature Conservancy, Boulder, Colorado, United States of America

Wen Zhuang
Key Laboratory of Coastal Environmental Processes and Ecological Remediation, Yantai Institute of Coastal Zone Research, Chinese Academy of Sciences, Yantai, Shandong, China
College of City and Architecture Engineering, Zaozhuang University, Zaozhuang, Shandong, China
University of Chinese Academy of Sciences, Beijing, China

Xuelu Gao
Key Laboratory of Coastal Environmental Processes and Ecological Remediation, Yantai Institute of Coastal Zone Research, Chinese Academy of Sciences, Yantai, Shandong, China

Hao Lian
State Key Laboratory of Freshwater Ecology and Biotechnology, Institute of Hydrobiology, The Chinese Academy of Sciences, Wuhan, China

University of Chinese Academy of Sciences, Beijing, China

Wei Hu, Rong Huang, Fukuan Du, Lanjie Liao, Zuoyan Zhu and Yaping Wang
State Key Laboratory of Freshwater Ecology and Biotechnology, Institute of Hydrobiology, The Chinese Academy of Sciences, Wuhan, China

Xiaoyu Li
State Key Laboratory of Desert and Oasis Ecology, Xinjiang Institute of Ecology and Geography, Chinese Academy of Sciences, Xinjiang, China
State Key Laboratory of Forest and Soil Ecology, Institute of Applied Ecology, Chinese Academy of Sciences, Liaoning, China

Lijuan Liu, Yugang Wang, Geping Luo and Xi Chen
State Key Laboratory of Desert and Oasis Ecology, Xinjiang Institute of Ecology and Geography, Chinese Academy of Sciences, Xinjiang, China

Xiaoliang Yang
College of Environmental Science and Forestry, State University of New York, Syracuse, New York, United States of America

Bin Gao
College of Resources Science and Technology, Beijing Normal University, Beijing, China

Xingyuan He
State Key Laboratory of Forest and Soil Ecology, Institute of Applied Ecology, Chinese Academy of Sciences, Liaoning, China

Kelly M. Searle, David L. Smith and Tamaki Kobayashi
Department of Epidemiology, Bloomberg School of Public Health, Johns Hopkins University, Baltimore, Maryland, United States of America

Timothy Shields and Gregory Glass
W. Harry Feinstone Department of Molecular Microbiology and Immunology, Bloomberg School of Public Health, Johns Hopkins University, Baltimore, Maryland, United States of America

Harry Hamapumbu and Philip E. Thuma
Macha Research Trust, Choma, Zambia

Sungano Mharakurwa
W. Harry Feinstone Department of Molecular Microbiology and Immunology, Bloomberg School of Public Health, Johns Hopkins University, Baltimore, Maryland, United States of America
Macha Research Trust, Choma, Zambia

William J. Moss
Department of Epidemiology, Bloomberg School of Public Health, Johns Hopkins University, Baltimore, Maryland, United States of America
W. Harry Feinstone Department of Molecular Microbiology and Immunology, Bloomberg School of Public Health, Johns Hopkins University, Baltimore, Maryland, United States of America

David A. Keith
Australian Wetlands Rivers and Landscapes Centre, University of New South Wales, Sydney, New South Wales, Australia
New South Wales Office of Environment and Heritage, Hurstville, New South Wales, Australia

Jon Paul Rodríguez
Centro de Ecología, Instituto Venezolano de Investigaciones Científicas, Caracas, Venezuela
Provita, Caracas, Venezuela
EcoHealth Alliance, New York, New York, United States of America
IUCN Commission on Ecosystem Management and IUCN Species Survival Commission, Gland, Switzerland

Kathryn M. Rodríguez-Clark
Centro de Ecología, Instituto Venezolano de Investigaciones Científicas, Caracas, Venezuela

Emily Nicholson
Centre of Excellence for Environmental Decisions, University of Melbourne, Victoria, Australia

Kaisu Aapala
Finnish Environment Institute, Helsinki, Finland

Alfonso Alonso
Smithsonian Conservation Biology Institute, National Zoological Park, Washington, D.C., United States of America

Marianne Asmussen
Centro de Ecología, Instituto Venezolano de Investigaciones Científicas, Caracas, Venezuela
EcoHealth Alliance, New York, New York, United States of America

Steven Bachman and Justin Moat
Royal Botanic Gardens, Kew, England

Alberto Basset
Department of Biological and Environmental Science, Ecotekne Center, University of Salento, Lecce, Italy

Edmund G. Barrow
IUCN Global Ecosystem Management Programme, Nairobi, Kenya

John S. Benson
Royal Botanic Gardens Trust, Sydney, New South Wales, Australia

Melanie J. Bishop
Department of Biological Sciences, Macquarie University, New South Wales, Australia

Ronald Bonifacio
Science Resource Centre, Department of Environment and Natural Resources, Adelaide, South Australia, Australia

Thomas M. Brooks
IUCN Commission on Ecosystem Management and IUCN Species Survival Commission, Gland, Switzerland
Nature Serve, Arlington, Virginia, United States of America

Mark A. Burgman
Australian Centre of Excellence for Risk Assessment, University of Melbourne, Victoria, Australia

Patrick Comer
NatureServe, Boulder, Colorado, United States of America

Francisco A. Comín
Pyrenean Institute of Ecology, Zaragoza. Spain

Franz Essl
Environment Agency Austria, Vienna, Austria
Department of Conservation Biology, Vegetation and Landscape Ecology, University of Vienna, Vienna, Austria

Don Faber-Langendoen
NatureServe, Arlington, Virginia, United States of America

Peter G. Fairweather
School of Biological Sciences, Flinders University, Adelaide, South Australia, Australia

Robert J. Holdaway
Landcare Research, Lincoln, New Zealand

Michael Jennings
Department of Geography, University of Idaho, Moscow, Idaho, United States of America

Richard T. Kingsford
Australian Wetlands Rivers and Landscapes Centre, University of New South Wales, Sydney, New South Wales, Australia

Rebecca E. Lester
School of Life and Environmental Sciences, Deakin University, Warnambool, Victoria, Australia

Ralph Mac Nally
Australian Centre for Biodiversity, School of Biological Sciences Monash University, Victoria, Australia

Michael A. McCarthy and Tracey J. Regan
Centre of Excellence for Environmental Decisions, University of Melbourne, Victoria, Australia

María A. Oliveira-Miranda
Provita, Caracas, Venezuela

Phil Pisanu
Science Resource Centre, Department of Environment and Natural Resources, Adelaide, South Australia, Australia

Brigitte Poulin
Tour du Valat Research Center, Arles, France

Uwe Riecken
German Federal Agency for Nature Conservation, Bonn, Germany

Mark D. Spalding
The Nature Conservancy and Conservation Science Group, Department of Zoology, University of Cambridge, Cambridge, England

Sergio Zambrano-Martínez
Centro de Ecología, Instituto Venezolano de Investigaciones Científicas, Caracas, Venezuela

Seyedeh Belin Tavakoly Sany
Institute of Biological Sciences University of Malaya, Kuala Lumpur, Malaysia
Food Science and Technology Research Institute, ACECR Mashhad Branch, Mashhad, Iran

Majid Rezayi
Food Science and Technology Research Institute, ACECR Mashhad Branch, Mashhad, Iran
Chemistry Department, Faculty of Science, University Malaya, Kuala Lumpur, Malaysia

Rosli Hashim and Aishah Salleh
Institute of Biological Sciences University of Malaya, Kuala Lumpur, Malaysia

Ali Mehdinia
Department of Marine Science, Marine Living Group, Iranian National Institute for Oceanography, Tehran, Iran

Omid Safari
Faculty of Natural Resources and Environment, Ferdowsi University of Mashhad, Mashhad, Iran

Index